Atomic No.	Symbol	Element	Atomic Weight	Density	Structure	a		
33	As	Arsenic	74.92	5.78	R	3.16		2.6
38	Sr	Strontium	87.62	2.58	FCC	4.302		
39	Y	Yttrium	88.91	4.48	CPH	3.554		
40	Zr	Zirconium	91.22	6.51	CPH	3.172	17	3.2
41	Nb	Niobium	92.91	8.58	BCC	2.858	13.7	4.06
42	Mo	Molybdenum	95.94	10.22	BCC	2.725		2.7
47	Ag	Silver	107.87	10.50	FCC	2.889	47	10.9
48	Cd	Cadmium	112.40	8.65	CPH	2.979	11	16.55
49	In	Indium	114.82	7.31	T	3.251	8	18
50	Sn	Tin	118.69	7.29†	T	3.022	1.57	13
51	Sb	Antimony	121.75	6.69	R	2.907	6–6.5	4.7–6
55	Cs	Cesium	132.91	1.91‡	BCC	5.235	11.3	54
56	Ba	Barium	137.33	3.59	BCC	4.347		
58	Ce	Cerium	140.12	6.77	FCC	3.649	6	4.44
60	Nd	Neodymium	144.24	7.00	H	3.658		3.33
73	Ta	Tantalum	180.95	16.67	BCC	2.860	27	3.6
74	W	Tungsten	183.85	19.25	BCC	2.741	50	2.55
78	Pt	Platinum	195.09	21.44	FCC	2.775	21.3	4.9
79	Au	Gold	196.97	19.28	FCC	2.884	11.6	7.9
80	Hg	Mercury	200.59	14.26§	R	3.005		
81	Tl	Thallium	204.37	11.87	CPH	3.457		16.3
82	Pb	Lead	207.20	11.34	FCC	3.500	2	7.4
83	Bi	Bismuth	208.98	9.80	R	3.071	4.6	
92	U	Uranium	238.03	19.05	OR	2.76	24	3.8–7.8

*R = rhombohedral; H = simple hexagonal; C = simple cubic; OR = orthorhombic; T = simple tetragonal.

*At 25°C.

†At −10°C.

‡At −46°C.

§At −46°C.

THE TECHNOLOGY
OF METALLURGY)

WILLIAM K. DALTON
Purdue University

Merrill, an imprint of
Macmillan Publishing Company
New York

Maxwell Macmillan Canada
Toronto

Maxwell Macmillan International
New York   Oxford   Singapore   Sydney

Editor: Stephen Helba
Production Editor: Stephen C. Robb
Art Coordinator: Peter A. Robison
Text Designer: Anne Flanagan
Cover Designer: Robert Vega
Production Buyer: Pamela D. Bennett
Illustrations: Accurate Art, Inc.

This book was set in Times Roman by Bi-Comp, Inc. and was printed and bound by R. R. Donnelley & Sons Company. The cover was printed by Phoenix Color Corp.

The Publisher offers discounts on this book when ordered in bulk quantities. For more information, write to: Special Sales Department, Macmillan Publishing Company, 445 Hutchinson Avenue, Columbus, OH 43235, or call 1-800-228-7854

Macmillan Publishing Company
866 Third Avenue
New York, NY 10022

Macmillan Publishing Company is part of the
Maxwell Communication Group of Companies.

Maxwell Macmillan Canada, Inc.
1200 Eglinton Avenue East, Suite 200
Don Mills, Ontario M3C 3N1

Library of Congress Cataloging-in-Publication Data
Dalton, William K.
    The technology of metallurgy / William K. Dalton.
        p.   cm.
    Includes index.
    ISBN 0-02-326900-6
    1. Metallurgy.   I. Title.
TN655.D35   1993
669—dc20   92-43577
                                                              CIP

Printing: 1 2 3 4 5 6 7 8 9   Year: 4 5 6 7

MERRILL'S INTERNATIONAL SERIES IN ENGINEERING TECHNOLOGY

INTRODUCTION TO ENGINEERING TECHNOLOGY

Pond, *Introduction to Engineering Technology, 2nd Edition*, 0-02-396031-0

ELECTRONICS TECHNOLOGY

Electronics Reference

Adamson, *The Electronics Dictionary for Technicians*, 0-02-300820-2

Berlin, *The Illustrated Electronics Dictionary*, 0-675-20451-8

Reis, *Becoming an Electronics Technician: Securing Your High-Tech Future*, 0-02-399231-X

DC/AC Circuits

Boylestad, *DC/AC: The Basics*, 0-675-20918-8

Boylestad, *Introductory Circuit Analysis, 7th Edition*, 0-02-313161-6

Ciccarelli, *Circuit Modeling: Exercises and Software, 2nd Edition*, 0-02-322455-X

Floyd, *Electric Circuits Fundamentals, 2nd Edition*, 0-675-21408-4

Floyd, *Electronics Fundamentals: Circuits, Devices, and Applications, 2nd Edition*, 0-675-21310-X

Floyd, *Principles of Electric Circuits, 4th Edition*, 0-02-338531-6

Floyd, *Principles of Electric Circuits: Electron Flow Version, 3rd Edition*, 0-02-338501-4

Keown, *PSpice and Circuit Analysis, 2nd Edition*, 0-02-363526-6

Monssen, *PSpice with Circuit Analysis*, 0-675-21376-2

Murphy, *DC Circuit Tutor: A Software Tutorial Using Animated Hypertext*, 0-02-385141-4

Murphy, *AC Circuit Tutor: A Software Tutorial Using Animated Hypertext*, 0-02-385144-9

Tocci, *Introduction to Electric Circuit Analysis, 2nd Edition*, 0-675-20002-4

Devices and Linear Circuits

Berlin & Getz, *Fundamentals of Operational Amplifiers and Linear Integrated Circuits*, 0-675-21002-X

Berube, *Electronic Devices and Circuits Using MICRO-CAP II*, 0-02-309160-6

Berube, *Electronic Devices and Circuits Using MICRO-CAP III*, 0-02-309151-7

Bogart, *Electronic Devices and Circuits, 3rd Edition*, 0-02-311701-X

Bogart, *Linear Electronics*, 0-02-311601-3

Floyd, *Basic Operational Amplifiers and Linear Integrated Circuits*, 0-02-338641-X

Floyd, *Electronic Devices, 3rd Edition*, 0-675-22170-6

FLoyd, *Electronic Devices: Electron Flow Version*, 0-02-338540-5

Floyd, *Fundamentals of Linear Circuits*, 0-02-338481-6

Schwartz, *Survey of Electronics, 3rd Edition*, 0-675-20162-4

Stanley, *Operational Amplifiers with Linear Integrated Circuits, 3rd Edition*, 0-02-415556-X

Tocci, *Electronic Devices: Conventional Flow Version, 3rd Edition*, 0-675-21150-6

Tocci & Oliver, *Fundamentals of Electronic Devices, 4th Edition*, 0-675-21259-6

Digital Electronics

Floyd, *Digital Fundamentals, 5th Edition*, 0-02-338502-2

Foster, *Sequential Logic Tutor: A Software Tutorial Using Animated Hypertext*, 0-02-338731-9

Foster, *Combinational Logic Tutor: A Software Tutorial Using Animated Hypertext*, 0-02-338735-1

McCalla, *Digital Logic and Computer Design*, 0-675-21170-0

Reis, *Digital Electronics through Project Analysis*, 0-675-21141-7

Tocci, *Fundamentals of Pulse and Digital Circuits, 3rd Edition*, 0-675-20033-4

Microprocessor Technology

Antonakos, *The 68000 Microprocessor: Hardware and Software Principles and Applications, 2nd Edition*, 0-02-303603-6

Antonakos, *An Introduction to the Intel Family of Microprocessors: A Hands-On Approach Utilizing the 8088 Microprocessor*, 0-675-22173-0

Brey, *8086/8066, 80286, 80386, and 80486 Assembly Language Programming*, 0-02-314247-2

Brey, *The Advanced Intel Microprocessor*, 0-02-314245-6

Brey, *The Intel Microprocessors: 8086/8088, 80186, 80286, 80386, and 80486: Architecture, Programming, and Interfacing, 3rd Edition*, 0-02-314250-2

Brey, *Microprocessors and Peripherals: Hardware, Software, Interfacing, and Applications, 2nd Edition*, 0-675-20884-X

Driscoll, Coughlin, & Villanucci, *Data Acquisition and Process Control with the MC68HC11 Microcontroller*, 0-02-330555-X

Gaonkar, *Microprocessor Architecture, Programming, and Applications with the 8085/8080A, 2nd Edition*, 0-675-20675-6

Gaonkar, *The Z80 Microprocessor: Architecture, Interfacing, Programming, and Design, 2nd Edition*, 0-02-340484-1

Goody, *Programming and Interfacing the 8086/8088 Microprocessor: A Product-Development Laboratory Process*, 0-675-21312-6

MacKenzie, *The 8051 Microcontroller*, 0-02-373650-X

Miller, *The 68000 Family of Microprocessors: Architecture, Programming, and Applications, 2nd Edition*, 0-02-381560-4

Quinn, *The 6800 Microprocessor*, 0-675-20515-8

Subbarao, *16/32 Bit Microprocessors: 68000/68010/68020 Software, Hardware, and Design Applications*, 0-675-21119-0

Electronic Communications

Monaco, *Introduction to Microwave Technology*, 0-675-21030-5

Monaco, *Preparing for the FCC Radio-Telephone Operator's License Examination*, 0-675-21313-4

Schoenbeck, *Electronic Communications: Modulation and Transmission, 2nd Edition*, 0-675-21311-8

Young, *Electronic Communication Techniques, 3rd Edition*, 0-02-431201-0

Zanger & Zanger, *Fiber Optics: Communication and Other Applications*, 0-675-20944-7

Microcomputer Servicing

Adamson, *Microcomputer Repair*, 0-02-300825-3

Asser, Stigliano, & Bahrenburg, *Microcomputer Servicing: Practical Systems and Troubleshooting, 2nd Edition*, 0-02-304241-9

Asser, Stigliano, & Bahrenburg, *Microcomputer Theory and Servicing, 2nd Edition*, 0-02-304231-1

Programming

Adamson, *Applied Pascal for Technology*, 0-675-20771-1

Adamson, *Structured BASIC Applied to Technology, 2nd Edition*, 0-02-300827-X

Adamson, *Structured C for Technology*, 0-675-20993-5

Adamson, *Structured C for Technology (with disk)*, 0-675-21289-8

Nashelsky & Boylestad, *BASIC Applied to Circuit Analysis*, 0-675-20161-6

Instrumentation and Measurement

Berlin & Getz, *Principles of Electronic Instrumentation and Measurement*, 0-675-20449-6

Buchla & McLachlan, *Applied Electronic Instrumentation and Measurement*, 0-675-21162-X

Gillies, *Instrumentation and Measurements for Electronic Technicians, 2nd Edition*, 0-02-343051-6

Transform Analysis

Kulathinal, *Transform Analysis and Electronic Networks with Applications*, 0-675-20765-7

Biomedical Equipment Technology

Aston, *Principles of Biomedical Instrumentation and Measurement*, 0-675-20943-9

Mathematics

Monaco, *Essential Mathematics for Electronics Technicians*, 0-675-21172-7

Davis, *Technical Mathematics*, 0-675-20338-4

Davis, *Technical Mathematics with Calculus*, 0-675-20965-X

INDUSTRIAL ELECTRONICS/INDUSTRIAL TECHNOLOGY

Bateson, *Introduction to Control System Technology, 4th Edition*, 0-02-306463-3

Fuller, *Robotics: Introduction, Programming, and Projects*, 0-675-21078-X

Goetsch, *Industrial Safety and Health: In the Age of High Technology*, 0-02-344207-7

Goetsch, *Industrial Supervision: In the Age of High Technology*, 0-675-22137-4

Geotsch, *Introduction to Total Quality: Quality, Productivity, and Competitiveness*, 0-02-344221-2

Horath, *Computer Numerical Control Programming of Machines*, 0-02-357201-9

Hubert, *Electric Machines: Theory, Operation, Applications, Adjustment, and Control*, 0-675-20765-7

Humphries, *Motors and Controls*, 0-675-20235-3

Hutchins, *Introduction to Quality: Management, Assurance, and Control*, 0-675-20896-3

Laviana, *Basic Computer Numerical Control Programming*, 0-675-21298-7

Pond, *Fundamentals of Statistical Quality Control*

Reis, *Electronic Project Design and Fabrication, 2nd Edition*, 0-02-399230-1

Rosenblatt & Friedman, *Direct and Alternating Current Machinery, 2nd Edition*, 0-675-20160-8

Smith, *Statistical Process Control and Quality Improvement*, 0-675-21160-3

Webb, *Programmable Logic Controllers: Principles and Applications, 2nd Edition*, 0-02-424970-X

Webb & Greshock, *Industrial Control Electronics, 2nd Edition*, 0-02-424864-9

MECHANICAL/CIVIL TECHNOLOGY

Dalton, *The Technology of Metallurgy*, 0-02-326900-6

Keyser, *Materials Science in Engineering, 4th Edition*, 0-675-20401-1

Kokernak, *Fluid Power Technology*, 0-02-305705-X

Kraut, *Fluid Mechanics for Technicians*, 0-675-21330-4

Mott, *Applied Fluid Mechanics, 4th Edition*, 0-02-384231-8

Mott, *Machine Elements in Mechanical Design, 2nd Edition*, 0-675-22289-3

Rolle, *Thermodynamics and Heat Power, 4th Edition*, 0-02-403201-8

Spiegel & Limbrunner, *Applied Statics and Strength of Materials, 2nd Edition*, 0-02-414961-6

Spiegel & Limbrunner, *Applied Strength of Materials*, 0-02-414970-5

Wolansky & Akers, *Modern Hydraulics: The Basics at Work*, 0-675-20987-0

Wolf, *Statics and Strength of Materials: A Parallel Approach to Understanding Structures*, 0-675-20622-7

DRAFTING TECHNOLOGY

Cooper, *Introduction to VersaCAD*, 0-675-21164-6

Ethier, *AutoCAD in 3 Dimensions*, 0-02-334232-3

Goetsch & Rickman, *Computer-Aided Drafting with AutoCAD*, 0-675-20915-3

Kirkpatrick & Kirkpatrick, *AutoCAD for Interior Design and Space Planning*, 0-02-364455-9

Kirkpatrick, *The AutoCAD Book: Drawing, Modeling, and Applications, 2nd Edition*, 0-675-22288-5

Kirkpatrick, *The AutoCAD Book: Drawing, Modeling, and Applications, Including Release 12, 3rd Edition*, 0-02-364440-0

Lamit & Lloyd, *Drafting for Electronics, 2nd Edition*, 0-02-367342-7

Lamit & Paige, *Computer-Aided Design and Drafting*, 0-675-20475-5

Maruggi, *Technical Graphics: Electronics Worktext, 2nd Edition*, 0-675-21378-9

Maruggi, *The Technology of Drafting*, 0-675-20762-2

Sell, *Basic Technical Drawing*, 0-675-21001-1

TECHNICAL WRITING

Croft, *Getting a Job: Resume Writing, Job Application Letters, and Interview Strategies*, 0-675-20917-X

Panares, *A Handbook of English for Technical Students*, 0-675-20650-2

Pfeiffer, *Proposal Writing: The Art of Friendly Persuausion*, 0-675-20988-9

Pfeiffer, *Technical Writing: A Practical Approach, 2nd Edition*, 0-02-395111-7

Roze, *Technical Communications: The Practical Craft, 2nd Edition*, 0-02-404171-8

Weisman, *Basic Technical Writing, 6th Edition*, 0-675-21256-1

P R E F A C E

Purpose and Scope of the Book. The major purpose of this book is to present the basics of metallurgy to students who are not majoring in metallurgy or materials science. The traditional study of materials and processes as covered in many other texts is important, but our graduates will be better able to develop and apply the technologies of today and tomorrow if they understand what is happening to metals when they are cold-worked, heat-treated, and alloyed, that is if they understand metallurgery.

Although the content of this book is based on scientific and engineering principles, it is not intended to be a book on metallurgical science. The word "technology" in its title is intended to indicate the focus of the book, not to differentiate between the disciplines of engineering and engineering technology.

This book was begun because I could not find a text suitable for the applied metallurgy course I teach. To meet my students' needs, the boundaries I have placed on the scope of this text have limited the mathematics to algebra and trigonometry, with minor reference to the concepts of calculus. I have also limited the material to what could be covered in reading assignments of 20 to 30 pages per week for a 15-week, one-semester course. Although the course I teach is based on a format of 30 fifty-minute classes and 12 or 13 two-hour labs, instructors can pick and choose from the material presented here to suit other schedules.

Organization of the Book. The text is divided into three major parts, each containing five chapters, and the appendixes. Part 1 deals with basic metallurgical principles such as what metals are, how mechanical properties are achieved, and how those properties are changed. Part 2 applies the basic metallurgical principles to the study of ferrous and nonferrous metals, and Part 3 discusses how metals fail and what can be done to reduce or eliminate failures. Appendix A contains photomicrographs and photographs, and Appendixes B through F provide information on mechanical testing and application of the test data. A glossary of important terms concludes the text.

Organization of the Chapters. Each chapter begins with a list of learning objectives, an introductory statement on the importance and application of the informa-

v

tion presented in the chapter, and an overview of the chapter content. The chapter then covers the information in approximately the same order used in setting forth the objectives and chapter overview. Each chapter is divided into numbered sections, and some sections are further divided into unnumbered subsections. Each chapter concludes with a summary section, review questions, and a list of sources for those interested in further study.

Special Features of the Book. To make the text "user-friendly" for both student and teacher, I've incorporated several pedagogical features.

Objectives. The learning objectives are written in terms of what the students should be able to do after studying the material in the chapter. Before reading a chapter students should use the objectives as a checklist of key concepts to look for as they read the material. After reading the chapter, students can review the text by reviewing the objectives. It should be apparent that the objectives and review questions both provide a source of questions for examinations.

New Terms. When a new term is introduced, it appears in boldface type and is defined and discussed in nearby text. These boldfaced terms are found in the glossary at the end of the book and can also be found in college-level dictionaries.

Data Tables. The scope of this book precludes the inclusion of large amounts of data. Rather, data for actual metals and alloys are used to illustrate or relate the metallurgical principles discussed. The sources of those data are given where appropriate; additional data are readily available in the handbooks published by such societies as the American Society for Metals (ASM), the Society of Automotive Engineers (SAE), and the American Society for Testing & Materials (ASTM).

Appendix A. Appendix A contains more than 40 photomicrographs and photographs; a brief introduction to the appendix describes the steps required to prepare specimens for examination by optical microscopy. Since the same photomicrograph or photograph may be referred to in widely separated portions of the text, I thought it best to collect all of them together in one place. In the text they are referred to as "Figure A," followed by a number and are placed in that numerical order in Appendix A.

Instructor's Resource Manual. This manual contains the following features: a sample course syllabus and schedule; solutions to the review questions contained in the text; and as a resource for users of this text, I am providing the details for 12 lab exercises. The information given with each lab exercise discusses the objectives of the exercise, the equipment required, preparation of test specimens, description of the outcome of the exercise, solutions to questions, and possible alternative methods for meeting the objectives.

Organization of Classes and Labs. The text can be used for both class and laboratory instruction; sample lab exercises are included as part of the Instructors Resource Manual. Part 1, Basic Metallurgical Basics Principles, can be augmented by laboratory exercises that involve such applications as tensile and hardness testing, cold working, annealing, development of freezing/cooling curves, generation of a phase diagram, and computer simulations of alloy freezing, and heat treatment. These laboratory activities can be enhanced in turn by the use of Appendixes B, C, and D.

Part 2, Ferrous and Nonferrous Metals, with the support of Appendix E, can be extended by work in the laboratory on heat treatment of steel and aluminum alloys, and by performing Jominy end-quench tests. Part 3, How Metals Fail, can be enhanced by performing impact tests (Appendix F), corrosion tests, and creep tests in the laboratory and by examining failed parts and components.

About the Author.   My interest in metals and the metals industry can be traced to the fact that I grew up in the Pittsburgh area. I worked in steel laboratories and mills during my summer vacations, and I studied metallurgy as a mechanical engineering student. For 32 enjoyable years I worked for the Aluminum Company of America, mostly in production management in fabricating plants, and for the last nine years I have taught about metals at Purdue University. I hope the text reflects the personal interest and practical involvement I have had with the subject.

Acknowledgments.   This book was begun as a master's degree project at Purdue University, and I wish to thank my committee for the encouragement they gave me. Professors Tom Hull, Jack Lillich, Ernie McDaniel, and Al Suess all supported a practical, useful project, as opposed to a study that would be "bound on all four edges and filed," one of Dr. Suess's often-voiced complaints about graduate-student projects.

A number of my colleagues deserve thanks for the help they gave me as the project expanded: Professors Hal Roach and Charles Thomas, and Deans Don Gentry and Fred Emshousen contributed to my understanding of how the teaching of metallurgy fit within the broad scope of engineering and technology education and helped to focus my attention on the most pertinent material for the text; Professors Bob English, Glenn Fritzlen, Ed Widener, Mileta Tomovic, and Ron Hosey added to my understanding of some of the basic metallurgical concepts; Professors Bruce Harding and Tom Kirk were invaluable tutors on the use of the computer, without which I could not have even begun this project.

I also thank the following reviewers, who provided helpful critiques: Fred Bouchard, Southeastern Louisiana University; Andrew Burke, Monroe County Community College; Wayne Coleman, Eastern Illinois University; Morris Ellenburg, Indiana Vocational Technical College; John Kray, Glendale College; Richard Kruppa, Bowling Green State University; C. J. Law, Western New Mexico University; Paul O'Leary, Montana College of Mineral Science and Technology; Jack Paige, New Hampshire Technical College; Walter Ryan, New Hampshire Vocational Technical College; Tilman Sorrell, Northeast Missouri State University; George Stanton, East Tennessee State University; and Ken Williams, Navarro College.

Those to whom I owe the largest debt are the countless people in the mills at Irvin, Homestead, and New Kensington, Pennsylvania, and Geelong, Australia,

Vernon, California, and Lafayette, Indiana, who contributed to my real education, i.e., the actual work of hot- and cold-rolling, structural shape rolling, tube drawing, foil rolling, ingot casting, extruding, and all the other heat-treating, tool-and-die making, and finishing operations that are required in the production processes. Some of these "teachers" were bosses, engineers, and metallurgists, but many of them were the people actually doing the work, the people who gave me a real education in what was going on.

I make no claim that this book contains original information. Rather, I have tried to take my own knowledge and experience plus the knowledge and experience of others—particularly one of my Cornell professors, Clyde W. Mason, from whose book (*Introductory Physical Metallurgy*, published by ASM International) most of the photomicrographs of Appendix A have come—and organize it in a way that seemed most readily understood by my students.

Since the beginning of the project my students have used portions of the text as I completed them. I thank them for the feedback and critiques they have knowingly and unknowingly given me. My best wishes go to the students, past, present, and future, who will ultimately determine whether this has indeed been a practical and useful project.

William K. Dalton

C O N T E N T S

BASIC METALLURGICAL PRINCIPLES

THE PROPERTIES OF METALS CAN BE CHANGED OR CONtrolled by three different processes: strain-hardening or cold-working; alloying; and heating or heat-treating. All three processes are influenced by and dependent on the crystalline nature of metals. So we begin Part 1 by studying the crystal or atom lattice structure of metals, lattice geometry, and how metals freeze as crystals.

In Chapters 2 and 3 we'll study opposing phenomena: the strengthening of metals by cold-working and the softening of metals by heating.

In Chapter 4 we begin the study of alloying. Here we'll look at its dependence on the mutual solubility or insolubility of metals, and then in Chapter 5 we'll develop the phase diagram, which provides the groundwork for the study of the heat-treating and alloying of the ferrous and nonferrous metals, which will be taken up in Part 2.

The material presented in these first five chapters can be supplemented by the use of materials found in Appendixes B, C, and D and through laboratory exercises and demonstrations.

C H A P T E R 1

Nature and Formation of Metal Crystals

OBJECTIVES

After studying this chapter, you should be able to

- describe the atom according to recent theory, incorporating some specific terms in your description:
  - electrons, protons, neutrons, and nucleus
  - electron waves and electron energy levels or states
  - transition elements
- explain how the number of electrons, protons, and neutrons; electron energy levels; and the location of transition elements are related on the periodic table
- explain how the characteristics of the elements are determined
- describe metallic bonding and how the current atomic theory explains it
- describe the "closest-approach" atom diameter and how it determines lattice size
- describe how atom diameter varies with position on the periodic table
- describe the geometry of and sketch the major lattice configurations (BCC, FCC, CPH), showing where the atoms touch
- explain the difference between FCC and CPH in terms of "stacking"
- define the packing factor and rank the lattices according to it
- use the Miller indices system to identify planes and directions in simple cubic and hexagonal systems
- describe the freezing process of a pure metal and draw its freezing curve, explaining as you do so
  - the difference between the potential and the kinetic energy contained in molten metal and how these energies are affected by freezing
  - the generation of dendrites and their dependence on the latent heat of fusion and thermal gradients (heat flow)

- the generation of grain boundaries and their dependence on the formation of the grain itself
- the nature of grain boundaries as discontinuities representing locations of energy
- the preferred growth direction(s) of lattices
- define and describe vacancies, interstitials, and dislocations

DID YOU KNOW?

Today the concept of the atomic structure of matter is universally accepted, but this was not always the case. It has taken us many centuries to come to this point. The atomists of ancient Greece held that matter was composed of small indivisible particles, an idea that opposed the popular philosophical view of the time, which held that matter was composed of five "elements": earth, air, fire, water, and the "quintessence." However, the atomic concept was virtually lost until Robert Boyle (1627–91) and others began working on the properties of gases. John Dalton developed it further in his work on the partial pressures of gases in the early 1800s, and throughout the nineteenth century many scientists contributed to an understanding of the atom's makeup.

Early in the twentieth century Niels Bohr developed a model of the hydrogen atom using Max Planck's concepts of quantum mechanics. Bohr's concept of this atom, which had a nucleus at its center and an electron moving about the nucleus in a circular orbit, explained some, but not all, of the observed behavior of atoms. It remained for others to show that the electron was not only a particle but also a wave, with an orbit of a length that is a multiple of the wavelength. With that addition, "modern physics" was able to explain and predict the behavior of atoms.

A PREVIEW

The purpose of this chapter is to show that metals, as used in most engineering applications, are crystals and to investigate those crystals and their formation.

Our study of metals begins by investigating their basic building block, the atom. The nature of the atom—its structure, characteristics, and properties—are related by the periodic table, which we'll look at in some detail. The arrangement of the electrons about the atom by increments of energy can be described by using the periodic table.

The method by which metal atoms bond together, the metallic bond, causes them to assume a crystalline form. The configuration of these crystals, or atom lattices, depends on the internal structure of the particular atom. Here we'll study the three most common lattices and some of their characteristics.

In studying the geometric atom lattices we'll use a method of identifying their planes and directions, the Miller indices system, and we'll use some examples to describe how the system works. We'll also discuss how metal atoms pass from the random orientation of a molten liquid to a crystalline solid, i.e., how they solidify and form dendrites. Understanding the formation of dendrites is crucial to an appreciation of how metal elements segregate and hence require homogenization treatments, a subject discussed later. Finally, we'll look at crystal imperfections, a subject central to the understanding of how metals can be deformed, the topic of the next chapter.

1.1   ATOMIC STRUCTURE

Everyday experiences in the physical world are adequately described by the physics of Isaac Newton, and examples of his laws are all about us—the acceleration of a car, the balancing of a check book (that is, what goes out must equal what comes in), the action of billiard balls. Our understanding of these laws is at the level of intuition. Unfortunately, the structure of the atom cannot be completely explained by Newton's physics and requires the use of quantum mechanics, which is *not* understood at an intuitive level.

Since the late 1800s the work of many scientists has slowly revealed the structure of the atom.* It is known that electrons behave as if they were both particle *and* wave, but the work is not complete. There is still much that is not known, particularly at the level of physical visualization. But the theory of quantum mechanics has given us many principles by which the behavior of the atom can be explained and predicted.

Atomic Particles.   The atom consists of a **nucleus** surrounded by negatively charged particles called **electrons.** The nucleus has a positive charge equal to the sum of the negative charge of the electrons, making the atom itself electrically neutral. The nuclei of most atoms consist of two types of particles—positively charged **protons** and neutral **neutrons.** (The hydrogen atom is the exception; its nucleus has only one proton and no neutrons.) The mass of the proton is almost 2000 times the mass of an electron. Let's consider the electron solely as a circulating *particle*, kept in orbit about the nucleus by a centripetal force. If an outside force accelerates the particle, an emission of radiation should result. But in this case, there is no emission, so the electron is obviously not behaving as a particle.

The Electron as a Wave.   In treating the electron as a wave, scientists wrote wave equations to account for the energy levels of the electrons in the various atoms. Let's look at some of the principles that derive from these studies.

---

*I refer here to the work of Plank, Rutherford, Bohr, Einstein, de Broglie, Schrodinger, Dirac, Pauli, Heisenberg, and many others.

- *As a wave the electron's energy is related to its frequency* (rather than to its mass and velocity). Figure 1-1 depicts one wavelength of a wave and shows that frequency and wavelength are inversely related.
- The wave of the electron around the nucleus must be in multiples of whole wavelengths in order that the wave not die out. Thus the electron's energy varies in increments related to the wavelength called **quanta.** That is, *an electron's energy does not vary continuously but is quantized.*
- There are three quantum numbers that account for the energy of an electron. The most significant of these, *n*, the principal quantum number, is the **orbit,** or **shell,** of the electron. The higher the *n*, the higher the energy level, or *energy state*.
- The maximum number of electrons that can have this principal quantum number of *n*, that is, the maximum number of electrons that can be in any shell, is $2n^2$.
- From the rule $2n^2$, the maximum number of electrons in each major shell is 2, 8, 18, 32, 50, etc. Only the first four shells fill completely; the outer three shells are never filled by the 100+ known elements.
- The other two quantum numbers also contribute energy so that within a shell there are **subshells** of electrons with a range of energies. The first subshell of any shell can contain a maximum of 2 electrons; the next can have a maximum of 6, the third 10, the fourth 14. The $2n^2$ rule still sets the limit on the total number of electrons in the shell.
- The maximum number of electrons that can have the *same* energy in an atom is two. Thus there are electron pairs that have the same energy but spin in opposite directions.
- *Electrons attempt to locate at the lowest energy level.* Because of the combination of three quantum numbers, there are instances where a higher shell (larger *n*) may have a lower energy level than a subshell from a lower shell. Thus as electrons are added in progressing through the periodic table of elements, the shells are not filled in strict order.

The Quantum Nature of Electron Energy. Two useful conclusions we can draw here are that (1) the energy of electrons does not vary on a continuous scale but is

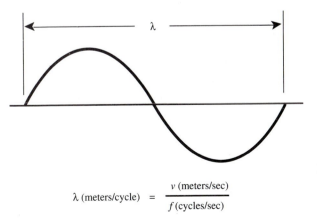

$$\lambda \text{ (meters/cycle)} = \frac{v \text{ (meters/sec)}}{f \text{ (cycles/sec)}}$$

FIGURE 1–1

A wave of wavelength λ, frequency f, and velocity v

quantized and (2) as electrons are added to the elements, they go into the shells or subshells with the lowest energy, not necessarily in the order of physical concentric rings.

1.2   PERIODIC TABLE

The periodic table of the elements shown in Table I-1 is arranged in the traditional manner, with the elements in the order of their atomic number—the number of protons or electrons in the element's atom. For example, hydrogen, with an atomic number of 1, has one electron and one proton in an atom of neutral charge; aluminum, atomic number 13, has thirteen of each; iron, atomic number 26, has twenty-six, etc.

Arrangement of Electron Shells or Periods. Electrons are arranged around the nucleus in shells that have definite patterns according to the quanta of energy they contain. In Table I-1 the shells are the same as the horizontal rows or **periods;** the arrangement of electrons in the shells is shown below the symbol for the element and the atomic number. For example, lithium (Li), atomic number 3, has two electrons in the first major shell, which fills it, and one electron in the next shell; aluminum (Al), atomic number 13, has two electrons in the first shell, eight in the next (which fills it), and three in the third shell; neodymium (Nd), atomic number 60 (in the Lanthanum Series), has complete shells of 2, 8, and 18 electrons, and incomplete shells of 22, 8, and 2 electrons for a total of 60.[†]

The first 18 atoms, from hydrogen to argon, fill their shells in an orderly fashion. Argon's first 10 electrons fill the first 2 shells, and the last 8 electrons fill 2 subshells of the third shell. By the $2n^2$ rule, the third shell can have a maximum of 18 electrons, or 10 more than argon. It would be expected that the next element, potassium (K), would continue to fill this shell with its nineteenth electron, but instead it starts the fourth shell. Calcium (Ca), the next element, also puts its twentieth electron in the fourth shell. This occurs because these two energy states are *lower* than the energy states remaining in the third shell, and one of the principles of quantum mechanics is that atoms will attempt to exist at the lowest possible energy state.

At the element scandium (Sc), the preferred (lower) energy location switches back to the third shell; from scandium to nickel (Ni), electrons are added to the third shell, not the fourth. Finally, at copper the third shell has its full complement of 18 electrons and from there to krypton (Kr) the electrons are added in consecutive fashion to the fourth shell.

The fifth, sixth, and seventh periods all begin as did the fourth. That is, instead of continuing to fill the shell of the prior inert gas, the first two elements in

---

[†]To conserve space in the table starting with potassium (K) the first 10 electrons, or two shells, are replaced with an X. Later, the first three shells, or 28 electrons, are replaced with a Y, and still later the first four shells, or 60 electrons, are replaced with a Z.

T A B L E I–1   The periodic table of the elements, showing element symbol, atomic number, number of electrons in each major shell, the principal atom lattice,* and the closest-approach distance, i.e., the hard-ball atom diameter†

Groups

Period	I-A	II-A	III-B	IV-B	V-B	VI-B	VII-B	VIII	VIII	VIII	I-B	II-B	III-A	IV-A	V-A	VI-A	VII-A	0
1	H 1 0.92 Å																	He 2 2
2	Li 3 2,1 BCC 3.039 Å	Be 4 2,2 CPH 2.225 Å											B 5 2,3 R 1.94 Å	C 6 2,4 R 1.42 Å	N 7 2,5 1.42 Å	O 8 2,6 1.20 Å	F 9 2,7	Ne 10 2,8
3	Na 11 2,8,1 BCC 3.716 Å	Mg 12 2,8,2 CPH 3.197 Å											Al 13 2,8,3 FCC 2.863 Å	Si 14 2,8,4 C 2.351 Å	P 15 2,8,5 OR 2.18 Å	S 16 2,8,6 2.32 Å	Cl 17 2,8,7	Ar 18 2,8,8
4	K 19 X,8,1 BCC 4.544 Å	Ca 20 X,8,2 FCC 3.947 Å	Sc 21 X,9,2 CPH 3.252 Å	Ti 22 X,10,2 CPH 2.890 Å	V 23 X,11,2 BCC 2.622 Å	Cr 24 X,13,1 BCC 2.498 Å	Mn 25 X,13,2 C 2.24 Å	Fe 26 X,14,2 BCC 2.482 Å	Co 27 X,15,2 CPH 2.497 Å	Ni 28 X,16,2 FCC 2.492 Å	Cu 29 X,18,1 FCC 2.556 Å	Zn 30 X,18,2 CPH 2.665 Å	Ga 31 X,18,3 OR 2.44 Å	Ge 32 X,18,4 C 2.450 Å	As 33 X,18,5 R 3.16 Å	Se 34 X,18,6 R 2.32 Å	Br 35 X,18,7	Kr 36 X,18,8
5	Rb 37 Y,8,1 BCC 4.858 Å	Sr 38 Y,8,2 FCC 4.302 Å	Y 39 Y,9,2 CPH 3.554 Å	Zr 40 Y,10,2 CPH 3.172 Å	Nb 41 Y,12,1 BCC 2.858 Å	Mo 42 Y,13,1 BCC 2.725 Å	Tc 43 Y,14,1 CPH 2.71 Å	Ru 44 Y,15,1 CPH 2.65 Å	Rh 45 Y,16,1 FCC 2.690 Å	Pd 46 Y,18 FCC 2.750 Å	Ag 47 Y,18,1 FCC 2.889 Å	Cd 48 Y,18,2 CPH 2.979 Å	In 49 Y,18,3 T 3.251 Å	Sn 50 Y,18,4 T 3.022 Å	Sb 51 Y,18,5 R 2.907 Å	Te 52 Y,18,6 H 2.864 Å	I 53 Y,18,7	Xe 54 Y,18,8
6	Cs 55 Y,18,8,1 BCC 5.235 Å	Ba 56 Y,18,8,2 BCC 4.347 Å	La 57 Y,18,9,2 H 3.733 Å	Hf 72 Z,10,2 CPH 3.127 Å	Ta 73 Z,11,2 BCC 2.860 Å	W 74 Z,12,2 BCC 2.741 Å	Re 75 Z,13,2 CPH 2.741 Å	Os 76 Z,14,2 CPH 2.675 Å	Ir 77 Z,14,2 FCC 2.714 Å	Pt 78 Z,17,1 FCC 2.775 Å	Au 79 Z,18,1 FCC 2.884 Å	Hg 80 Z,18,2 R 3.005 Å	Tl 81 Z,18,3 CPH 3.457 Å	Pb 82 Z,18,4 FCC 3.500 Å	Bi 83 Z,18,5 R 3.071 Å	Po 84 Z,18,6 C 3.338 Å	At 85 Z,18,7	Rn 86 Z,18,8
7	Fr 87 Z,18,8,1 5.235 Å	Ra 88 Z,18,8,2	Ac 89 Y,18,9,2 FCC 3.755 Å	Ku 104 Z,32,10,2	Ha 105 Z,32,11,2	106 Z,32,12,2												

X = the first 10 electrons contained in the first two completed shells (2 + 8 = 10)
Y = the first 28 electrons contained in the first three completed shells (2 + 8 + 18 = 28)
Z = the first 60 electrons contained in the first four completed shells (2 + 8 + 18 + 32 = 60)

The Lanthanum Series of 14 elements fits between lanthanum, number 57, and hafnium, number 72.

Ce 58 Y,20,8,2	Pr 59 Y,21,8,2	Nd 60 Y,22,8,2	Pm 61 Y,23,8,2	Sm 62 Y,24,8,2	Eu 63 Y,25,8,2	Gd 64 Y,25,9,2	Tb 65 Y,27,8,2	Dy 66 Y,28,8,2	Ho 67 Y,29,8,2	Er 68 Y,30,8,2	Tm 69 Y,31,8,2	Yb 70 Y,32,8,2	Lu 71 Y,32,9,2

The Actinium Series of 14 elements fits between actinium, number 89, and number 104.

Th 90 Z,18,10,2	Pa 91 Z,20,9,2	U 92 Z,21,9,2	Np 93 Z,22,9,2	Pu 94 Z,24,8,2	Am 95 Z,25,8,2	Cm 96 Z,25,9,2	Bk 97 Z,27,8,2	Cf 98 Z,28,8,2	Es 99 Z,29,8,2	Fm 100 Z,30,8,2	Md 101 Z,31,8,2	No 102 Z,32,8,2	Lr 103 Z,32,9,2

*BCC = body-centered cubic, FCC = face-centered cubic, CPH = close-packed hexagonal, R = rhombohedral, C = cubic, T = tetragonal, H = hexagonal, OR = orthorhombic

† 1 angstrom unit (Å) = 10^{-8} cm or 10^{-10} m.

Source: Charles Barrett and T. B. Massalski. *Structure of metals* (3d rev. ed.). Oxford: Pergamon, 1980. table A–6, 626 and 235.

the row start a new shell, then the electrons are added to the prior shell. Thus *there are a large number of elements in the center of the table where electrons are always being added to prior shells.*

The sixth and seventh periods are unusual in that each has a series of 14 elements that add electrons to even earlier shells. For example, in the sixth period, lanthanum (La) adds an electron to the incomplete *fifth* shell, as does the next element, hafnium (Hf), and so on across the period. But between lanthanum and hafnium are 14 elements, the Lanthanum Series, which all add electrons to the incomplete *fourth* shell. The Actinium Series in the seventh period is similar in that those elements all add electrons to the incomplete *fifth* shell.

Relationship of Groups.   Elements in any group or column tend to have similar properties. For example, the very active metals lithium (Li), sodium (Na), and potassium (K) are in group I-A; the very active gases fluorine (F), chlorine (Cl), and bromine (Br) are in group VII-A; and the inert gases helium (He), neon (Ne), and argon (Ar) are in group 0. Group I-B is an example of some familiar metals that have similar properties.

The first 18 elements in the table are separated, placed above the rest of the table, and appear in groups headed by a Roman numeral followed by an A. Reading across the periods, you can see that the Roman numeral for the A columns is equal to the number of outermost electrons, or **valence electrons.**

This is not the case for the large body of elements in the center of the table (group III-B across to II-B). With few exceptions, these B elements all have one or two valence electrons and thus do not fit with the pattern of the A groups, where electrons are added to the outermost positions as the atomic number goes up. Groups I-B and II-B have a pattern that leads into the III-A group, but all of the elements in the center of the table are in a "transition" from adding electrons to inner shells to adding them to outer shells; these elements are known as **transition elements.**

Atom lattice and atom diameter.   Atom lattice and atom diameter will be discussed shortly, but we note here that *in any column the elements tend to have the same atom lattice and their diameters increase going from top to bottom*; this is especially valid for the transition elements. The atom diameter among the transition elements is very consistent in any period, although the diameter increases from right to left among the other elements.

1.3 ━━━━━━   **ATOMIC BONDING**

The world would not exist as we know it if the only materials available were atomic forms of the elements. Salts, gases, polymers, and metals are formed by the **bonding** together of the atoms of the elements in four different ways. It is evident from the periodic table that the elements add electrons to fill the outer shell to a maximum of eight electrons. Three of the atomic-bonding methods—

ionic, covalent, and **metallic** bonding—rely on this property. Obviously, it is the metallic bond that is of major interest here. The fourth method, van der Waals, or molecular, bonding, generates a weak attraction between inert gases and other molecules with completed outer shells that typically have a large total number of electrons. This method contributes very little to an understanding of the metallic bond, and we'll not discuss it further.

Ionic Bond. The ionic bond is typical of the union of elements from opposite sides of the periodic table; elements in these groups have few electrons or spaces for electrons in their outermost shells. For example, sodium has one outermost electron and chlorine lacks one to complete its outermost shell. These two elements take advantage of the opposite signs of their valences; the sodium *donates* its valence electron to the chlorine, and common salt, NaCl, is formed. Although the compound NaCl is electrically neutral, the two elements exist in the compound as ions of opposite electrical charge and are therefore attracted to one another. The materials formed by this type of bonding typically have a high melting temperature, are hard and brittle, exist in a three-dimensional **crystal** (crystals will be discussed shortly), and easily dissociate into ionic forms; for example, NaCl in water forms Na^+ and Cl^- ions.

Covalent Bond. In a covalent bond, the element *shares* its valence electrons to complete the outer shell. Covalent bonding is typically found in gaseous molecules and compounds and in the formation of long-chained polymers; it also bonds those elements that have four or more valence electrons. For example, the hydrogen atom has one electron in a shell that can hold only two. The hydrogen molecule, H_2, is two hydrogen atoms combining to use each other's one electron to complete that shell. The molecules of other noninert gases like oxygen, O_2, other gaseous compounds like ammonia, NH_3, and liquid compounds like water, H_2O, are formed in the same way.

In forming the ammonia (NH_3) and water (H_2O) molecules, the vacant sites in the outer shells of the nitrogen and oxygen are completed by sharing electrons with the appropriate number of hydrogen atoms; the electrons from each hydrogen atom pair with one of the electrons from the other element. Thus in ammonia there are three bonded pairs; in water, there are two.

Carbon, with four electrons in the outer shell (and lacking four to complete the shell), uses the covalent bond when it forms the gases carbon dioxide (CO_2), methane (CH_4), and acetylene (C_2H_2); a solid diamond is formed by covalently bonded carbon atoms.

Metallic Bond. An inspection of Table I-1 shows that the metallic elements have only a few electrons in their outermost shell. Compared with electrons closer to the nucleus, these outer electrons are more loosely held and can be attracted to other nuclei. In this way, these valence electrons form a cloud to which numerous nuclei are attracted. Figure 1-2 is the schematic representation of the atoms in a metallic solid. The black circles are the nuclei; the dots represent the electron cloud.

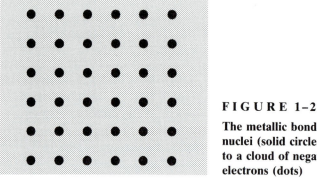

F I G U R E 1–2

**The metallic bond: Positive
nuclei (solid circles) attracted
to a cloud of negative
electrons (dots)**

Thus *metals are an arrangement of nuclei and inner electron shells having a
net positive charge that are attracted to a cloud of negative valence electrons.*
Although the positive center portions tend to repel each other, the attraction to
the negative electron cloud is greater, and a net cohesive force bonds the metal
atoms together. By observation, we arrive at the logical conclusion that metal
atoms do, in fact, prefer to exist in combination (bonded together) since they do
not exist in nature as isolated atoms.

The other bonding methods require the valence electrons to remain in position
relative to the nucleus to satisfy the electrical attraction-repulsion forces. Since
the electrical and thermal characteristics of metals require the movement of electrons, it is obvious that the other bonding methods cannot apply to metals. That is,
the metallic bond is useful in explaining the ability of metals to conduct heat and
electricity. Also, since the valence electrons are the ones involved in the cloud,
metallic bonding accounts for the ability of atoms with differing numbers of total
electrons to exist side by side in alloys, as long as the atoms are not too different in
diameter. (The importance of this is discussed in Chapter 4.)

1.4   ATOMIC DIAMETER

Internal Energy of an Atom.   One isolated atom is an electrically neutral body.
The repulsive forces between the like-charged electrons are offset by their attraction to the positive nucleus, and the internal energy of the atom is said to be at a
reference point of zero. Suppose two such atoms of the same metal are a great
distance apart. At this distance, each atom would have very little influence upon
the other, and their internal energies would still be zero.

Now suppose that the two atoms are brought closer and closer together so
that the electrical balance in each atom begins to be influenced by the other. The
like-charged particles cause a repulsive force, but the unlike charges are attracted.
The result is the metallic bond described previously.

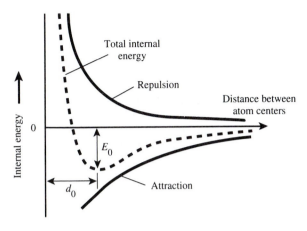

Internal energy →

0

E_0

d_0

Total internal energy

Repulsion

Distance between atom centers

Attraction

FIGURE 1-3

The internal energy of an atom (the resultant of repulsion and attraction forces between two atoms) vs. the distance between the centers of the atoms

Closest-Approach Distance. Figure 1-3 depicts the effect on the internal energy of the atoms as the distance between the centers of the nuclei is reduced: The repulsive force increases in the positive direction, and the attractive force also increases, but in the opposite or negative direction. As the distance between the atoms decreases, the resultant of these two forces is a net attractive force that is at an energy level below that of an isolated atom. The net attractive force increases (negatively) to the point of minimum energy, E_0, which occurs at the **closest-approach distance,** d_0, between the atoms. Since atoms prefer to exist in as low an energy state as possible, they tend to reside at this optimum, but relatively close, distance from their neighbor(s) rather than at some greater or lesser distance.

The distance illustrated in Figure 1-3 is the distance between the centers of the nuclei. If the nuclei (with the inner shells of electrons) are thought of as hard balls, the distance between their centers would be the diameter of the atoms. This closest-approach distance, or **hard-ball atom diameter,** has been measured by **X-ray diffraction.** In this technique, high-frequency X rays are beamed through thin specimens of metal; as the rays penetrate the metal, they are diffracted in relation to the distance between the atoms. The angle of the diffracted X rays is used to calculate the distance between the centers of the atoms and thus their closest-approach distance, or hard-ball atom diameter.

Atom Diameter and the Periodic Table. The experimentally determined atom diameters are shown in Table I-1. An inspection of the table reveals generalities about how atom diameters are related by the periodic table:

• The diameters of most common metals are between 2.4 and 3.5 Å[‡].
• Going from the top of the table down any group, with few exceptions, the diameters get *larger*; that is, as electrons are added, the diameters get bigger.

[‡]One angstrom (Å) equals 10^{-8} cm, or 10^{-10} m.

- Looking at the A groups in any row or period, with few exceptions, as electrons are added (i.e., moving from left to right) the atom diameters get *smaller*. That is the more active elements on the left, with fewer valence electrons, have larger hard-ball diameters than the elements farther along in that period.
- *In the transition elements*, the B groups in the center of the table, *the atom diameters do not vary consistently with the number of electrons*. In groups III-B through VII-B the electrons are added to inner electron shells, closer to the nucleus with its strong attraction, and the diameter decreases consistently. In periods 4, 5, and 6, following across groups VIII through II-B, the diameter generally increases as the 18-electron shell is completed and electrons are added to the next outer shell; then in group III-A the diameter begins to decrease again, and although it is sometimes larger than the diameter in II-B, III-A is always smaller than III-B.

Summary.   At this point a brief summary may be helpful.

- The atom has a nucleus with electrons moving about it in orbits that vary in increments, or quanta, of energy.
- The energy levels of the electrons determine the characteristics and properties of the elements.
- The periodic table orders the elements by the number of their electrons or protons, and certain characteristics and properties of elements in the same row or column tend to be similar.
- From the periodic table the metallic atoms are seen to have just a few electrons in their outermost or valence shells; these atoms undergo *metallic bonding*, which is different from the bonding in other types of elements.
- Because of the net attraction forces between metallic atoms, these atoms reside next to each other at a characteristic distance that is their equivalent hard-ball atom diameter, a distance dependent on the number and configuration of their electrons.

These concepts are the background for our study of metals as they are actually observed and used—as crystals.

1.5   CRYSTALLINE NATURE OF METALS

The solid state is distinguished from the gaseous and liquid states in that solids are self-contained, i.e., they have a definite volume and they retain their shape. But solid metals differ from other solids because the metallic bonding of their atoms forms well-defined structures, or crystals. Metals are said to be **crystalline;** they exist in **crystal lattices** or **atom lattices.** Although all metals are subject to metallic bonding, they do not all bond with the same pattern or configuration in their crystals. Table I-1 identifies the crystal lattices of the metallic elements, usually at approximately room temperature.

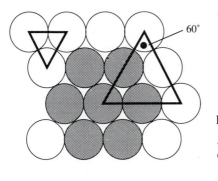

FIGURE 1–4

Atoms on a plane in the densest packing possible

FIGURE 1–5

A second-layer atom of a densest-packed array nestled in the saddle point

There are fourteen basic crystal lattices plus numerous variations, but many of them occur at temperatures other than room temperature and in compounds rather than elements.[§] An inspection of Table I-1 shows that, with few exceptions, the more common metals exist in the form of close-packed hexagonal (CPH) lattices, face-centered cubic (FCC) lattices, or body-centered cubic (BCC) lattices. These are the three lattice structures that will be dealt with here.

Close-Packed Patterns. If a number of hard-ball atoms are arrayed in a plane in the densest possible packing, they will adopt a close-packed pattern, or array, like that shown in Figure 1-4. Note that within that array are numerous triangular and hexagonal patterns.

Atom Stacking. When three hard-ball atoms are clustered in a triangle, they create a low cradle or saddle; any hard ball placed on top of them will tend to settle or nestle into that lowest point. Figure 1-5 shows an upper atom "in the saddle"; the lower layer of three atoms is shaded and the upper atom is an open circle so the atoms below can be seen.

If the three lightly-shaded atoms of Figure 1-5 are now placed on top of the more heavily shaded atoms of Figure 1-4 and aligned with the saddle points, Figure 1-6 is created. There are now three layers of atoms, but note that the atom in the center of the seven darkly shaded atoms in the bottom layer can be seen through the open-circled atom in the top layer. That is, these two atoms are in line vertically. The layers are identified as *A* for the bottom, *B* for the second layer,

---

[§]Some elements, the most notable being iron, change their atom lattice with temperature. This change or transformation is termed *allotropy* and is dealt with in more detail in Chapter 6.

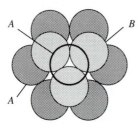

F I G U R E   1–6

Three layers of atoms in a densest-packed array

F I G U R E   1–7   Three layers of close-packed atoms, showing the difference between *ABAB* and *ABCABC* stacking

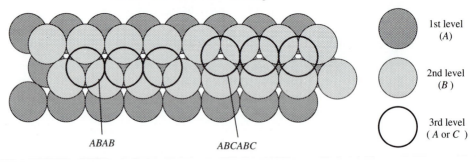

and *A* for the third layer, since it is in line with the first. This stacking of atoms is termed *ABAB*.

Figure 1-7 depicts a larger array of closely packed atoms stacked three high. Note that in the second, or *B*, layer of atoms there are two sets of saddle points where the third layer can nestle. In the left portion of the figure the top layer of atoms is located over the bottom layer; i.e., it is the *ABAB* stacking just described. In the right portion of the figure, the (white) plane *under* the lowest *A* level can be seen through the open circle of the top atom; i.e., the atoms in the first layer are not under the atoms in the third layer. Instead, the third-layer atoms are in a different *C* pattern, so this atom stacking is *ABCABC*.

If Figure 1-7 were expanded to include many layers of atoms with many atoms in each layer, we would have a more realistic view of a metallic crystal. Planes or sheets of atoms are stacked one on top of another. If the atoms in each sheet are close-packed, the pattern shown in Figure 1-4, then the atoms can be visualized as a series of stacked hexagons.

Close-Packed Hexagonal and Face-Centered Cubic Lattices.   Figures 1-8 and 1-9‖ show how the difference between *ABAB* and *ABCABC* stacking results in two different crystal lattices. The small circles used to locate the centers of the hard-

---

‖In Figures 1-8 through 1-14 the atoms are not shown as hard balls because this would hide the relationships of the atoms inside the lattice.

FIGURE 1-8 Atoms in the close-packed hexagonal lattice, showing alternate *A* layers in vertical alignment, or *ABAB* stacking

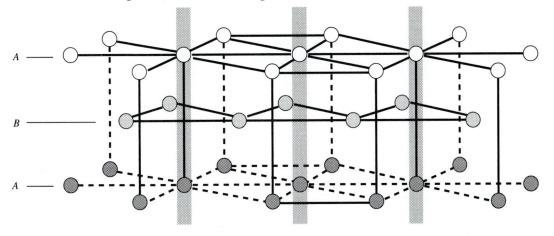

ball atoms are shaded to indicate the levels shown in Figure 1-7. The shape of the lattice is shown in light lines.

Figure 1-8 shows the *ABAB* stacking. The hexagonal configuration in the two *A* planes is evident in the center of the drawing; the resulting **close-packed hexagonal, or CPH, lattice** appears as a six-sided box, with seven atoms on its top and bottom and three suspended between. The large shaded lines show that the atom stacking repeats *every other* layer, or *ABAB* stacking. Although not evident in the figure, the three atoms in the *B* layer are part of similar hexagonal patterns and would match planes of atoms above and below the two *A* planes shown.

FIGURE 1-9 Atoms in the face-centered cubic lattice, showing atoms aligned in every third layer, or *ABCABC* stacking

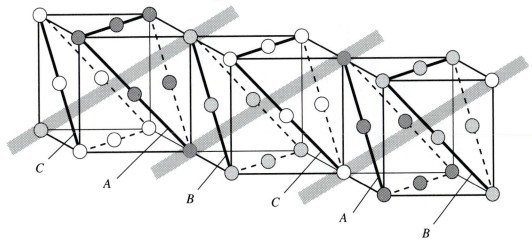

Figure 1-9 shows a series of parallel triangular planes in heavy outline that have atoms in the dense-packed configuration of Figure 1-4. The large shaded lines show that the atoms repeat in their stacking every third layer, or the *ABCABC* stacking. But, as shown by the light lines, the shape of the lattice is now cubic rather than hexagonal. If the terminology used with the hexagonal pattern were used here, this lattice would be described as "close-packed cubic." Although this would be more descriptive of the structure, because of the obvious location of atoms in the center of each face, it is commonly known as **face-centered cubic,** or **FCC, lattice.**

The only difference between the close-packed hexagonal and cubic structures is in the location of the atoms in the third layer. If the atoms repeat every other layer, the lattice is CPH; if they repeat every third layer, the lattice is FCC. It's all a matter of which saddle point is used by the atoms in the third layer. Chapter 2 points out that this small difference in lattices is responsible for the fact that metals with FCC lattices are generally more ductile than those with CPH lattices.

Body-Centered Cubic Lattice.   The **body-centered cubic,** or **BCC, lattice** is the lattice of iron at room temperature; it is commonly encountered. Table I-1 shows that, excluding the reactive metals in group I-A, most of the other BCC metals are clustered under the transition metals vanadium and chromium.

The atoms are located in the lattice as the name implies—one atom in the center of the body surrounded by eight atoms at the corners of a cube. Assuming hard-ball atoms, the corner atoms contact the atom in the center but do not touch each other.

In a single crystal of a BCC metal, which would contain many individual lattices, the atoms lie in parallel planes arranged in squares such that they don't touch each other. Immediately above and below each plane are similar planes of atoms arranged in similar squares, but offset so that each atom nestles in the saddles created between four atoms in the plane above and four atoms in the plane below. Thus every atom is the corner of one cube and, at the same time, a center

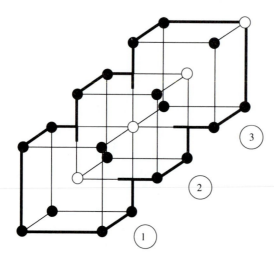

F I G U R E   1–10

Atoms in the body-centered cubic lattice, showing that the center atom ◯ of one cube is the corner atom of another

atom in another. Figure 1-10 shows how the cubes interrelate. The center atom of the first cube is actually the corner atom of the second cube, and in like fashion the center atom of the second cube is the corner atom for the third cube, and so forth

1.6   GEOMETRY OF LATTICES

The previous section dealt with the general arrangement of many atoms in the three most common lattice types. Now we look at the relationships of the atoms within the individual lattice.

The size of the lattice is indicated by the lattice parameters. The **lattice parameters** (*a* and *c* in the following discussion) are the length(s) along an axis or edge of the lattice and are usually measured in angstrom (Å) units.

CPH Lattice. Figure 1-11 depicts a single CPH lattice. The atoms in all three layers are in the close-packed configuration. The three atoms in the intermediate layer are each nestled in the saddle point of the layers above and below. In each horizontal plane the atoms are touching, so the distance between the centers of the atoms is equal to their diameter, and the centers are at 60° angles to each other. In the CPH lattice the lattice parameter *a* is equal to the atom diameter *d*.

Figure 1-11 shows another lattice parameter *c*, the distance between the top and bottom planes. The relationship of an atom in the intermediate plane to the three atoms in the bottom plane is shown in Figure 1-5. Lines drawn between the centers of these four atoms form a three-sided pyramid, with sides equal to the atom diameter or lattice parameter *a*. The vertical height of this pyramid is one-

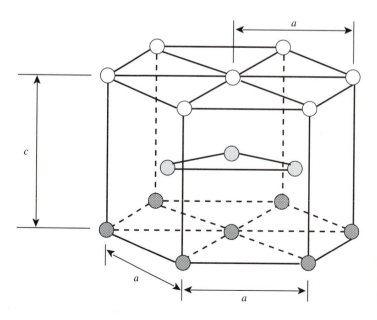

FIGURE 1–11

The CPH lattice, showing the lattice parameters *a* and *c*

half the height of the lattice, or one-half c. Using the Pythagorean theorem and trigonometry, it can be shown that this half height is $c/2 = a\sqrt{(2/3)}$, or $c = 2a\sqrt{(2/3)}$, or $1.633a$, or the ratio of $c/a = 1.633$.

This ratio assumes that the atoms are spherical hard balls. Table I-2 shows experimental data for the lattices of some common metals. For the CPH metals the c/a ratio is included; note that the ratios do not equal the theoretical value. This means the atoms are not behaving as spheres but as footballs. For a lattice oriented as in Figure 1-11, metals with ratios greater than 1.633 have atoms standing on end, and those with ratios less than 1.633 have atoms lying on their side.

FCC Lattice.   The geometry of the FCC lattice can be seen in Figure 1-12. Since the edge of the close-packed plane intersects the face of the cube on its diagonal, the three atoms touch each other, or the diagonal distance from one corner to the other is 2 times the atom diameter. Since the face is a square with sides equal to the lattice parameter a, by the Pythagorean theorem the atom diameter and lattice parameter are related by the expression

$$a^2 + a^2 = (2d)^2$$

which simplifies to

$$2a^2 = 4d^2$$
$$a^2 = 2d^2$$
$$a = d\sqrt{2}, \text{ or } a/d = \sqrt{2}$$

T A B L E I–2 Experimentally determined atom lattice data, by lattice type*

Close-Packed Hexagonal					Face-Centered Cubic			Body-Centered Cubic		
Element	Atom Diam- eter	Lattice Parameters a	c	Ratio c/a	Element	Atom Diam- eter	Lattice Param- eter a	Element	Atom Diam- eter	Lattice Param- eter a
Be	2.225	2.286	3.583	1.568	Al	2.863	4.0496	Cr	2.498	2.8846
Cd	2.979	2.979	5.617	1.886	Cu	2.556	3.6147	Fe	2.482	2.8664
Co	2.497	2.506	4.069	1.624	Au	2.884	4.0788	Mo	2.725	3.1468
Mg	3.197	3.209	5.211	1.624	Ni	2.492	3.5236	W	2.741	3.1650
Ti	2.890	2.951	4.679	1.586	Pt	2.775	3.9239	V	2.622	3.0282
Zn	2.665	2.665	4.947	1.856	Ag	2.889	4.0857	C 1.42		
								Carbon is hex, not BCC		

*Units are angstroms: Å $= 10^{-8}$ cm or 10^{-10} m.

Source: Charles Barrett and T. B. Massalski. *Structure of metals* (3d. rev. ed.). Oxford: Pergamon, 1980, table A-6, 626.

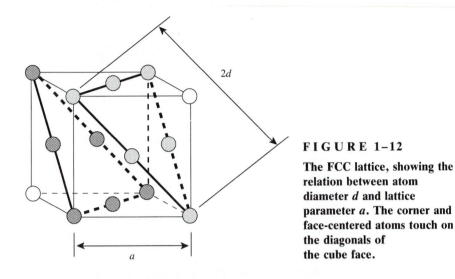

FIGURE 1-12

The FCC lattice, showing the relation between atom diameter d and lattice parameter a. The corner and face-centered atoms touch on the diagonals of the cube face.

Analysis of the data of Table I-2 for the FCC lattice shows that the ratios of the experimental values for lattice parameter and atom diameter, a/d, are almost exactly $\sqrt{2}$.

BCC Lattice. Figure 1-13 illustrates the BCC lattice. Referring to the previous description of this lattice, atom contact occurs between the atoms at the corners and the one atom nestled between them. Thus atoms are not in contact on a diagonal of the cube face, but on a diagonal of the cube—the diagonal that runs between opposite corners of the cube. This diagonal lies on a plane of dimensions a and (by Pythagorean theorem) $\sqrt{(2a^2)}$. Since this diagonal also equals 2 times

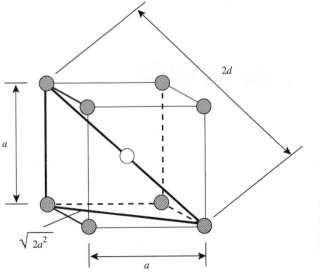

FIGURE 1-13

The BCC lattice, showing the relation between atom diameter d and lattice parameter a. The corner and body-centered atoms touch on the diagonals of the cube.

the atom diameter, the following relationship exists:

$$a^2 + [\sqrt{2a^2}]^2 = (2d)^2$$

This can be simplified to

$$a^2 + 2a^2 = 4d^2$$
$$3a^2 = 4d^2$$
$$a = d\sqrt{(4/3)} = 1.155d, \text{ or } a/d = 1.155$$

Analysis of the data for BCC metals in Table I-2 shows this relationship to exist between the experimental values of lattice parameter and atom diameter.

1.7   PACKING FACTOR

The **packing factor** is a measure of how densely packed the atoms are in an atom lattice. It is defined as the *decimal ratio of the volume of the atoms in the unit cell, or lattice, to the volume of the unit cell itself,* or

$$\text{Packing factor } pf = \frac{\text{Volume of atoms}}{\text{Volume of lattice}} = \frac{\text{No. atoms per unit cell} \times (4/3)\pi(d/2)^3}{\text{Volume of lattice}}$$

In this calculation, the atoms are assumed to be hard balls with a radius of $d/2$. By making use of the relationships between the lattice parameter(s) and the atom diameter developed previously, we can determine a dimensionless ratio for the three common lattice types being considered here. Before proceeding, however, we must determine the number of atoms in a single lattice, i.e., the **unit cell.**

Unit Cell. From Figures 1-11, 1-12, and 1-13 the number of atoms in a single lattice can be determined. CPH has seven in the top layer, seven in the bottom, and three in the intermediate layer, for a total of seventeen; FCC has eight corner atoms and one in the center of each of the six sides for a total of fourteen; BCC has eight corner atoms and one in the center of the cube for a total of nine. But these numbers are misleading; they attribute too many atoms to each lattice.

In the FCC lattice a corner atom is shared with seven other lattices that surround it, or one lattice has only one-eighth of each of the corner atoms. In a similar way, each face atom is shared with an adjacent lattice, or only one-half—three—of each of the six face atoms belong to a lattice. Thus one lattice has only $1 + 3 = 4$ atoms. That is, the *FCC lattice has four atoms per unit cell.*

A similar analysis shows that the *BCC lattice has two atoms in its unit cell.* The combined eight corner atoms contribute the equivalent of only one atom, plus the one atom in the center of the cube, for a total of two atoms per unit cell.

In the CPH lattice it is more difficult to identify the unit cell. Figure 1-14 shows two views of the lattice. The perspective view is similar to Figure 1-11, except that part of another lattice is shown on the left and a six-sided **prism** is outlined in heavy lines. This prism has atoms at each of its eight corners plus one

FIGURE 1-14 Two views of the CPH lattice, showing that the lattice repeats as a 60–120° prism with a unit cell of two atoms

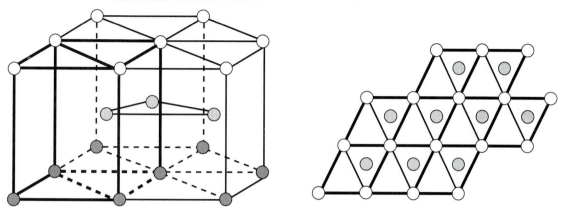

Perspective view   Top view

internal atom. The internal atom is not in the geometric center of the prism (as in the BCC) but nestled in the saddle created by three of the four atoms in the upper layer and three of the four atoms in the lower layer. (Remember, the hard-ball atoms in any layer are actually touching; in these views only the atom centers are shown.)

The top view shows that although this lattice is termed hexagonal, and the hexagons are evident, the *lattice actually repeats as a prism* with internal angles of 60° and 120°. For this prism-shaped unit cell, the number of atoms is determined as in the BCC: Each of eight corner atoms contributes one-eighth to an equivalent of one atom, plus the internal atom; the *CPH lattice has a total of two atoms per unit cell.*

With the ability to determine the number of atoms per unit cell, and using the relationships between a and d developed in Section 1.6, we can now determine the packing factor of the lattices. The previous definition,

$$\text{Packing factor } pf = \frac{\text{Volume of atoms}}{\text{Volume of lattice}} = \frac{\text{No. atoms per unit cell} \times (4/3)\pi(d/2)^3}{\text{Volume of lattice}}$$

can now be applied to the three lattices being studied.

FCC Packing Factor. The FCC lattice has four atoms per unit cell, and its volume is the lattice parameter cubed, or a^3; since $a = d\sqrt{2}$, the FCC packing factor is 0.74:

$$pf = \frac{4 \times (4/3)\pi(d/2)^3}{a^3} = \frac{4 \times (4/3)\pi(d/2)^3}{(d\sqrt{2})^3} = \frac{4 \times 4\pi d^3}{2\sqrt{2}\, d^3 \times 3 \times 8} = \frac{\pi}{3\sqrt{2}} = 0.74$$

CPH Packing Factor. The unit cell for the CPH lattice contains two atoms, but it is a prism with a rhombohedral base with sides equal to the lattice parameter a, which also equals the atom diameter d. In addition, the height of the cell is related to the lattice parameter or atom diameter by $c = 1.633d$. It can be shown that the area of a rhombus of sides d is $d^2(\sqrt{3})/2$. The packing factor for CPH is 0.74, assuming spherical atoms:

$$pf = \frac{2 \times (4/3)\pi(d/2)^3}{1.633d \times d^2(\sqrt{3})/2} = \frac{2 \times 4 \times 2\pi d^3}{1.633d^3 \times 3\sqrt{3} \times 8} = \frac{2\pi}{3\sqrt{3} \times 1.633} = 0.74$$

Note that this is the same as for the FCC lattice. Since the CPH and FCC lattices are generated from stacks of closely packed atoms, it is logical that their packing factors are the same. However, these calculations are based on spherical hard-ball atoms, and as shown in Table I-2, this is not a good assumption for all the metals with the CPH lattice.

BCC Packing Factor. The BCC lattice has only two atoms per unit cell; its volume is a^3, and $a = 1.155d$. The BCC packing factor is 0.68:

$$pf = \frac{2 \times (4/3)\pi(d/2)^3}{a^3} = \frac{2 \times (4/3)\pi(d/2)^3}{(1.155d)^3} = \frac{2 \times 4\pi d^3}{(1.155)^3 d^3 3 \times 8} = \frac{\pi}{3(1.155)^3} = 0.68$$

Because of the limited amount of atoms nesting in the BCC lattice, it is not surprising that its density is the lowest of the three.

1.8   MILLER INDICES

From the previous discussion it is evident that atoms are located on certain planes and diagonals and it is often necessary to describe the locations. For example, most metals freeze on the plane that is the face of their cubic or hexagonal lattice, and some metals are most easily deformed when a force is applied along the

F I G U R E 1–15 Axes used with the Miller indices system

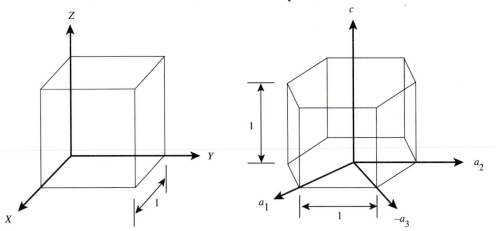

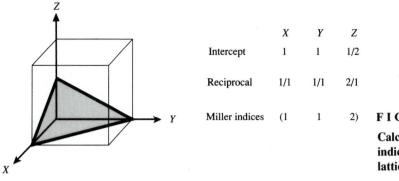

	X	Y	Z
Intercept	1	1	1/2
Reciprocal	1/1	1/1	2/1
Miller indices	(1	1	2)

FIGURE 1–16

Calculation of the Miller indices for a plane in a cubic lattice

diagonal of the cube. To make such references more precise, the Miller index system is used.

The system makes use of notations for the axes of the lattices, as shown in Figure 1-15. The cubic lattice uses three axes, X, Y, and Z, each axis coinciding with an edge of the cube. In the hexagonal lattice, four axes are used, three in the plane that forms the base of the figure—a_1, a_2, a_3—and one for the vertical axis c. The lattice parameters (length of the sides) of the cube and hexagonal prism are assumed to be one unit in length.

Planes in Cubic Lattices. To specify a plane in the lattice, it is necessary to determine where the plane intercepts the axes, take the reciprocal of these intercepts, and multiply out the common denominator, i.e., resolve any fractions. The Miller indices of that plane are the resulting integers, which are enclosed in parentheses.

Figure 1-16 shows a plane at an angle in a cubic lattice. The calculations are shown in a table structured to reinforce the discipline needed for the Miller system; the use of such a table will help avoid errors.

FIGURE 1–17 Calculation of the Miller indices for a plane in a hexagonal lattice

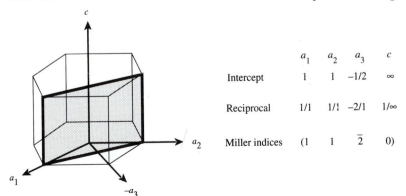

	a_1	a_2	a_3	c
Intercept	1	1	–1/2	∞
Reciprocal	1/1	1/1	–2/1	1/∞
Miller indices	(1	1	$\bar{2}$	0)

The plane of Figure 1-16 intercepts the *X* and *Y* axes at the extremes of the cube, or at values of 1; it intercepts the *Z* axis halfway to the edge of the cube, or at a value of ½. The reciprocals of these are taken; ¹/₁ is of course 1, and 1 over ½ is 2, so the resulting Miller indices is 112. The parentheses ( ) indicate that the number represents a plane. The Miller indices are written without any commas, hence (112).

Planes in Hexagonal Lattices.   Figure 1-17 shows a plane in a hexagonal lattice. The system is used as in the cubic lattice except that there are four axes instead of three. This plane intersects the a_1 and a_2 axes at the edge of the figure, or at values of 1. Note that the third axis is labeled $-a_3$. This is because the three base-plane axes are separated by 120°, which puts the $+a_3$ axis going away from the front of the diagram. Any plane that intersects the a_1 and a_2 axes cannot reach the $+a_3$ axis, so the $-a_3$ axis is used on the front side of the diagram. In this example, the plane intersects the $-a_3$ axis halfway between the origin and the edge of the figure, or at $-½$.

The fourth intercept is a special case, but not an unusual one. The plane does not intersect the *c* axis but is parallel to it; that is, they meet at infinity: intercept = ∞. Taking the reciprocals of 1 and $-½$ is straightforward, but the reciprocal of infinity is zero; the resulting number is (11$\bar{2}$0). (Note that the negative sign is written *above* the number.)

General Rules and Special Planes.   In the Miller system *the reciprocal of infinity is zero, and*—vice versa—*the reciprocal of zero is infinity.* Therefore, if a plane went through the origin of the axes, its intercept would be zero and its reciprocal infinity, which is not a valid Miller index. Thus one of the rules of the Miller system is that *planes cannot go through the origin,* or zero point, of the axes system. This also means that *the presence of a zero in the Miller indices of a plane indicates the plane is parallel to that axis.*

Figures 1-18 and 1-19 show two very common planes that are parallel to axes. In Figure 1-18 the intercepts are 1, 1, and ∞, and the reciprocals are the Miller

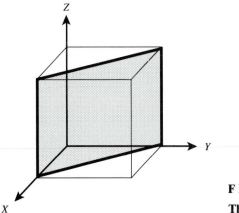

FIGURE 1–18

The (110) plane

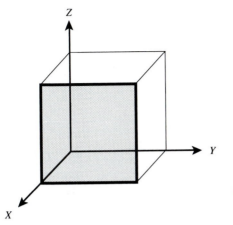

FIGURE 1-19

The (100) plane

indices of (110). Figure 1-19 is similar, except there are two intercepts of ∞, and thus the resulting Miller indices is (100). (It is a common error of students to shortcut the discipline of the Miller system and identify the parallel condition as having an intercept of zero, which leads to an incorrect Miller index of ∞.)

The (100) plane of Figure 1-19 is an important plane, since it is the plane that makes up the **cube face.** The other faces of the cube would all have the same three digits, but they would be in a different order, e.g., (010) and (001). The cube faces that are on the *X, Y,* and *Z* axes cannot be described with the same set of axes, since this would violate the rule that planes do not go through the origin. To describe these planes, the origin is moved to another corner and the directions measured back, down, or to the left in a negative direction.

It is often desirable to talk about all the planes that have a certain orientation in a lattice, e.g., the 100 planes that make up the faces of a cube. To do this, the Miller system uses {} brackets. Thus the six face planes of a cube (Figure 1-19) are described by {100}, and the six diagonal planes of a cubic lattice (Figure 1-18) are the {110} planes.

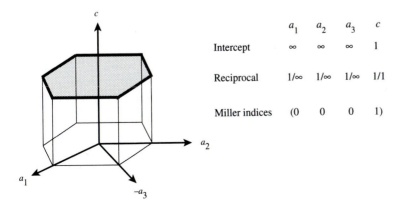

	a_1	a_2	a_3	c
Intercept	∞	∞	∞	1
Reciprocal	1/∞	1/∞	1/∞	1/1
Miller indices	(0	0	0	1)

FIGURE 1-20

The (0001), or basal, plane

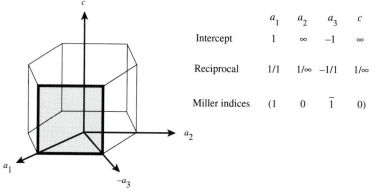

	a_1	a_2	a_3	c
Intercept	1	∞	−1	∞
Reciprocal	1/1	1/∞	−1/1	1/∞
Miller indices	(1	0	$\bar{1}$	0)

FIGURE 1–21

The ($10\bar{1}0$), or prism-face, plane, parallel to two axes

The hexagonal system requires two sets of planes to describe the outer surfaces of the lattice, as illustrated in Figures 1-20 and 1-21. The (0001) plane is parallel to the three axes that lie in the base of the lattice and is referred to as the **basal plane.** It is the close-packed plane discussed earlier.

The ($10\bar{1}0$) plane in Figure 1-21 is an outer face of the prism; note that any prism face can be described with the same four digits, but that one of the 1s must always be negative, and the sequence of the digits changes. Collectively the **prism faces** are the {$1\bar{1}00$} planes.

Two examples of planes that are at angles with the axes are shown in Figures 1-22 and 1-23. The plane in Figure 1-22 intercepts the axes at values of 1; the reciprocals are simply 1, so the plane is (111). In Figure 1-23 the plane intercepts the a_1 and a_3 axes at values of 1 and −1, and the c axis at ½, but it is parallel to the a_2 axis. When we take the reciprocals, we find that the Miller indices are ($10\bar{1}2$).

Directions in Lattices. The Miller system can also be used to identify directions within an atom lattice. Different from the identification of planes, all directions must start at the origin. Figure 1-24 shows such a direction. To determine the

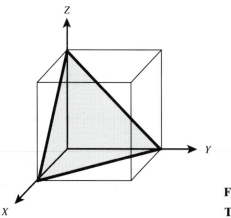

FIGURE 1–22

The (111) plane

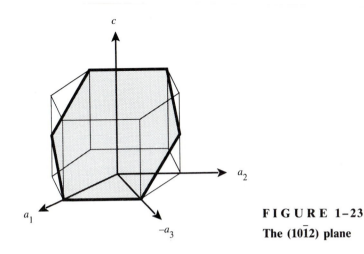

FIGURE 1–23

The (10$\bar{1}$2) plane

Miller indices of this direction, assume that the cube is a box with X, Y, and Z coordinates. In this example, the arrow pierces (intercepts) the box in the middle of one of its sides (a cube face). The Miller indices of the direction are the coordinates of the hole made by the arrow, with fractions resolved.

In Figure 1-24 the intercept is in a face that is one unit along the Y axis from the origin, so the Y coordinate is 1. The hole is in the middle of the face, so it is a half unit out the X axis and a half unit up on the Z axis, so those coordinates are both ½. In XYZ order the intercepts are ½, 1, and ½. In determining directions *reciprocals are not taken,* but fractions are resolved, in this case by multiplying by 2. The resulting number is [121]; to distinguish directions from planes, square brackets [ ] are used. If all similar directions are being considered, pointed brackets $\langle \rangle$ are used.

Figures 1-25 and 1-26 show the same planes as Figures 1-18 and 1-19, but they include directions *of the same number.* The direction shown in Figure 1-25 lies in the plane of the XY axes, so its Z value is zero. It pierces the cube at the corner where $X = 1$ and $Y = 1$, so the direction is [110]. In Figure 1-26 the direction lies on the X axis; it never intercepts the Y or Z axes, so their intercepts are zero. It pierces the cube where $X = 1$, so the direction is [100].

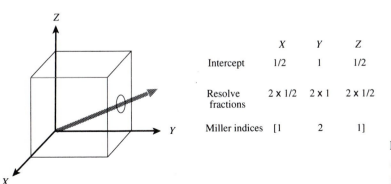

	X	Y	Z
Intercept	1/2	1	1/2
Resolve fractions	2 x 1/2	2 x 1	2 x 1/2
Miller indices	[1	2	1]

FIGURE 1–24

Calculation of the Miller indices for the [121] direction

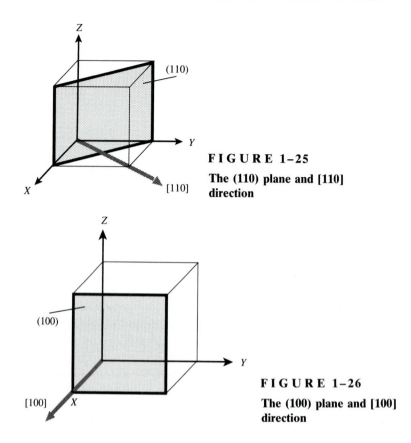

FIGURE 1–25

The (110) plane and [110] direction

FIGURE 1–26

The (100) plane and [100] direction

Note that planes and the directions with the same number are perpendicular. This occurs because the reciprocals of intercepts are used to name the planes, but the directions are read directly. This is a general principle: In *the Miller system planes and directions with the same number are perpendicular.*

In the hexagonal lattice there are three axes in the basal plane, but there are normally only two dimensions in a plane; this makes determining directions difficult. However, if the plane perpendicular to a direction can be determined, its direction is known. This method is suitable for simpler directions and is illustrated in Figure 1-27. To find a direction in the hexagonal lattice, find the plane perpendicular to it and change the brackets.

Using the Miller Indices System.   In the above examples, a plane or direction was shown in a lattice and the Miller indices then determined. In many cases planes and directions will be described by just the number and you will be expected to visualize the plane, the atom location, and the direction. This requires that the process described above be used in reverse order.

First, the brackets tell whether the indices represent a plane or a direction; then the number of digits tells whether the lattice is cubic or hexagonal. If the indices are for a *plane,* the reciprocal of the numbers is taken; if any of those digits

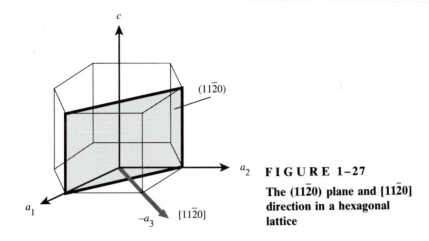

FIGURE 1-27

The $(11\bar{2}0)$ plane and $[11\bar{2}0]$ direction in a hexagonal lattice

is greater than 1, divide by the largest digit so measurements will be no greater than 1, and these are the intercepts. To determine intercepts for a *direction*, divide the indices themselves (not their reciprocals) by the largest index. Given the intercepts, the plane or direction can then be visualized, and sketched on paper if necessary.

Summary. The Miller indices are a system of notations used to identify planes and directions in atomic lattices. The system is founded on 11 basic principles:

- A lattice is assumed to be one unit long from the origin to its edge.
- To define a plane or a direction requires three digits in a cubic lattice, four digits in a hexagonal lattice.
- By definition a plane *cannot* go through the origin of the axes.
- By definition a direction *must* go through the origin of the axes.
- A plane and a direction with the same numbers in their Miller indices will be perpendicular to each other.
- To determine the Miller indices of a plane, take the reciprocal of the intercepts and resolve any fractions in order to find a number made up of integers.
- Planes parallel to axes have intercepts of infinity and an index of zero.
- In the cubic system a direction is defined by the coordinates of the point where the direction arrow leaves the unit cube.
- In the cubic lattices, directions that lie on an axis have an intercept and index of 1 for that axis and intercepts and indices of 0 for the other two axes.
- In the hexagonal lattice, directions can be determined from the number of the plane to which they are perpendicular; a more straightforward method of determining directions was not provided here.
- A system of parenthetical symbols identifies the Miller indices type:
  () denote a specific, individual plane
  {} denote all planes of a similar orientation
  [] denote a specific, individual direction
  ⟨⟩ denote all directions of a similar orientation

1.9   SOLIDIFICATION, OR CRYSTALLIZATION, PROCESS

Liquid and Solid States. In the **liquid state** atoms of a metal are free to move about in a random manner; the atoms are not in any particular order or pattern. At this high temperature—"high" in the sense that it is above the melting temperature—the atoms are high in energy. This energy is of two kinds: **kinetic energy** because the atoms are in vigorous motion and **potential energy** because they are farther from their neighbors than their closest-approach distance. Figure 1-3 shows that atoms that are farther apart than the closest approach will be at a higher energy level than the minimum possible.

In the **solid state** a metal's atoms are in their space lattice, i.e., they are crystalline, and thus the process of freezing is termed **crystallization.** In the solid state the atoms are at a lower energy level since they are at a lower temperature and at their closest-approach distance from neighbors.

Crystallization Mechanism. Crystallization occurs in two stages. First, a nucleus is established on which a crystal (or **grain**) can form, and second, the crystal spreads until it meets another crystal, or the surface containing the liquid metal. The **nucleating sites** are usually associated with a low-energy location in the melt. That is, the nucleating sites are located in areas that offer an opportunity for the higher-energy molten atoms to lose energy. Usually the nucleating sites are where heat is being removed, e.g., a mold wall. Nucleation also can be triggered by oxides or other particles about which molten atoms begin to form their lattice.

Crystallization progresses by a multiplication of the metal's lattice, generally the atoms being added on the less-dense planes. Consider that in the hexagonal lattice the {0001} plane is the dense, close-packed plane; in the FCC lattice it is {111}. Atoms attempting to freeze on these dense planes would have to fit exactly into their "place." By comparison, the prism-face or cube-face planes, {10$\bar{1}$0} and {100}, have more room to fit new atoms and are the preferred freezing planes. Thus the lattice systems have **preferred directions** in which growth usually takes place. In the hexagonal lattice it is ⟨10$\bar{1}$0⟩, or the directions perpendicular to the prism faces. For the cubic lattices, it is ⟨100⟩, or the directions perpendicular to the cube faces.

Removal of Energy and the Formation of Dendrites. Figure 1-28 shows the temperature-time relationship (i.e., cooling curve) as a pure metal loses energy and cools to room temperature. Between *A* and *B* the metal cools as a liquid; between *D* and *E* it cools as a solid. The conversion of the metal from a liquid to a solid, its crystallization, occurs between *B* and *D*.

If heat could be extracted at no temperature difference, *and* if there were no energy given off as the metal changed states, the freezing could take place at one temperature, i.e., path *B–D*, the theoretical freezing temperature of the metal. This is not the case, however; some **undercooling** is required, and cooling follows the dotted path through point *C* in order to become solid at *D*.

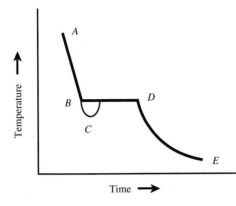

FIGURE 1–28

Cooling curve for a pure metal, showing the undercooling at C required to remove the latent heat of fusion

In addition to the practical temperature difference ΔT required to cause heat to flow, the potential energy mentioned above must also be removed. As the atoms relocate to their closest-approach distance, the latent heat of fusion must be removed. Since the **latent heat of fusion** is generated from the actual freezing process, it is logical that the heat would occur at the **interface,** or junction of the solid and liquid. That is, the latent heat of fusion can be thought of as heating the solid-liquid interface. This is depicted in Figure 1-29.

Notice that the solid-liquid interface is warmer than the surrounding metal; it has been heated by the potential energy (latent heat of fusion) given off as freezing atoms assume their lattice position. That is, "solid" atoms at the interface are slightly higher in temperature than some of the nearby "liquid" atoms, but they are still at a lower energy level, since they are frozen, and are "attractive" to the liquid atoms. Thus narrow arms of frozen metal are formed in this lower-temperature area of the liquid metal. The initial formation is called a **primary arm.** See Figure 1-30.

As the primary arm is created, i.e., freezes, it becomes a new interface with the molten metal (Figure 1-29) and forms new **secondary arms.** Secondary arms are a new interface with the molten metal; they can create tertiary arms by the same process by which they were created. The resulting structure or formation

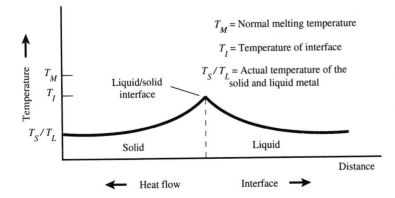

T_M = Normal melting temperature

T_I = Temperature of interface

T_S / T_L = Actual temperature of the solid and liquid metal

FIGURE 1–29

Temperature at the solid-liquid interface

Source: Adapted from T. H. Sanders, Jr. Glossary of terms. Course notes for MSE 536 Solidification, Purdue University.

F I G U R E 1–30 **Formation of dendrites at the solid-liquid interface**

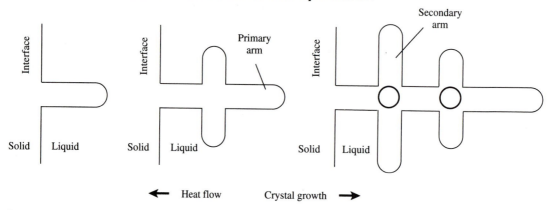

Source: Adapted from T. H. Sanders, Jr. Glossary of terms. Course notes for MSE 536 Solidification, Purdue University.

has the appearance of a tree and is therefore called a **dendrite.**[#] Figure 1-30 illustrates the formation of dendrites.

The dendrite arms are part of the lattice of the metal being frozen, and will be at a 90 or 120° angle to the primary arm, depending on whether the lattice is cubic or hexagonal. As cooling continues, the remaining molten metal freezes between the dendrite arms. In Figure 1-30 the heat flow is shown going to the left; therefore the dendrites and the interface move to the right. (See the photomicrographs in Figures A-1, A-2, A-5, A-25, and A-26 of Appendix A, which show dendritic structures or evidence that the metals have frozen by forming dendrites. The introduction to Appendix A describes how photomicrographs are made.)

In order to form, dendrites require a temperature gradient (T_I—T_S of Figure 1-29) to generate this flow of heat, or **directional cooling.** The lower the temperature gradient, the slower the cooling rate (temperature change per time) and the farther apart the dendrite arms are formed. As will be discussed (Section 5.8), in commercial practice it is normally desirable to have the dendrite arm spacing as *small* as possible, since dendrites cause **segregation** of alloying elements. This problem is corrected by diffusion (see the next section), which involves a heating process called **homogenization.** Homogenization can be accomplished more easily if the arms are closer together. Slow cooling, on the other hand, will create large dendrites with wide arm spacings that will require expensive higher temperatures and longer heating cycles in the homogenization process. (See Figures A-2 through A-6.)

Columnar and Equiaxed Grains. Directional cooling tends to create crystals, often called *grains*, that grow in the direction opposite to the heat flow (see Figure 1-30). If the temperature gradient is great enough, these crystals can become quite

---

[#]The word is derived from *dendron,* the Greek word for tree.

FIGURE 1-31 Schematic of a section through frozen ingots. Ingot *A* has columnar crystals that meet at 45° angles. Ingot *B* has columnar crystals near the mold wall, but the center has frozen with equiaxed crystals.

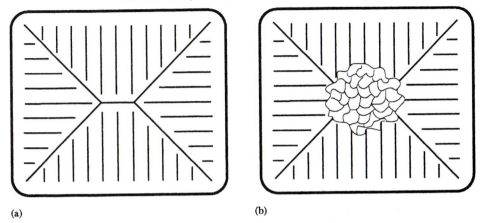

(a)   (b)

long and thus appear as columns, or **columnar grains.** These are shown schematically in ingot *A* of Figure 1-31. As shown there, these grains can create a plane of weakness where they meet at the 45° angle.

If there is not a decided temperature gradient, then nucleation will begin in a more random fashion. Heat may be extracted to the air through the surface of the molten metal, or oxides or other impurities may act as nuclei to begin the energy-removal process. Rather than forming crystals with a definite directional orientation, the grains will be **equiaxed,** taking a shape of approximately equal dimensions on the three axes.

Ingot *B* of Figure 1-31 illustrates the generation of equiaxed grains. It started to form columnar grains but the cooling rate was not rapid enough to continue the columnar growth all the way to the center. Before the columnar grains could reach the center, nucleation sites there initiated the freezing of equiaxed grains.

Techniques have been perfected that enable the casting of turbine blades with columnar grains that are as long as the blade, and further refinement of the process now enables casting the blades as one grain. This is done by passing the crystals that first solidify through a maze that permits only one crystal orientation to emerge, and further solidification aligns along this orientation. The alloys used for this application are usually nickel-based FCC metals, so the solidification normally proceeds on the {100} faces in the ⟨100⟩ direction.

Shrinkage Cracks. If heat were removed from a solidifying ingot in one direction, e.g., through the walls of the mold, then there would always be a reservoir of molten metal to feed any cavities caused by the contraction of the freezing metal. However, in the commercial production of pigs and ingots this is often not the case. Instead, heat is removed in a number of directions, which causes the molten metal reservoir to freeze too soon; this starves the solidifying metal and a shrinkage crack, or **pipe,** forms. If this crack is open to the surface, it creates a severe safety hazard when liquids somehow enter, say by storage out-of-doors, and remain there. If the ingot is later charged in a furnace and becomes submerged below the level of the molten metal, the moisture converts to steam so rapidly that the ingot essentially explodes. To avoid this problem, ingots must be protected from the elements or dried in a furnace before charging in a melting furnace.

Crystal or Grain Size and Grain Boundaries.   The size of the crystal or grain that results from the crystallization process is determined when it freezes into the next grain. That is, the freezing process itself determines the grain size.** If many crystals are freezing, the resulting grain size will be small; if just a few are freezing, the grain size will be large. Thus eventually the size of the frozen grain is determined by the number of nucleating sites available to the molten liquid. *The more nucleating sites there are, the more grains will be generated and therefore the smaller the resulting grains will be.*

The nucleating sites can be located according to temperature gradients, as described above, or they can be located according to the presence of insoluble particles about which the metal will begin to freeze. In many commercial operations, such "grain refiners" are purposely added to the molten metal. In casting aluminum ingot, boron or titanium-boron is added to achieve a smaller grain size.

Grain boundaries are areas where the atom lattice of one grain does not match with the lattice of its neighbor(s); these boundaries are areas of discontinuity, or areas where the atoms are not at their closest-approach distance and thus are at higher energy levels. The grain boundary can be considered an irregular noncrystalline structure that is stronger and less plastic than the metal grain.

Just because the grain boundary is created by the last metal to freeze does not necessarily mean that it is the first metal to become liquid when heated. Tests have been reported[††] where a metal was slowly heated to within less than 1°C of the melting temperature, while being observed under an electron microscope, without melting the grain boundaries.

1.10   CRYSTAL IMPERFECTIONS

Even in relatively small metal specimens, the crystallization or freezing process requires the alignment of millions of atoms into a precise location in the crystal lattice. For example, assume a BCC metal with a lattice parameter of 3.33 Å. From Table I-2 it can be seen that this is a representative size. If this metal freezes

---

**ASTM Standard E112 covers the measurement of grain size. The ASTM grain size number, which goes from 1 to 10, is the *number* of grains in a standard area. A small number labels a large grain size; a large number labels a small grain size.

[††]Reported by Robert W. Balluffi in the 10th Annual Peter G. Winchell Memorial Lecture, "Grain Boundary Phase Transitions," given at Purdue University, March 26, 1991.

into a square bar that is 1.0 cm on a side, the number of atoms on each 1.0-cm side would be

$$\frac{1 \text{ cm}}{3.33 \text{ Å/atom}} \times \frac{1}{10^{-8} \text{ cm/Å}} = 30 \times 10^6 \text{ atoms}$$

and the total number that would have to crystallize *in each layer*, 1.0 cm by 1.0 cm, would be this number squared, or 900 million million!

The crystals created by freezing such huge numbers of atoms are not all perfect. Some crystal lattices are created with too many atoms, some without enough; others become contaminated with other particles of a different size. In some cases the imperfections become aligned and create a defect many thousands of atom diameters long, and in other cases the lattices within a crystal become misaligned.

The most common imperfections are *vacancies, interstitials,* and *dislocations,* all of which strain the atom lattice, leaving nearby atoms at higher than minimum energy levels. Vacancies and dislocations are of interest because of their usefulness in explaining the relationship between deformation and strength (work-hardening) in metals (discussed further in Chapter 2); interstitials are useful in understanding the alloying of steel (discussed in Chapters 4 and 6). The presence of vacancies and interstitials are also useful in explaining how atoms of one metal can move or diffuse through the lattice of another to reduce segregation of alloying elements. (Diffusion is discussed further in Section 5.8, and the concept is used frequently in the text.)

Vacancies. As shown in Figure 1-32, a vacancy is merely an *unoccupied atom site,* but because atom location is a matter of electrical balance (see Figure 1-3), the absence of an atom has an effect that goes beyond just the adjacent atoms. Because of the vacancy, each nearby atom must come to a new separating distance from its neighbors that minimizes its energy, but that energy level will still be higher than it would be at the closest-approach distance. Figure 1-32 attempts to illustrate the effect by showing the lattice gradually caving toward the vacancy; the disturbance would probably be spread over more atoms than shown.

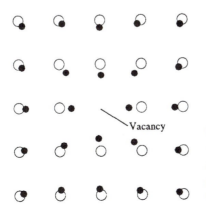

FIGURE 1-32

The vacancy in the center causes atoms to relocate to positions represented by the black dots

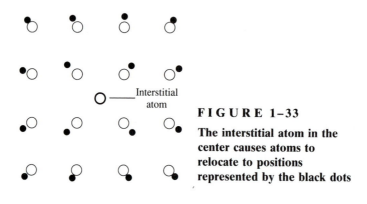

Interstitial atom

FIGURE 1–33

The interstitial atom in the center causes atoms to relocate to positions represented by the black dots

Interstitials. The term *interstitial* means *between the spaces.* Figure 1-33 shows such an interstitial atom between the atoms of a lattice. This atom isn't a part of the lattice itself, but fits between the spaces of the lattice. The most common interstitials occur when metals that have vastly different atom diameters are alloyed. Table I-2 lists the diameter of the carbon atom because it is often an interstitial alloying element with BCC metals. The interstitial relationship between carbon and iron, the basis of steel alloys, occurs because of this large difference in diameter.

The disruptive effect of an interstitial atom is depicted in Figure 1-33. The interstitial atom or particle tends to spread the atom lattice, which disturbs the electrical balance of the atoms. The atoms will seek compromise positions that minimize their energy, but their energy will still be higher than if they were at their closest-approach distance.

Dislocations. A dislocation is a *region of misaligned atoms existing between otherwise properly aligned atoms.* In the lattice of Figure 1-34 the atoms at the top are in their proper lattice, as are those at the bottom, but the two are not in alignment. The plane of shaded atoms does not permit the two portions to align. Instead, an irregular tapered region has been created where the atoms are dislocated; they are not in their proper locations.

From the view of the atom lattice at the bottom, the shaded plane of atoms is an ''extra'' plane of atoms; from the view of the lattice at the top, the bottom portion ''lacks'' a plane. Thus it can't be known if this dislocation is due to too many or too few planes of atoms. It is merely a place where the atoms should line up but don't. Because of its sharp corner or edge, it is termed an **edge dislocation;** it might extend many atom diameters across the crystal.

Dislocations need not be in uniform increments of the lattice parameter. That is, the mismatch can taper, causing the dislocated area to progress through the

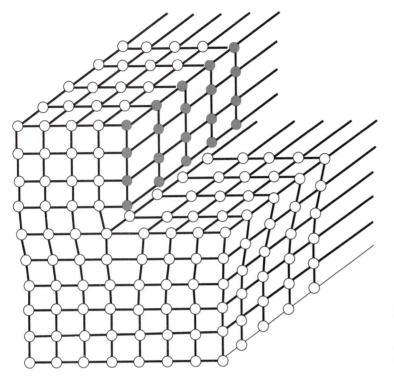

FIGURE 1–34

An edge dislocation: The row of shaded atoms creates an edge through the crystal

lattices of atoms in a spiral or screw-thread fashion. Such dislocations are known as **screw dislocations** and can have either right- or left-hand rotation.

SUMMARY

1. The characteristics and properties of the elements are determined by the energy of their electrons.

2. The energy of electrons does not vary continuously but in steps; the energy is quantized.

3. Metals have few *valence* electrons and bond together in the metallic bond— positively charged centers attracted to a cloud of negatively charged electrons.

4. The net resultant of the attractive and repulsive forces between two atoms is a reduced energy level of both atoms that occurs at the closest-approach distance, which is termed the *hard-ball atom diameter*. The most commonly used metals have atom diameters of 2.4 to 3.5 Å.

5. The periodic table shows the amazing relationships that exist between the

elements; it orders the elements by atom diameter, lattice type, and other characteristics and properties.

6. As metal atoms take up their closest-approach distance, they do so in a crystal lattice.

7. Close-packed atoms generate a hexagonal lattice (CPH) if stacked *ABAB,* or a cubic lattice (FCC) if stacked *ABCABC.*

8. In *ABAB* stacking, close-packed planes are aligned every other layer. In *ABCABC* stacking, close-packed planes don't repeat until the fourth layer.

9. The BCC lattice is not a close-packed lattice. Its packing factor, or atom density, is less than that of the close-packed lattices.

10. Lattices repeat in metal crystals by their geometric unit cell: CPH is a two-atom prism, FCC is a four-atom cube, and BCC is a two-atom cube.

11. The size of a lattice is dependent on the size of its atoms. The relationships between the lattice parameters and the atom diameters are constants for the BCC and FCC lattices and approach the theoretical values for CPH.

12. The Miller indices system uses integers to describe planes and directions in lattices. A complete summary is included at the end of Section 1.8, but some system principles are repeated here:
    a. Planes cannot go through the origin of the axes.
    b. Directions must go through the origin of the axes.
    c. Planes and directions with the same number are perpendicular to each other.
    d. Special brackets are used to identify planes and directions.
    e. To identify a direction in the hexagonal lattice, find the indices of the perpendicular plane (rule c above) and use the direction brackets.

13. Solidification is the process of changing atoms from the random orientation of molten metal to a (solid) crystal lattice and thus is also termed *crystallization.*

14. Solidification begins at nucleation sites where energy is removed from the molten metal. Both kinetic and potential energy must be removed to enable the metal to freeze, i.e., crystallize.

15. The removal of the latent heat of fusion during the final solidification process tends to warm the molten-solid interface, causing freezing to take place in the form of dendrites.

16. The grains solidify in the direction opposite to the direction of the heat removal and may become "columnar" if the temperature gradient is large in a specific direction.

17. Temperature gradients cause nucleation, but the number of nuclei created depends on the rate of cooling; fast cooling or freezing rates generate more small grains, and slow freezing rates generate fewer and larger grains.

18. Nucleation can be encouraged by the presence of insoluble particles; sometimes grain refiners of this type are purposely used.

19. Lack of significant directional cooling permits nucleation such that the resulting grains may lack any well-defined freezing direction; these grains are then said to be "equiaxed."

20. Grain boundaries are created by the freezing together of adjacent grains or crystals; these boundaries are irregular, noncrystalline regions.

21. The hexagonal and cubic crystals freeze in the direction perpendicular to their lattice faces, that is, $\langle 10\bar{1}0 \rangle$ and $\langle 100 \rangle$.

22. The freezing of even small crystals involves aligning many atoms into their lattice. Crystals are not perfectly formed and contain vacancies, interstitials, and dislocations. These crystal imperfections are useful in explaining work-hardening, alloying, and diffusion.

REVIEW QUESTIONS

1. Describe the overall format or organization of the periodic table as it relates to the numbers of electrons, electron shells, atom diameters, and crystal lattices. Where are the transition metals? Why are they different?

2. In determining the energy of an electron, what is the significance of the electron's behaving as a wave?

3. How does the metallic bond differ from the ionic and covalent bonds? Describe how the closest-approach distance relates to the metallic bond.

4. Explain the relationship between the hard-ball atom diameter and the lattice parameter(s) for the CPH, FCC, and BCC lattices.

5. What is atom stacking? What differences are created in atom lattices?

6. What is meant by the packing factor?

7. What is the packing factor of a lattice that has atoms at the eight corners of a (90°) cube? Each atom touches three other atoms in the cube.

8. What is the packing factor for a lattice that is a body-centered rectangle whose three dimensions are a, a, and $1.2a$? The corner atoms do not touch each other, but the atoms do touch on the diagonal of the cube.

9. What are the Miller indices of the following?

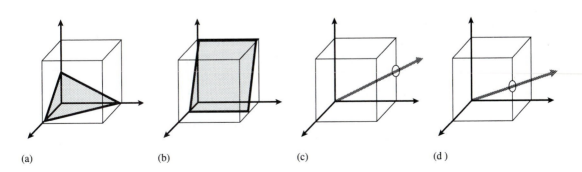

(a)   (b)   (c)   (d)

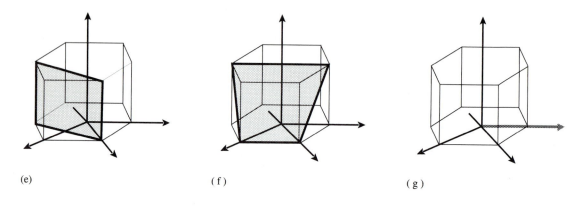

(e)   (f)   (g)

10. For figures (b) and (f), what are the Miller indices of the directions perpendicular to these planes?

11. Sketch the following:
    a. (111)   d. (112̄2)
    b. [0001]   e. (231)
    c. [11̄2]   f. [231]

12. What is allotropy?

13. Describe the crystallization process of a pure metal. Use its cooling curve in your explanation.

14. Explain how the latent heat of fusion causes dendrites to form. Describe a vacancy, an interstitial, and a dislocation. What is the significance of each?

15. How are grain boundaries created? How are they similar to dislocations?

16. How are diffusion and segregation related?

FOR FURTHER STUDY

Avner, Sidney H. *Introduction to physical metallurgy*. New York: McGraw-Hill, 1974, chap. 2.

Barrett, Charles, and T. B. Massalski. *Structure of metals* (3d rev. ed.). Oxford: Pergamon, 1980. Chapter 1 discusses the various atom lattices.

Beiser, Arthur. *Modern technical physics* (5th ed.). Reading, MA: Addison-Wesley, 1987. Chapters 28 and 29 are helpful in understanding atom structure and atomic bonding.

DeGarmo, E. Paul, J. Temple Black, and Ronald A. Kohser. *Materials and processes in manufacturing* (7th ed.). New York: Macmillan, 1988, 63–71.

Flinn, Richard A., and Paul K. Trojan. *Engineering materials and their applications* (3d ed.). Boston: Houghton Mifflin, 1986, secs. 1.4 and 2.1–2.8.

Metals handbook, desk edition. Metals Park, OH: American Society for Metals, 1985. Pages 1-44 to 1-48 contain data on the physical properties of the elements.

Reed-Hill, Robert E., and Reza Abbaschian. *Physical metallurgy principles* (3d ed.). Boston: PWS-Kent, 1992. Chapters 1 (The structure of metals) and 14 (Solidification) are especially helpful.

C H A P T E R 2

Cold-Working
of Metals

After studying this chapter (in conjunction with Appendixes B, C, and D) and doing some relevant lab work, you should be able to

- describe and differentiate between the modes of deformation, slip, twinning, and cleavage and determine the plane on which slip is most likely to occur in the common lattice types

- differentiate between ductile deformation and brittle failure or cleavage and determine the angle of the plane, relative to the applied tensile load, on which deformation and failure will occur

- explain the nature and characteristics of cold work

- define critical resolved shear stress (CRS) and explain how it affects the workability of a metal and how its value is changed by alloying

- describe the effect of cold work on mechanical properties

- explain how polycrystalline deformation is different from single-crystal deformation and why, for ductile metals, the engineering stress-strain curve goes to a maximum and then falls off as rupture approaches

- use dislocations to explain the mechanisms of slip and fracture

- describe how having different-size atoms in a lattice can strengthen a metal

DID YOU KNOW?

Many of the uses of metals depend on their ability to be formed and fabricated through many different metalworking methods. Metals have a unique ability to be formed and strengthened at the same time; i.e., they work-harden when cold-worked.

When metals are cold-worked, much of the energy expended shows up as heat; e.g., high-speed drawing or rolling can make metals too hot to touch. However, not all the energy goes off as heat; some is trapped in the atom lattice. This trapped energy is capable of initiating recrystallization during reheating of the metal (as will be discussed in Chapter 3).

A PREVIEW

To understand the mechanism of cold-working, we'll first investigate it by looking at *single* crystals of metals. We'll look at what happens on atomic slip planes during loading to cause elastic and plastic deformation and then consider what happens within the atom lattices and how the three major lattices react differently. Lastly, we'll discuss the cold-working of polycrystalline (many-crystal) metals.

The information in this chapter and in Appendix D, Calculation of Cold Work, is relevant to the field of metalworking—for example, bending, extruding, rolling, and drawing.

2.1   DEFORMATION

Forces Applied to Atom Planes.   When a tensile or compressive force is applied to a single metal crystal—i.e., a number of atom lattices with common alignment—the atoms still attempt to retain their proper spacing. If the forces are perpendicular to a densely packed plane, e.g., $\langle 0001 \rangle$, so that the atoms are only pulled apart or pushed together, any movement that occurs will be of a rather short distance; the forces that attempt to keep the atoms at their closest-approach distance (Figure 1-3) will be in opposition. If the force is large enough, the lattice will be literally ripped apart or cleaved.

However, if the force is applied at a certain angle, e.g., $\langle 11\bar{2}0 \rangle$, the atoms can slide over one another. Figure 2-1 shows two adjacent rows of atoms, depicted as "hard balls," in a single crystal of a metal with a force applied. This force f_s is a shearing force that tries to make these two rows slide past each other. The opposition to the force is that the atoms in the upper row are nestled in saddles and must slide up and over their neighbors below. As f_s is increased and the atoms are displaced, their internal energies go up and it becomes difficult to move them farther.

In Figure 2-2 the force has increased to f_{se}, a force great enough that the atoms are sitting at a crest. If f_{se} is reduced, the atoms will return to their original position and the lattice will return to normal. But if this force is increased to some higher value, say f_{sp}, then the atoms will slide, or slip, over the crest and take up new locations.

f_{se} is the limit of elastic loading; if any force equal to or less than f_{se} is applied and then removed, the atoms will return to their original positions. Any movement of the atoms in this range of loading constitutes *elastic deformation*.

Any force greater than f_{se}, say f_{sp}, will permanently displace the atoms and result in what is termed *plastic deformation*, or *slip*. When f_{sp} is removed, the

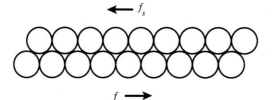

F I G U R E 2–1

Adjacent rows of atoms with a shearing force applied

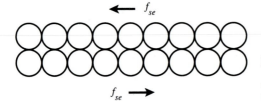

F I G U R E 2–2

Adjacent rows of atoms at the limit of elastic loading

F I G U R E 2–3 Slip planes in a single crystal

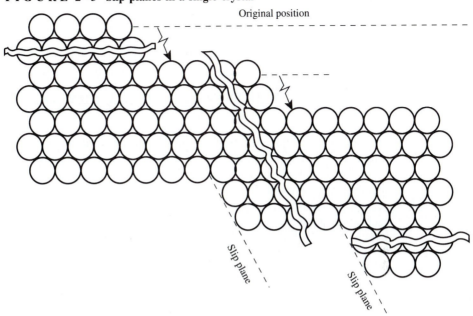

Source: Clyde W. Mason. *Introductory physical metallurgy*. Materials Park, OH: ASM International, 1947, fig. 15, 29. Used by permission.

atoms will have taken up new locations with new neighbors. That is, the lattice does not return to its original condition; it has been plastically deformed, or cold-worked, and is at a higher energy level than before.

Elastic Deformation. In **elastic deformation** the atoms of a lattice can be pictured as slightly displaced from their original lattice location, but still retaining the original lattice design. Each atom has the same neighbor(s) and its same relative position in a particular lattice; e. g., a corner atom is still a corner atom, though a cubic lattice will become a parallelepiped* under elastic loading. When the load is removed, the atoms return to their original locations; i.e., the metal springs back.

Plastic Deformation, or Slip. With loads greater than f_{se} the atoms are strained too far from their normal positions to recover on release of the stress, that is, **plastic deformation** has occurred. That is, from their position in Figure 2-2 they slide, or slip, over into the next depression. However, once a plane of atoms begins slipping, it moves much more than just into the next depression. Figure 2-3 is a representation of how slip would appear in a single crystal of a metal. Note

*A solid with each of its six sides a parallelogram.

that *during slip atoms do change neighbors and usually move many atom diameters from their original location*. Also note that the distance from one plane of slip to the next is many atom diameters. That is, crystals do not slip on immediately adjacent planes.

2.2   PLASTIC DEFORMATION OF A SINGLE CRYSTAL

Figure 2-4 is a representation of the tensile deformation that takes place in a single-crystal specimen with one of its slip planes oriented as shown. In Figure 2-4(a) the tensile force is applied but slip has not started. In Figure 2-4(b) the specimen has slipped, but the forces have moved sideways so the deformation occurs like the sliding of a deck of cards. In this case the atoms have slipped past each other on the slip planes, and although they have new neighbors, the lattice arrangement and distances between atoms remain largely undisturbed.

Figure 2-4(c) is a more realistic view of what actually happens during tensile loading. Since the forces do not move sideways but stay aligned, the individual "layers" rotate and/or bend. Eventually enough of the planes will not be in an orientation favorable for slip and the piece will fracture.

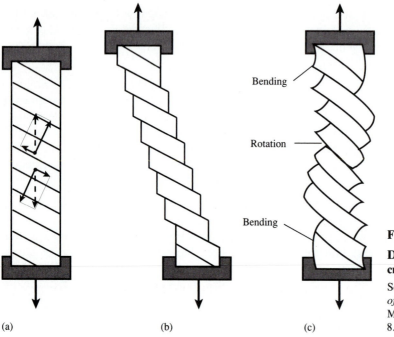

Bending

Rotation

Bending

(a)   (b)   (c)

FIGURE 2–4

Deformation of a single crystal by tensile loading

Source: B. D. Cullity. *Elements of X-ray diffraction*. Reading, MA: Addison-Wesley, 1956, fig. 8.25. Used by permission.

FIGURE 2–5

Slip planes on the surface of a polished brass specimen deformed in tension

Source: R. M. Brick and A. Phillips. *Structure and properties of alloys* (2d ed.). New York: McGraw-Hill, 1949, 85. Used by permission.

The result of the slipping process is a series of stepwise movements within the crystal. Figure 2-5 shows a single crystal that had a line scribed on its polished face and was then deformed in tension. The set of diagonal lines shows the edges of atomic planes that have slipped. Note that the vertical line is bent, showing that the upper part of the specimen is to the right of the lower part.

2.3   MECHANISM OF SLIP

In Figures 2-2 and 2-3 slip was pictured as the sliding of many atoms in one layer over many atoms in another layer. All the atoms moved in unison. This description is helpful, but it cannot depict what actually occurs in the metal crystal. The sum of the interatomic forces that would have to be overcome to slide one complete layer of atoms over another would require a strength of metals many times greater than is known to exist.

The fact that metals slip at stresses far below these levels is explained by *dislocations*. Because of dislocations, rows of absent or *dis*located atoms in the lattice, the *atoms don't have to move all at once; they can move one dislocation at a time*. Figure 2-6(a) is a portion of the edge dislocation shown in Figure 1-34; the views in Figure 2-6(b) through (e) track the movement of the dislocation through the lattice when force f_{sp} is applied.

As the force f_{sp} shears the atoms, the dislocation moves to the right so that by the final view in Figure 2-6 it has moved to the extreme right and the metal has been displaced one lattice parameter. Slip should not be thought of as happening instantaneously, but rather as a *progressive movement of dislocations across the lattices*.

As the dislocations are used, the force necessary to continue deformation increases. That is, the metal gets stronger; it work-hardens or is cold worked. *Failure or cleavage occurs when the dislocations are all used.*

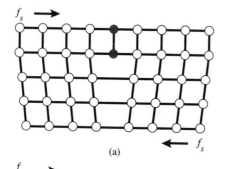

(a)

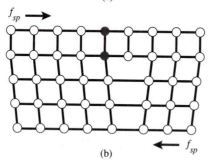

(b)

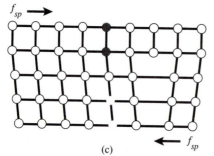

(c)

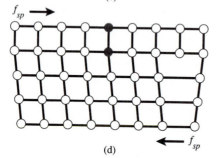

(d)

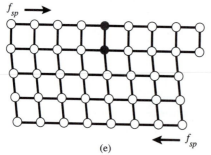

(e)

FIGURE 2–6

The moving of a dislocation through a crystal lattice

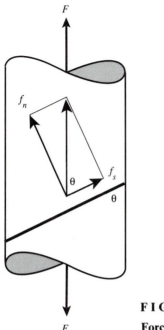

FIGURE 2–7

Forces on a slip plane

2.4   FORCES IN LOADING

Our objective now is to study the **slip planes** in the three major lattices, but that requires an understanding of the forces that act on a lattice. Figure 2-7 shows tensile loading on a specimen. The force F is applied axially to an area A that is normal (perpendicular) to the axis.

Assume a plane through the specimen at any angle θ. The area of this plane, A_θ, is

$$A_\theta = \frac{A}{\sin \theta}$$

The force F can be vectored into two components, a shearing force f_s lying along the plane, and another, f_n, perpendicular to the plane.

From Figure 2-7, $\cos \theta = f_s/F$, and solving for stress,[†]

$$\sigma_s = \frac{f_s}{A_\theta} = \frac{F \cos \theta}{A/\sin \theta} = \frac{F}{A} \sin \theta \cos \theta = \frac{1}{2}\frac{F}{A} \sin 2\theta$$

This expression is maximized when the angle of 2θ is 90°, or when θ is 45°. That is, deformation by shear is maximized on planes that are 45° to the applied force.

[†]For a definition and discussion of the terms *stress* and *strain* see Appendix B.

Although this analysis is done here for a tensile force, the result is the same for a compressive force.

Also from Figure 2-7, $\sin \theta = f_n/F$ and $f_n = F \sin \theta$. Solving for the normal stress, we have

$$\sigma_n = \frac{f_n}{A_\theta} = \frac{F \sin \theta}{A/\sin \theta} = \frac{F}{A} \sin^2 \theta$$

This expression is maximized when the angle θ is maximized, that is, when θ is 90°. Thus the forces that tend to separate the specimen—pull it apart rather than cause it to deform by shear—occur on planes that are at 90° to the applied force.

These calculations confirm what is generally observed in tension testing: Ductile metals fail by shear, and their failure area is distorted, having planes at angles of 45°; brittle metals fail by normal forces, have little distortion, and their failure area is at 90° to the force.

2.5   PREFERRED SLIP PLANES

FCC Lattice. Figure 2-8 shows the slip process in a single FCC lattice. The atoms are not shown but can be visualized at the corners of the cube, plus one in the center of each face; atoms a and c are indicated at the lower corners of the cube. The force F is applied perpendicular to the edge of the lattice. From the discussion of Section 2.4, the maximum shearing force is the vector of this force, f_s at 45° to force F, which goes along the edge of the (111) plane. This causes slip in the [$\bar{1}$10] direction in multiples of half the lattice diagonal.

This is the ideal **slip system** (a combination of slip plane and direction) for the FCC lattice. There are eight {111} planes in the FCC lattice, one related to each of the corners; they are thus known as the *octahedral planes*. In addition, these planes are the dense, close-packed planes (see Figure 1–4). This means that the

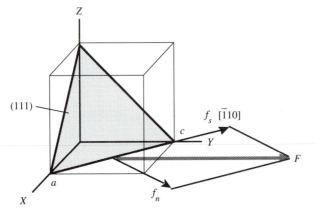

FIGURE 2–8

Slip in an FCC lattice, showing the shear and normal components of an applied force

distance between parallel {111} planes will be greater than the distance between any other set of parallel planes. That is, the atoms will have the least interference as they try to "slip" over each other.

Figure 2-9 is a plan view of this process, showing the movement of corner atoms a and c of Figure 2-8 to their new locations a' and c'; the atoms move up and to the right in the $[\bar{1}10]$ direction along the line where the (111) plane intersects the XY plane. Of course, as discussed previously, once slip begins, the atoms actually move many more diameters than shown here. (The atoms in the center of the cube faces are not shown in Figure 2-9.)

BCC Lattice.   The {110} plane is the most densely packed plane of the BCC lattice, with five atoms on a rectangular plane (see Figure 1-13), but they are not in the close-packed configuration. The preferred direction with the BCC lattice is the cube diagonal, ⟨111⟩.

Figure 2-10 illustrates why the FCC and BCC lattices do not have the same preferred slip planes. In Figure 2-10(a), just a few {1$\bar{1}$1} planes (in heavy lines) in an FCC lattice can contain all but the two atoms in the extreme corners, and they would be picked up by parallel planes in the adjoining lattices. Figure 2-10(b) shows, however, that if {1$\bar{1}$0} planes were to be used for the FCC lattice, additional planes (light lines) would have to be used in order to include all the atoms. This would put the planes so close together that the atoms would protrude into the adjacent planes and slip would be prevented; thus {111} planes in the ⟨110⟩ direction form the preferred slip system for the FCC lattice.

Figure 2-10(c) shows that in the BCC lattice a series of just a few {1$\bar{1}$0} planes would contain all the atoms except those at the extreme corners, which would be picked up by parallel planes in the adjoining lattices. Figure 2-10(d) shows that if, instead, {1$\bar{1}$1} planes were used, many more planes would have to be used so they

FIGURE 2–9   Atoms slipping in an FCC lattice

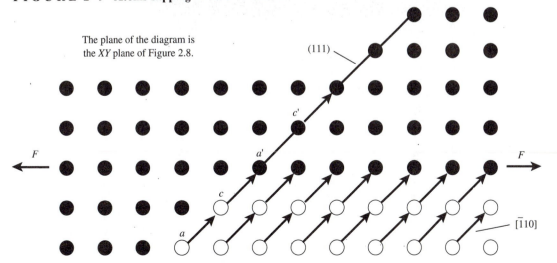

The plane of the diagram is the XY plane of Figure 2.8.

F I G U R E 2–10 **Preferred slip planes in cubic lattices**

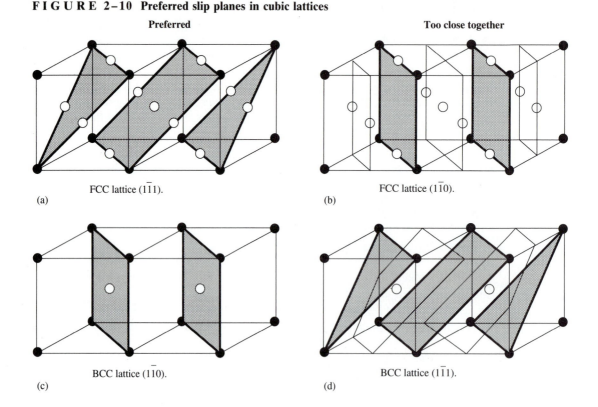

Preferred   Too close together

(a)   FCC lattice $(1\bar{1}1)$.   (b)   FCC lattice $(1\bar{1}0)$.

(c)   BCC lattice $(1\bar{1}0)$.   (d)   BCC lattice $(1\bar{1}1)$.

would be too close together to be effective; thus the preferred slip system for the BCC lattice is the {110} plane in the ⟨111⟩ direction.

CPH Lattice.   The close-packed plane of the CPH lattice is the same as that for the FCC (see Figure 1-4) except there is only one (0001) plane in each lattice. The preferred slip direction is normally the diagonal of the hexagon, ⟨11$\bar{2}$0⟩ (see Figure 1-27). The preferred slip system is {0001} and ⟨11$\bar{2}$0⟩.

2.6   CRITICAL RESOLVED SHEAR STRESS

Tests have been conducted on single crystals of metals, measuring the shearing stress (the f_s of Figure 2-8 divided by the area) required to just initiate plastic deformation, or **slip.** Since this is a threshold value, it is referred to as *critical,* and since it is a component of the applied force or stress, it is said to be *resolved;* that is, the **critical resolved shear stress.**

Table II-1 has values for a few metals and shows the plane and direction on which they were measured. The values are for specific metals, and although they do tend to fall into lattice groups, it is important to recognize that the *critical resolved shear stress is a physical constant for a given metal, not for a lattice*

T A B L E II–1 Values of critical resolved shear stress

Metal	Lattice	Slip Plane	Slip Direction	Critical Resolved Shear Stress, psi
Silver	FCC	{111}	⟨1$\bar{1}$0⟩	54
Copper	FCC	{111}	⟨1$\bar{1}$0⟩	71
Aluminum	FCC	{111}	⟨1$\bar{1}$0⟩	114
Magnesium	CPH	{0001}	⟨11$\bar{2}$0⟩	64
Cobalt	CPH	{0001}	⟨11$\bar{2}$0⟩	960
Titanium	CPH	{10$\bar{1}$0}	⟨11$\bar{2}$0⟩	1990
Iron	BCC	{110}	⟨$\bar{1}$11⟩	3980
Niobium (Cb)	BCC	{110}	⟨$\bar{1}$11⟩	4840
Molybdenum	BCC	{110}	⟨$\bar{1}$11⟩	10400

Source: W. J. M. Tegart. *Elements of mechanical metallurgy.* New York: Macmillan, 1966, 106. Used by permission.

type. Also, the table gives the stress needed to initiate slip, not continue slip; the *true stress,*[‡] which continues slip, will increase as the deformation progresses.

The FCC lattices tend to deform easily because they have densely packed {111} planes, as mentioned earlier. Although the CPH group has only one close-packed plane, (0001), the BCC lattice has none. Thus the FCC and CPH tend to be more easily deformed than the BCC.

The fact that a CPH metal has a low critical resolved shear stress, determined from a single crystal, does not necessarily mean that a polycrystalline specimen would deform as easily. In polycrystalline metals the slip planes will not necessarily be in a position favorable for slipping to occur. Since the CPH has only one close-packed plane, in the polycrystalline form it is not as easily deformed as Table II-1 would suggest.

The values of the table are for the metals themselves, not alloys with other metals. Other metals have different atom diameters, and combining atom sizes in a lattice changes the critical resolved shear stress. We'll discuss alloying and its effects in Chapter 4. It's sufficient here to point out that, in general, a combination of atoms in a lattice increases the strength and hardness of metals and decreases their ductility and plasticity. (Visualize Figures 2-1 and 2-2 with combinations of atoms of different diameters.)

2.7 TWINNING

When shearing stresses are applied to some metals, their atom lattices bend somewhat rather than slipping. This is termed *twinning* because the resulting distorted

[‡]True stress is the load divided by the actual area. See Appendix B, Section B.4.

lattice is a *mirror image,* or *twin,* of the original structure. Such an image would be viewed if a mirror were placed along the edge of the twinned region of Figure 2-11.

Twinning changes the relative position or orientation of atoms in their lattice. Atoms that were a lattice parameter apart become separated by half a diagonal, and vice versa. The result is that the atoms don't move far and have the same neighboring atoms; they stay in the same lattice, but their location in it changes.

Consider the (110) planes of FCC lattices shown in Figure 2-11. The heavy lines outline (110) planes, the light vertical lines represent the half-diagonal points, and the circles are the centers of the atoms. The horizontal distances between adjacent solid-black and heavily outlined atoms is half the diagonal, or two spaces equals $a\sqrt{2}$. The vertical distance between the atoms is the lattice parameter a. When twinning occurs (the shaded atoms), the distance between the atoms changes from half the diagonal to the full lattice parameter, i.e., the shaded atoms move down so that they are now a to the right from their (approximately) horizontal neighbor. Their distance from the other atom in the other direction is now half the diagonal.

Figure 2-11 shows only a few atom lattices involved in the twin; in reality there will be many more and the twin regions are large enough to be seen quite readily in the microscope. In microstructures, twins appear as parallel stripes or patterns across grains; the "stripes" going across the grains of the photomicro-

F I G U R E   2–11   Twinning in a lattice

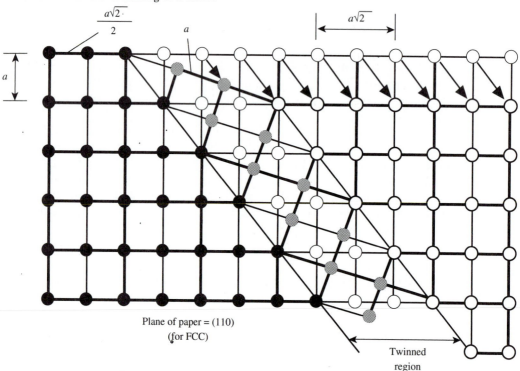

Plane of paper = (110)
(for FCC)

Twinned region

graphs in Figures A-7, A-8, A-10, A-11, and A-12 are twins. Because twins are so geometrically precise, they are usually easy to distinguish from the more irregularly shaped grain boundaries.

Twinning occurs in FCC metals but is probably most common in CPH metals, which inherently have few favorable slip planes. Although the example above used the FCC lattice, the same kind of relationship can exist in a CPH lattice where atoms that are close-packed can move to a lattice parameter c apart.

Although twinning is caused by shear stresses, there is no threshold value comparable to the critical resolved shear stress in slip. Compared with slip, twinning does not cause large amounts of deformation since the atoms move only a portion of a lattice parameter. For some metals, twinning is important, not for the amount of deformation it causes, but for rotating lattice planes into positions where they can slip.

2.8 POLYCRYSTALLINE ELASTIC DEFORMATION

Most of the discussion up to now has dealt with deformation of single metal crystals. Although more and more applications are being made of metals in this form, most metals are still used in a polycrystalline form, as accumulations of crystal lattices that are not aligned. Also, most of the preceding information has dealt with slip and plastic deformation, but before deforming plastically, metals go through an elastic region. This is an important region since it contains the range of stresses within which structures, vehicles, etc. are designed to operate.

Modulus of Elasticity. Some metals deform elastically more easily than others; this difference in behavior is expressed as the modulus of elasticity, or E, and is calculated as the ratio of stress to *elastic* strain.[§] That is, E relates stress and strain only when their relationship is linear; E *cannot* be used to relate stress to strain beyond the elastic region.

The modulus of elasticity, or **stiffness,** is determined by the basic atomic forces of the metal and is a constant for that metal. It is altered very little by alloying, cold-working, or heat treatment; however, the range of stress and strain over which the metal will be elastic can be changed greatly by each of these three methods. Practically, the modulus of elasticity is one of the determinants in how much sagging or flexing occurs in machines, bridges, airplane wings, and other structures where rigidity or stiffness is essential. Springback, so important in cold-rolling and drawing operations, depends directly on the modulus of elasticity.

Magnitude of Deformation. During elastic deformation, the modulus of elasticity E relates stress and strain. Because the modulus is so large for most common

[§]The principle of the modulus of elasticity was first identified by Robert Hooke and is associated with his name, but it is also known as Young's modulus. Since E is stress/strain, the units are psi/inch/inch, or simply psi in the English system. For most engineering metals, the value of E is in the millions and is often written as a number $\times 10^6$ psi.

metals (between 5 and 50×10^6 psi) the amount of *elastic* strain possible is very small. Metals such as steel, aluminum, and brass have maximum elastic strains measured in thousandths of an inch per inch. By contrast, during plastic deformation, there is no simple relationship between "engineering" stress and strain, and the magnitude of strain possible can be very large. The 2-in. gage length of a tensile specimen of a ductile metal such as 70-30 brass may stretch to over 3 in., that is, have a strain of over 50%.

2.9   POLYCRYSTALLINE PLASTIC DEFORMATION

Polycrystalline Deformation Process. In polycrystalline metals the lattices of adjoining crystals or grains are not aligned, and thus their slip systems are not aligned. However, among all the many crystals there will be some planes in an orientation favorable for slip; i.e., their critical resolved shear stress can be exceeded by an applied force.

Because slip takes place at angles, and because grains are in intimate contact with one another, slip in one grain affects its neighboring grains, twisting and distorting them. This repositions grains that would not slip initially into favorable orientations. The deformation proceeds throughout the metal until all the dislocations are used and grains are at angles not favorable for slip, at which point local deformation (necking) begins, the area decreases (load-carrying ability goes down), and fracture occurs.

Such a scenario helps explain why in polycrystalline metals, FCC metals are so ductile compared with some CPH metals having lower critical resolved shear stresses. Randomly oriented FCC lattices have an inherently larger number of close-packed {111} planes and many more slip systems available than the CPH lattice, which has only one {0001} plane in any given crystal.

Change in Mechanical Properties. Polycrystalline metals generally go through some permanent plastic deformation or slip before they fracture. This is a valuable property of metals, for without it all rolling, bending, drawing, extruding, and forging operations would be impossible and metals could only be cast or cut to shape and size. Also, as metals slip, they resist further slip; i.e., their strength against yielding goes up. This provides a factor of safety between yielding and breaking and allows unevenly distributed stresses to be equalized. Extremely brittle metals, on the other hand, break apart without appreciable slip and do not provide these advantages.

As seen in Figure 2-3, *slip occurs within the grain(s)*, along a particular slip plane, rather than *between* the grains; i.e., it does not occur along grain boundaries. Generally, to strengthen metals, their resistance to slip has to be improved. Thus, in most cases, when metals are strengthened by cold-working, alloying, or heat-treating, the strengthening occurs within the grain, not in the grain boundaries.

Since slip occurs within the grains, the larger the grain the more opportunity there is for deformation. Grain boundaries, on the other hand, interrupt the slip process. Thus the generality is that larger-grain metals are more ductile and smaller-grain metals are stronger.

As metals are plastically deformed, they get stronger and harder but lose ductility or plasticity. However, the influence of plastic deformation or cold work is not uniform: The initial cold work done on an annealed metal increases the hardness and yield strength much more than it increases the tensile strength, and when the amount of cold work becomes significant, about 30 to 50%, the percentage change in properties by further cold work is greatly reduced. (See Appendix D and Figure D-1.)

The cold-working of metals requires an expenditure of energy. When metals are cold-worked extensively, they become warm,[‖] but not all the energy goes off as heat; some small amount is stored in the lattice. In Chapter 3, where the heating of metals is studied, this small amount of stored energy is identified as the key to initiating the phenomenon of recrystallization.

SUMMARY

1. In elastic deformation the atoms of a lattice are slightly displaced but return to their original locations when the load is removed.

2. In plastic deformation, or slip, atoms move many atom diameters or lattice parameters from their original locations, change neighbors, and do not return to their original positions on release of the stress.

3. Atoms slip on planes that are many atom diameters apart; i.e., crystals do not slip on immediately adjacent planes.

4. Slip is a series of stepwise movements of atoms within the crystal on planes that are favorably oriented.

5. When there are no planes favorably oriented for slip the specimen fractures or cleaves.

6. All the atoms in a plane can't move in unison; too high a force would be required to offset the forces between all those atoms. It would mean that the tensile and yield strengths of metals would be many times greater than they are.

7. The reason metals do not have higher strength is because the atoms don't have to move all at once but can move one dislocation at a time.

8. As the dislocations are used, the force necessary to continue deformation increases; i.e., the metal is cold-worked or work-hardened and gets stronger. Fracture or cleavage occurs when the dislocation are all used.

‖Rapid cold-rolling or cold-drawing can raise metal temperatures enough to easily burn the skin.

9. Deformation by shear is maximized on planes that are 45° to the applied force.

10. Normal or separating forces that tend to pull the specimen apart occur on planes that are at 90° to the applied force.

11. Preferred slip systems (combinations of slip plane and direction) for lattices occur on planes with the densest packing of atoms. These planes will have the greatest distance between similar parallel planes, and thus the atoms will have the least interference as they "slip" over each other.

12. The preferred slip systems are, for FCC, {111} planes in the ⟨110⟩ directions; for BCC, {110} planes in the ⟨111⟩ directions; for CPH, {0001} planes in the ⟨11$\bar{2}$0⟩ directions.

13. The threshold shearing stress required to just initiate plastic deformation, or slip, is the critical resolved shear stress.

14. The critical resolved shear stress is a physical constant for a given metal, not for a lattice type. Also, it is the stress to initiate slip, not continue it. Once slip is initiated, the stress required to continue it will increase as the deformation progresses.

15. Having more than one size of atom in a lattice will change the critical resolved shear stress.

16. Twinning is a geometric change in the relative orientation of atoms in their lattice. Atoms that were a lattice parameter apart may come to be half a diagonal apart, and vice versa. The atoms don't move very far, and their lattice stays intact, but their locations in it have changed.

17. In photomicrographs, twins appear as parallel stripes or patterns across grains.

18. Twinning is caused by shear stresses, but there is no threshold value comparable to the critical resolved shear stress in slip.

19. Twinning does not cause large amounts of deformation but may rotate lattice planes into positions where they can slip.

20. The modulus of elasticity, or E, is calculated as the ratio of stress to *elastic* strain.

21. The modulus of elasticity, or stiffness, is a constant for a metal. Alloying, cold-working, and heat treatment change the range over which the metal will be elastic, but they change E very little.

22. Most metals—e.g., steel, aluminum, and brass—have maximum elastic strains of a few thousandths of an inch per inch.

23. In polycrystalline metals, slip in one grain twists and distorts neighboring grains, putting them into orientations favorable for slip. Deformation continues until all the dislocations are used and grains are no longer at angles favorable for slip; then local deformation (necking) begins, the area decreases (load-carrying ability goes down), and fracture (cleavage) occurs.

24. Slip occurs *within grains*, along a particular plane, rather than between the grains. Generally, to strengthen metals, their resistance to slip has to be

improved. So, strengthening by cold-working, alloying, or heat-treating occurs within the grain, not in the grain boundaries.

25. As metals are plastically deformed, they get stronger and harder but also less ductile or **malleable.**

26. The influence of cold work is not uniform; the initial cold-working of an annealed metal increases the hardness and yield strength more than it increases the tensile strength, and when working gets to a significant level, additional cold-working changes properties very little.

27. Small amounts of cold work are stored in the deformed lattices. If the stored energy is greater than a critical value, the deformed lattices recrystallize, when heated to a high enough temperature. (The phenomenon of recrystallization is the major topic of Chapter 3.)

REVIEW QUESTIONS

1. Describe the difference in the behavior of atoms subjected to elastic and plastic deformation.
2. What are the key differences between slip, twinning, and cleavage?
3. Using the concept of dislocation, describe plastic deformation including localized necking and fracture; see section 3.4.
4. Use sketches of the three atom-lattice types to illustrate their preferred slip systems.
5. What is critical resolved shear stress?
6. Describe the difference between plastic deformation in single crystals and in polycrystalline metals.
7. Discuss the role played by grain boundaries in attempts to strengthen metals by cold-working, alloying, or heat-treating.
8. Discuss the factors determining or influencing the modulus of elasticity.

FOR FURTHER STUDY

Alexander, W. O., G. J. Davies, K. A. Reynolds, and E. J. Bradbury. *Essential metallurgy for engineers.* Wokingham, U.K.: Van Nostrand Reinhold, 1985, secs. 2.3–2.6.

Avner, Sidney H. *Introduction to physical metallurgy* (2d ed.). New York: McGraw-Hill, 1974, chaps. 2 and 3.

DeGarmo, E. Paul, J. Temple Black, and Ronald A. Kohser. *Materials and processes in manufacturing* (7th ed.). New York: Macmillan, 1988, 71–76, chap. 18, Cold-working processes.

Dieter, George E. *Mechanical metallurgy* (3d. ed.). New York: McGraw-Hill, 1986, chap. 1.

Flinn, Richard A., and Paul K. Trojan. *Engineering materials and their applications* (3d ed.). Boston: Houghton Mifflin, 1986, secs. 3.1–3.10.

Harris, J. N. *Mechanical working of metals.* New York: Pergamon, 1983, chap. 1.

Reed-Hill, Robert E., and Reza Abbaschian. *Physical metallurgy principles* (3d ed.). Boston: PWS-Kent, 1992, chap. 4.

C H A P T E R   3

Heating and Hot-Working of Metals

OBJECTIVES

After studying this chapter, you should be able to

- define the three stages of annealing and the uses of each
- describe the different temperatures at which the three stages occur, using graphs of
  - energy reclaimed versus annealing temperature
  - electrical conductivity versus annealing temperature
- describe the recrystallization process, including the roles played by critical cold work, incubation, and nucleation
- predict the effect on annealing results by a change in
  - prior cold work
  - annealing temperature
  - annealing time
  - purity or alloying level
  - insolubles
  - prior grain size
- explain how the combination of temperature and strain rate determines whether deformation is hot-working or cold-working
- explain the economics behind the choice of hot- or cold-working methods

DID YOU KNOW?

It is difficult to think of an application for a metal that does not require that it be heated during either its manufacture or its use. The heat can be used for annealing, to soften the metal; during hot-working operations; during heat-treating or tempering of the metal; during welding or brazing operations; or in a high-temperature application. Annealing and hot-working, as discussed in this chapter, are important topics whose principles apply any time a metal with a comparable amount of cold work is heated.

A PREVIEW

We'll first describe the heating process and then briefly define the three stages of annealing. Each of the stages will be investigated in detail, though we'll concentrate on recrystallization, the second stage. The major factors influencing recrystallization—percent cold work, heating temperature, and heating time—will be investigated first and then factors of less significance will be studied. An incorrect understanding of the peculiarities of recrystallization can lead to erroneous conclusions and incorrect solutions to processing and product problems. The dividing line between hot-working and cold-working involves recrystallization, and this is the last subject considered. At that point, we use an example to illustrate the importance of the recrystallization process.

3.1   THREE STAGES OF ANNEALING

As a polycrystalline metal is heated, it passes through three stages, if the conditions are suitable. If **residual stresses** exist and the metal is heated to a comparatively low temperature, it will go into the **recovery stage.** If the metal has enough cold work and is heated to a higher temperature, it will enter the **recrystallization stage.** Heating the metal to a still-higher temperature and holding it there for a relatively long period of time will cause it to experience **grain growth.** Commercial treatments to achieve the results just described are usually called *anneals*; however, any heating process that causes the metal to progress through the stage(s) described can have the same effect whether or not it is termed ''annealing.''

Recovery Anneals.   Recovery anneals are usually termed **stress-relief anneals;** they reduce or remove the residual stresses caused by operations such as rapid quenching, welding, or machining.

Recrystallization Anneals.   Since *crystallization* is the process of freezing randomly oriented atoms of molten metal into its atom lattice, *recrystallization* suggests the re-forming, or *returning of the metal to its original unstrained atom lattice*. Recrystallization occurs when heat energy, combined with energy trapped in a deformed lattice, is sufficient to initiate the formation of a new, unstrained atom lattice. If recrystallization is carried to completion (that is, a high enough temperature is applied for a long enough time), there will be no cold work remaining in the metal. Cold-worked metals are usually given recrystallization anneals to soften them so that additional cold work can be done. These anneals are often termed **process anneals,** or **intermediate anneals,** either because they are part of the cold-working process or because the anneals were *not* used to achieve the final mechanical properties.

Grain-Growth Anneals.   If its temperature is raised above that required for recrystallization, the metal will enter the grain-growth stage. In this stage the grains become larger and, as a consequence, fewer in number; there will also be fewer grain boundaries. The reduction of the energy contained in the grain boundaries seems to be the force that drives the growth of the grains.

3.2   STORED ENERGY

Cold-Work Energy Stored.   It was mentioned in Chapter 2 that when metal is cold-worked much of the energy used goes off as heat, but that some cold work is trapped and stored in the deformed lattice. Figure 3-1 relates the energy stored during a tensile test of high-purity copper to the cold work, as measured by percent strain. The *stored-energy curve* (solid line) shows the *amount* of heat, in calories per mole of the copper specimen, that was not given off as heat. The

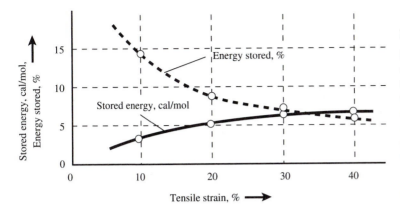

FIGURE 3–1

Stored energy and percent of energy stored vs. tensile strain (as a measure of cold work) for high-purity copper

Source: Adapted from the data of P. Gordon and used with permission from *Transactions of the Metallurgical Society*, 1955, *203*, 1043, a publication of The Minerals, Metals, & Materials Society, Warrendale, PA 15086.

energy-stored curve (dashed line) shows this energy as a *percentage* of all the energy used to deform the copper.

There are a number of important relationships shown by these data. First *the amount of stored energy* (solid line) *increases as the amount of strain is increased.* That is, the more cold work done, the more energy stored, and the stored-energy curve will always have a positive slope. Second, *the percentage of the energy stored* (dashed line) *goes down as the total energy expanded goes up.* That is, a large percentage of a small amount of cold work is stored, and/or a small percentage of a large amount of cold work is stored.

Return of Stored Energy.   Figure 3-2 shows that the stored energy is returned during later heating of cold-worked metal. The data for this figure was obtained by comparing the energy required to heat two 99.97% copper specimens, one cold-worked, one uncold-worked, to the same temperature. The cold-worked specimen took less energy than the specimen that had not been cold-worked. That is, in heating to any temperature, the amount of energy read from the graph is the energy the cold-worked specimen provided from its deformed lattice (Figure 3-1) that did not have to be supplied from the furnace.

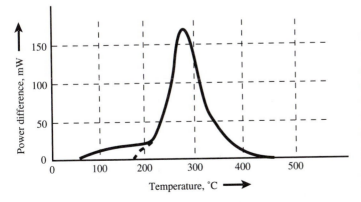

FIGURE 3–2

The difference in power required to heat two copper specimens. One specimen was cold-worked; the other was not.

Source: L. M. Clarebrough, M. E. Hargreaves, and G. W. West. *Proceedings of the Royal Society*, London, 1955, *232A*, 252. Used by permission.

Notice that the power returned from the cold-worked specimen can be divided into three distinct portions—a small amount up to about 200°C (390°F), the major contribution between 230 and 350°C (445 and 660°F), and another small amount above 360°C (680°F). These three divisions coincide with the three stages of annealing for copper.

Recovery stage. Below about 200°C (390°F) the copper is primarily undergoing recovery. Here the energy due to *elastic* stresses trapped in the deformed lattices of the cold-worked specimen is given off and reduces the power required to heat the specimen. Note that the slope of the curve is positive; the higher the temperature, the more energy released. In this stage the lattices are not re-formed, since the temperature and the level of deformation are not sufficient to initiate recrystallization, so *mechanical properties change very little.* Residual stresses are merely reduced or removed.

Recrystallization stage. Recrystallization occurs between approximately 250 and 340°C (480 and 640°F) where the large reduction of power occurs, i.e., where the strained lattice makes the largest contribution to the heating. If the amount of cold work is greater than the critical cold work, then heating in this temperature range removes the cold work and re-forms the lattice into a completely strain-free lattice, and the *maximum change in mechanical properties occurs.* In recrystallization the new grains are formed from the deformed portions of the old lattice. The size of the new grains is dependent on a number of factors that will be discussed later.

Grain-growth stage. Above about 360°C (680°F) the metal is in the grain-growth stage. By this temperature most if not all of the strain energy of the cold work is removed, and it appears that the grain-boundary energy drives the enlargement of the grains. In general, the grains that are larger in the beginning consume the smaller ones. Since the volume of the heated metal is fixed, the number of grains must decrease as the grain size goes up.

The changes in mechanical properties that occur because of grain growth are the same as would be expected from a larger grain size. The presence of fewer grain boundaries increases ductility and lowers strength, so percent elongation goes up and the tensile and yield strengths come down, but the *changes in properties are small compared with those due to recrystallization.* Hardness is affected by the reduced number of grain boundaries and, with some indenters, continues to decrease as grain size goes up.

Temperature Overlap. In the above discussion, the temperatures for each stage were given as a range, but there were gaps between the ranges. Actually, the three stages of annealing overlap in temperature. This is because the level of deformation or cold work of individual grains varies considerably throughout a polycrystalline metal. Some grains are oriented so that extensive slip can occur; others are not. Thus at any given temperature some grains have enough internal energy to completely recrystallize, while others may only be going through recovery. From Figure 3-2 it is evident that most of the recrystallization occurs at about 300°C (570°F), but some grains may begin recrystallizing at temperatures as low as 175°C

(347°F), while others may not begin until a temperature of 400°C (750°F) is reached. By similar reasoning, some grains may go through recovery at temperatures as high as 250°C (480°F), and some grains may begin grain growth at temperatures as low as 325°C (615°F).

3.3   RECOVERY STAGE OF ANNEALING

Elastic Stresses.   When the load that caused the cold work or plastic deformation in a polycrystalline metal is removed, the applied stress goes to zero, but that does not mean the internal stresses have gone to zero. The varying orientation of the slip planes causes the grains to be plastically worked in different amounts, and they are no longer the same size and shape as they were originally. Thus, while the stress is being reduced through the elastic region and one misshapened grain presses against another, the amount and direction of the elastic strains generated varies throughout the structure. As a result, elastic, or residual, stresses are created.

It is the elastic nature of these stresses that causes problems. Residual stresses are analogous to stretched rubber bands and compressed springs just waiting to be released. So it is that a machining operation or a hostile environment will release a residual stress in a metal, resulting in distortion, stress corrosion, or some other defect.

Removal of Stresses.   The data for Figure 3-3 were generated the same way as they were for Figure 3-2, except in this case they are for nickel. The power curve of Figure 3-3, the lowest curve, shows two peaks of energy return—one at 600°C

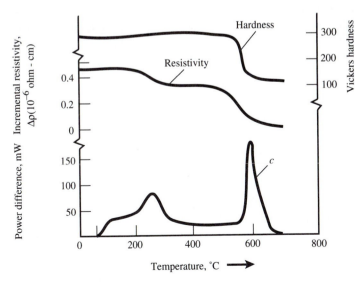

FIGURE 3–3

The difference in power to heat cold-worked and un-cold-worked nickel to various temperatures; resistivity and hardness vs. annealing temperature for cold-worked nickel

Source: L. M. Clarebrough, M. E. Hargreaves, and G. W. West. *Proceedings of the Royal Society*, London, 1955, *232A*, 252. Used by permission.

(1110°F), which is in the middle of the recrystallization range, and another at 250°C (480°F), where recovery occurs. Note that resistivity shows a major change at 250°C (480°F), but that the hardness is unaffected. As previously stated, recovery anneals reduce the residual elastic stresses and thus reduce resistivity (i.e., improve conductivity) but do not appreciably change mechanical properties; this is shown by the hardness curve. On the other hand, the hardness curve confirms that in the recrystallization stage, about 600°C, the mechanical properties change significantly.

3.4   RECRYSTALLIZATION STAGE OF ANNEALING

Mechanism.   *For recrystallization to occur, the energy from lattice strain plus the energy provided by the heat source (furnace) must be greater than a certain threshold value.* That is, first there is a period during which the strain and heat energy accumulate, or **incubate**. If the energy level is sufficient, nucleating sites begin to form in the deformed lattice. These nuclei form first in those areas of the latice that have the highest strain energy, and recrystallization begins there. This can be seen in the photomicrograph of Figures A-7 through A-9. Figure A-7 shows the evidence of the slip planes after the brass was cold-worked about 30%. Figure A-8 shows that, after a short period of heating, nuclei begin to show along those slip planes. In Figure A-9, heating for 5 sec longer permits those nuclei to spread over the whole specimen, establishing new, very small, recrystallized grains.

Time Required.   Figure 3-4 shows how the recrystallization process proceeds with time. The figure relates the percent of the microstructure recrystallized to the time a cold-worked specimen is held at a given temperature. The process begins slowly; as more and more nucleation sites are formed, the rate of recrystallization (slope of the curve) increases. When about half of the structure is recrystallized, the rate begins to slow. Completely recrystallizing the last of the structure, which is low in strain energy and thus lacks nucleating sites, takes time, and the process finishes at a slower rate.

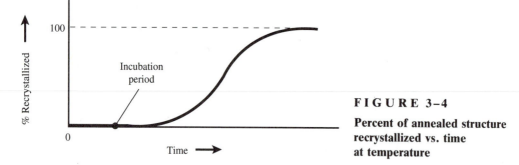

F I G U R E 3–4

Percent of annealed structure recrystallized vs. time at temperature

Nucleating Sites and Grain Size.   If the general level of strain energy is low, then at comparable temperatures fewer nuclei will be initiated and a few large grains will be formed.

Since the number of nucleating sites is directly related to the amount of cold work, and since the grain size is inversely related to the number of nuclei, the *recrystallized grain size will be inversely related to the amount of cold work*. This is shown in Figure 3-5. Beyond the critical cold work, the higher percent the cold work, the smaller will be the recrystallized grain size.

Major Factors in Recrystallization.   In order for recrystallization to begin, *time* is required for *incubation* and then time is required for the formation of *nucleation sites*. Once the process is underway, additional time is required as new grains spread out from the nucleating sites and form lattices with atoms in their proper locations. That is, the atoms must relocate, while in the solid state, to their lowest energy level, their closest-approach distances. This does not happen instantaneously; time is required for the process to come to completion.

Thus, as we can see from what has been discussed so far, *recrystallization requires cold work, temperature, and time*. Figure 3-6 shows the relationship of these three variables. The data are based on specimens of zirconium rods swaged (cold-forged) to two different levels of cold work (one approximately 4 times the other) and then annealed at combinations of temperature and time until completely recrystallized. To point out the relative importance of the three variables, two sets of data points are compared. Note that the time axis varies by multiples of 10.

At a constant temperature of 560°C (1040°F), the specimen that was cold-worked 51% completely recrystallizes in about 1.1 h; at that same temperature the specimen worked 13% takes approximately 50 h to recrystallize. However, if the time is held constant at 1 h, the 51% cold-worked specimen will recrystallize at about 570°C (1058°F), but the 13% cold-worked specimen has to be heated to about 630°C (1165°F).

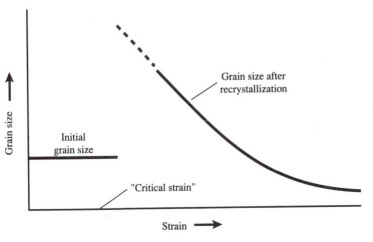

Grain size ↑

Grain size

Initial grain size

Grain size after recrystallization

"Critical strain"

Strain ➡

FIGURE 3–5

Grain size vs. percent cold work after a recrystallization anneal

Source: Clyde W. Mason. *Introductory physical metallurgy*. Materials Park, OH: ASM International, 1947, fig. 26, 47. Used by permission.

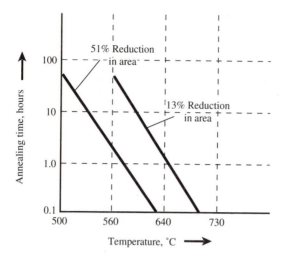

FIGURE 3–6

Annealing time vs. temperature to recrystallize a zirconium rod swaged to two levels of cold work

Source: R. M. Treco. *Proceedings of AIME regional conference on reactive metals,* 1956, 136. Used with the permission of American Institute of Mining, Metallurgical, and Petroleum Engineers, Inc., 345 East 47th St., New York, NY 10017.

The point of the comparisons is that at the same temperature the specimen with the lower cold work has to be heated about 45 times longer, but it can be recrystallized in the same length of time by raising the temperature only 60°C (108°F). Thus we can conclude that *the amount of cold work is a very important factor in how readily recrystallization occurs, and temperature is a more important factor than time.*

Critical cold work. In reaching these conclusions, we have to remember that cold work must be above some minimum level of cold work or recrystallization will not occur. This can be seen in Figure 3-7, which shows how the required level of cold work varies with percent cold work and temperature for recrystallized specimens of single-phase (alpha) brass. Note that at the higher levels of cold work, i.e., over 15%, recrystallization can occur when the metal is annealed at the lowest temperatures, 400 and 450°C (752 and 842°F). However, with a small amount of cold

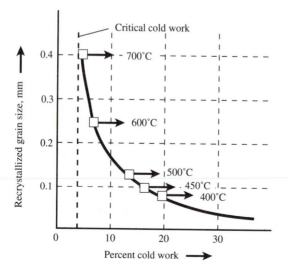

FIGURE 3–7

Recrystallized grain size vs. percent cold work for alpha brass annealed at various temperatures

work, recrystallization only occurs at the highest temperature, 700°C (1292°F), and below about 3% cold work, the critical cold work for this metal, recrystallization does not occur at all.

Following the temperatures from the high cold-work end and moving to the left, we see that the temperature required for recrystallization increases. That is, the combination of the heat and cold-work energies is not high enough for nucleation to occur, so only the highest temperature is sufficient to cause recrystallization at the low cold-work end.

Figure 3-7 also makes the point that the *recrystallized grain size is dependent on the number of nuclei*, i.e., the amount of cold work; *for a given amount of cold work, the recrystallized grain size is the same, regardless of the temperature.*

Summary. Figure 3-8 provides a good summary of the discussion so far. The relationship between the recrystallized grain size, percent cold work (or percent reduction in thickness), and temperature is all shown. Note the following factors:

- As the annealing temperature goes up, recrystallization begins at lower and lower values of cold work; i.e., the minimum required cold work goes down.
- At some temperatures no recrystallization takes place, regardless of the amount of cold work.

FIGURE 3-8

Recrystallized grain size vs. percent reduction in thickness vs. annealing temperature, for rolled "Armco" iron

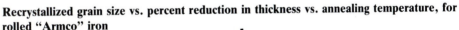
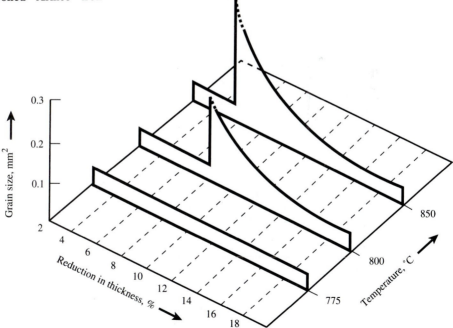

Source: Clyde W. Mason. *Introductory physical metallurgy*. Materials Park, OH: ASM International, 1947, fig. 28, 49. Based on work by H. F. Kaiser and H. F. Taylor; used by permission.

- For very low values of cold work no recrystallization occurs, regardless of annealing temperature; i.e., the critical cold work was not reached.
- For any given amount of cold work, the recrystallized grain size is the same, regardless of annealing temperature.
- The higher the percent cold work, the smaller the recrystallized grain size.

Minor Factors in Recrystallization.   There are three minor factors that influence a metal's ability to recrystallize—its purity, the amount of insolubles present, and the grain size before cold work.

Purity. The **purity** of a metal is dependent upon whether atoms of other metals are present; i.e., a 100% pure metal has atoms of only that metal in its atom lattice. In commercially produced metals of even the highest purity, this is very rare. The atoms of some other metal or metals usually are present as an unwanted impurity or as a purposely added alloying element. These foreign atoms, or **solute atoms,** reside in the atom lattice of the original metal and in effect substitute for its atoms.* Their effect is to raise the recrystallization temperature of the alloy above that of the pure metal.

Table III-1 shows recrystallization temperatures for a few pure metals and alloys. For the aluminum data, note the difference between 99.999% pure and 99.0% pure; the presence of a small amount of other atoms raises the temperature 209°C (375°F). The much larger amounts of alloying elements added to make up the aluminum alloys increase the temperature only another 28°C (50°F). The data on the copper alloys shows that the effect of alloying on the recrystallization temperature varies with the element and that a small amount of one element can have more effect than larger amounts of another.

From the discussion of slip in Chapter 2 you can see that the insertion of an atom of a different diameter into a lattice would inhibit the ability of the lattice to slip on that plane; i.e., the addition would prevent a dislocation from moving. In a somewhat similar way, the substituted atoms hinder the formation of a new lattice during recrystallization. The *solute atoms seem to pin the slip planes and grain boundaries and inhibit their movement, increasing the temperature required for recrystallization.*

In general, the initial alloying addition has the most effect on increasing the recrystallization temperature; as alloying amounts increase, they have less and less effect on the temperature.

Table III-1 indicates that the recrystallization temperatures for zinc, tin, and lead are below room temperature. This does not mean that the metals cannot be cold-worked; if they are deformed, slip occurs and their properties change, but in time, at room temperature, they will recrystallize.

Insoluble impurities. **Insoluble impurities,** such as oxides and gases, do not become part of the metal's atom lattice; that is, these impurities are, in fact, *insoluble*. Such particles do not change the recrystallization temperature appreciably, but they do encourage nucleation and thus form more numerous and therefore

*As we'll see in Chapter 4, the alloys formed by these substituting atoms are termed *substitutional alloys*.

TABLE III-1 Approximate recrystallization temperatures for several metals and alloys

Material	Recrystallization Temperature	
	°F	°C
Copper (99.999%)	250	121
Copper (OFHC)	400	204
Copper-5% zinc	600	316
Copper-5% aluminum	550	288
Copper-2% beryllium	700	371
Aluminum (99.999%)	175	79
Aluminum (99.0%+)	550	288
Aluminum alloys	600	316
Nickel (99.99%)	700	371
Nickel (99.4%)	1100	593
Monel metal (Nickel + 30% copper)	1100	593
Iron (electrolytic)	750	399
Low-carbon steel	1000	538
Magnesium (99.99%)	150	66
Magnesium alloys	450	232
Zinc	50	10
Tin	25	−4
Lead	25	−4

Source: Albert G. Guy. *Physical metallurgy for engineers*. Reading, MA: Addison-Wesley, 1962, table 8-4. Reprinted with the permission of the publisher.

smaller grains. Commercially, insoluble particles are sometimes purposely added to metals to encourage a small-grained structure after heating.

Grain size before cold work. Grain boundaries have been previously identified as discontinuities in the atom spacing that are locations of higher energy. This energy can also contribute to the energy required for recrystallization. A metal specimen that has a smaller prior grain size, i.e., a smaller grain size before cold-working, will have more grain boundaries and more energy than a specimen of a larger grain size cold-worked the same percentage. The smaller-grained specimen will tend to recrystallize more easily, i.e., at lower temperature and/or in less time. Compared with other factors that affect recrystallization, the influence is slight, but the prior grain size of the metal being heated is a factor in recrystallization.

3.5   GRAIN-GROWTH STAGE OF ANNEALING

If a cold-worked metal is heated at a high enough temperature for a long enough period of time, it will completely recrystallize and all evidence of the cold work, i.e., the slip lines, will disappear. If heating is continued, and particularly if the

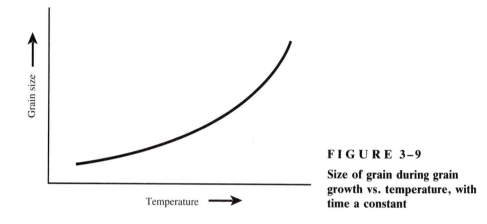

FIGURE 3–9

Size of grain during grain growth vs. temperature, with time a constant

temperature is increased, the grain size increases as shown in Figure 3-9; the metal experiences grain growth. In grain growth the larger grains expand and move into and absorb the adjacent smaller grains; the grains increase in size and decrease in number.[†] Theoretically, if a metal was held long enough at a high enough temperature, it would become one grain. In practice this is difficult to accomplish.

The force that seems to drive this growth process is the elimination of the grain boundaries themselves. Atoms at grain boundaries are not at their closest-approach distance and thus are not at their lowest energy level; as grains grow, the reduction in the extent of their boundaries enables the overall energy level of the atoms to decrease.

Since grain growth depends on the presence of grain boundaries, it can occur in un-cold-worked metals such as castings as well as cold-worked metals that have recrystallized.

Most annealing treatments are not carried into the grain-growth stage because (1) grain growth does not lower mechanical strength enough to significantly aid in the continuation of cold work, (2) costly fuel is consumed, and (3) expensive equipment is tied up. Also, tensile forces applied to metals with large grains twist and distort the surface grains, causing the surface to become rough, a condition termed "orange peel."

3.6   HOT-WORKING

Cold Work vs. Hot Work.   In Appendix D, Calculation of Cold Work, the reasons for using cold-working operations are discussed in detail. Basically, cold-working is used *to achieve a good surface finish, hold dimensions to closer tolerances, and/or achieve a desired level of properties.*

[†]This process has been studied by observing soap bubbles; larger bubbles seek to reduce the curvature of their boundaries by consuming the smaller bubbles.

The major disadvantage of cold-working operations is their limited ability to deform the metal. Heating the metal reduces its flow stress and enables deformation that would be impossible if done at room temperature; it also reduces the size or capacity of the equipment required. An example is the hot-rolling of metal slabs a foot or more thick to thicknesses of a fraction of an inch in just one trip through a 10-stand hot-strip mill.‡

However, there is a price to pay for these advantages. Hot-working operations present difficulties because of the high temperatures required. The equipment is expensive and deteriorates rapidly because of the heat; refractory-lined equipment can help but has its own costs and operating problems. Not the least of the costs of hot-working is the fuel expended in heating the metal.

Some products involve only hot-working, but many metal manufacturing processes begin with hot-working operations and finish with cold-working. Once the metal is reduced to a reasonable size or thickness, the work is done cold in order to achieve one or more of the purposes cited above: to get a good finish, control tolerances, and achieve certain properties or levels of properties. The proportion of the reduction done hot versus cold is basically an *economic decision* that must take into account the capital needed for equipment and facilities and the operating expenses required to manufacture a product of the desired quality at the necessary profit.

Differences Between Hot- and Cold-Working. Distinguishing between hot- and cold-working can be done by comparing the effect each has on mechanical properties. *Hot-working is done at temperatures that do not provide strengthening to the metal, while cold working does impart a strengthening.* In applying this distinction, you must realize that there are many operations that are done "hot," i.e., above room temperature, that work-harden the metal and leave it stronger than it was before it was deformed. These are therefore actually "cold-working" operations.

The effect of the rate of working. For deformation to be classified as "hot-working," the strengthening effects of the deformation must be removed by recrystallization. Since recrystallization involves incubation and nucleation, both of which require temperature *and* time, part of the distinction between hot- and cold-working involves the time aspect of the application of the forces doing the working. If temperatures are high enough—for instance, in the recrystallization range—and working slow enough, the metal will have time to recrystallize and no work-hardening will take place. The metal will have been "hot-worked." Conversely, if the rate of metalworking is increased so that the metal does not have time to recrystallize, the metal will be "cold-worked," even though it was done at an elevated temperature.

In distinguishing between hot- and cold-working, it may be helpful to recall the discussion of the factors that affect recrystallization: Recrystallization occurs at various temperatures, depending on the amount of deformation (percent cold

---

‡These mills consist of 10 stands of rolls arranged in tandem, i.e., one right after another. Some mills are a half-mile long.

work) and the amount of time available. So, at a *temperature* in the recrystallization range, if the rate of *deformation* is high enough, recrystallization will occur if sufficient *time* is available.

An example. The hot extrusion of metals is an example of an operation where the rate of deformation can determine whether the deformation imparts cold work or not. The extrusion of metals is not unlike pushing toothpaste out of a tube—the harder you squeeze, the faster it comes out. If working is done in the recrystallization temperature range and the rate of deformation is slow, it will be a hot-working operation. If the rate is increased enough, the part will be cold-worked. This may not sound serious, but the presence of cold work in the part means it has the capability of recrystallizing when heat-treated.

Some portions of the part may not have received the critical cold work and will not recrystallize during heating; their structures will consist of the original grains as deformed by the extrusion process. In those portions of the part that are heavily cold-worked, recrystallization will occur and small grains will result. But, in those portions where the cold work *barely* exceeds the critical amount, the recrystallized grain size might be very large.

If the cold work varies across the piece, or from end to end, its grain size will vary inversely. Thus it is possible to produce a wing spar or some other critical part that does not have a uniform grain size, which might keep the part from meeting the minimum strength requirements. The normal solution to this dilemma is to control temperature and rate of working such that the cold work imparted is either less than or *much* greater than the critical amount. This ensures that, if reheated, the part will be a single-type structure—unrecrystallized or recrystallized—and eliminates the possibility of creating the large grains that are symptomatic of a very small amount of cold work. Another remedy is to add alloy elements to retard recrystallization.

SUMMARY

1. There are three stages in the annealing, or heating, of metals:
   a. The recovery stage, which
      - occurs at lower temperatures
      - accomplishes relief of elastic, or residual, stresses
      - changes mechanical properties very little
      - improves electrical conductivity
   b. The recrystallization stage, which
      - occurs in the middle temperature range
      - requires a critical amount of cold work that triggers nucleation sites
      - promotes therefore the spread of new grains
      - causes major changes in mechanical properties
   c. The grain-growth stage, which
      - occurs after metals are heated to high temperatures for long periods of time
      - encourages some grains to enlarge at the expense of others

- is usually not a desirable process since large grains may yield poor surface quality in a product and it's an expensive process for the amount of reduction in properties achieved

2. Grain size after recrystallization is primarily determined by the number of nucleating sites, which is determined by the amount of prior cold work.

3. For a given level of cold work, mechanical properties will vary with heating temperature in the general pattern shown in Figure 3-10. Recrystallization (region II) occurs at temperatures where the curves have their steepest slopes, i.e., where the properties change the most with little change in temperature; recovery (region I) is at temperatures below that range, and grain growth (region III) is at temperatures above the range.

4. The energy causing recrystallization is a combination of the cold-work (internal) energy of the metal plus the energy from the heating source.
   a. If more than the critical cold work is present, whatever energy isn't provided by one source can be made up by the other.
   b. The heating source may contribute its heat at a high temperature for a short period of time or at a low temperature for a long period of time.

5. Grain boundaries and insoluble particles tend to provide energy to the cold-worked grains and make it easier for recrystallization to occur—"easier" meaning at a lower temperature or in less time; i.e., less energy has to be provided from the outside source.

6. Solute atoms, i.e., alloying elements present in the atomic lattice, tend to require additional energy for recrystallization to occur; their presence raises the temperature required for recrystallization.

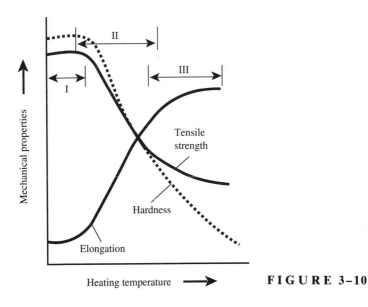

FIGURE 3-10

7. Compared with hot-working, cold-working has the advantage of contributing mechanical properties to the metal, providing a better surface, and controlling dimensional tolerances better. Hot-working can accomplish much larger amounts of reduction than cold-working, because the heat reduces the flow stress of the metal, but the high temperatures involved make a hot-working process expensive to operate and maintain.

8. The decision to combine hot- and cold-working is an economic one basic to the way the metalworking operation is established.

9. Metalworking operations performed at high temperatures can become cold-working operations if they are performed rapidly enough. Both the temperature and the rate of working enter into determining whether cold work will be done. This can be important if the part is to be heated again; if enough cold work has been imparted to the piece, then it will recrystallize, which may or may not be desirable.

REVIEW QUESTIONS

1. What are the major distinguishing characteristics of each of the three stages of annealing or heating?
2. What are the uses of each of these three stages?
3. Would castings be given recrystallization anneals? Explain.
4. Describe the factors that should be considered in deciding whether a cold-worked metal part will or will not recrystallize upon reheating.
5. Explain the changes that occur in the three curves of Figure 3-3 as the temperature goes up.
6. If a cold-worked part is given a recrystallization anneal,
   a. what happens to the time necessary to accomplish recrystallization if the temperature is increased?
   b. what happens to the time necessary for recrystallization if the cold work is reduced (and the temperature stays the same)?
7. Explain what happens to the recrystallized grain size of a cold-worked part if the amount of cold work is increased.
8. What are the factors that can cause an increase in the temperature required to recrystallize a cold-worked metal part?
9. Explain how grain growth occurs and what effect it has on a metal's properties.
10. Explain how metalworking operations done at elevated temperatures can be either hot-working or cold-working.
11. What are the advantages and disadvantages of (a) hot-working and (b) cold-working operations?
12. Explain how the decision would be made as to whether metalworking operations should be hot-working, cold-working, or a combination of both.
13. A 1-in.-thick bar of annealed metal is bent 90°, heated to its recrystallization range for 30 min and cooled to room temperature. After suitable preparation, the 1-in. face of

the bend area is inspected under a microscope. Which part of the bend area will have the largest grains? Which part will have the smallest?

14. A tapered ¼-in.-thick bar of annealed copper of the dimensions shown in the figure is pulled (stretched) so that the 8-in. part becomes 9-in. long. It is then given a recrystallization anneal. Draw a graph showing how the grain size would compare along the 9-in. length.

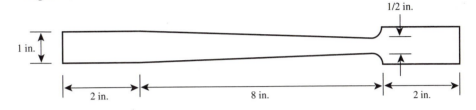

15. Discuss how the curves in Figure 3-10 would shift if the percentage of cold working before heating was increased.

FOR FURTHER STUDY

Alexander, W. O., G. J. Davies, K. A. Reynolds, and E. J. Bradbury. *Essential metallurgy for engineers.* Wokingham, U.K.: Van Nostrand Reinhold, 1985, secs. 2.11 and 2.12.

Avner, Sidney H. *Introduction to physical metallurgy* (2d. ed.). New York: McGraw-Hill, 1974, chap. 4, Annealing and hot working.

DeGarmo, E. Paul, J. Temple Black, and Ronald A. Kohser. *Materials and processes in manufacturing* (7th ed.). New York: Macmillan, 1988, chap. 17, Hot working processes, 76–80.

Reed-Hill, Robert E., and Reza Abbaschian. *Physical metallurgy principles* (3d. ed.). Boston: PWS-Kent, 1992, chap. 7, Annealing.

CHAPTER 4

Alloys

OBJECTIVES

After studying this chapter, you should be able to

- identify and describe the three metallic single phases and sketch their cooling curves
- explain what it means to form a metallic solid solution and what the conditions of saturation are
- describe and distinguish between the two types of solid solutions and state their characteristics and properties and the rules that govern their formation
- describe and distinguish between the two types of compounds and state their characteristics and properties and the rules that govern their formation

Part of the metallic world we experience is made of pure metals, but most of it is not. Jewelry, electrical conductors, and aluminum foil are some of the uses met by pure, or almost pure, metals, but most of our world, and especially that portion requiring high mechanical strength, is made of combinations of metals. Some of these combinations, or alloys, are solutions of metals, some are compounds, and many are mixtures.

In some alloys the amounts of other metals may be very large and in others the "alloying element" may be present in a very small quantity. Monel metal is an approximately 70-30 combination of nickel and copper; most steels contain less than 1% carbon.

A PREVIEW

An alloy is a combination of a metallic element with another metallic or nonmetallic element. We begin our study by describing the difference between pure metals and alloys. We then discuss the uses of alloys. We'll examine and describe each of the forms in which alloys exist—solid solutions, compounds, and mixtures.

The solubility of one metal in another (a solute atom dissolved in a solvent lattice) occurs at the atom-lattice level. Here we look at what solid solutions are, how they form, and what happens when saturation occurs and solubility is exceeded. Saturation has three conditions, and we'll discuss these. We'll define and describe the two types of solid solutions—substitutional and interstitial—and the two types of compounds—intermetallic and interstitial.

We'll examine the concept of phases and the single-phase forms of metals. The homogeneous forms of metals and alloys—pure metals, solid solutions, compounds—are single phases; mixture alloys are combinations of these homogeneous forms. When a metal changes states (e.g., liquid to solid) a change of phase occurs. These concepts are very important to an understanding of both phase diagrams and heat-treating, which are topics of later chapters. Finally, we'll look at the effect on mechanical properties of solute atoms entering the lattice of another metal.

This chapter deals only with binary alloys, alloys of two metals. You should realize that there are many applications in the commercial world where three, four, or more elements are combined in an alloy. The influence of each element on the whole must then be evaluated. Later, especially in the chapters on steels, we'll discuss situations where we have to account for the influence of more than one additional element in an alloy.

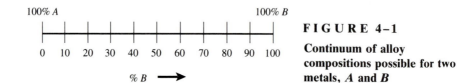

FIGURE 4–1

Continuum of alloy compositions possible for two metals, A and B

4.1   WHY ALLOYS?

One of the disadvantages of pure metals is that their mechanical strength can only be changed by annealing or by cold-working. Alloying enables metals to be strengthened two ways: through solid-solution strengthening (combining metals with different-size atoms, like monel metal) or by making a metal heat-treatable, e.g., adding carbon to iron or putting a small amount of copper in aluminum. In general, alloying elements are added to metals to

- improve (or change) various mechanical properties—e.g., tensile strength, ductility, or strength at specific temperatures
- improve (or change) various physical properties—corrosion resistance, electrical and/or magnetic characteristics, modulus of elasticity
- influence the microstructure of the metal—e.g., control the grain size in cast, heat-treated, or annealed products
- enable metals to be heat treated

The binary alloys possible from combining two pure metals can be thought of as a continuum containing all possible combinations of the two metals; this is shown in Figure 4-1. For discussion purposes, it is common practice to refer to these as metals A and B, but we could be talking of copper and nickel or many other combinations. In Figure 4-1 the left end of the line is 100% A, 0% B; the right end is 0% A, 100% B. Because we read from left to right, only the percentage of the metal at the right (B) is tracked. Obviously an alloy of 70% B also contains 30% A.

4.2   PURE METALS

A crucible of 100% pure molten metal,* nickel, for example, is a sea of randomly oriented atoms, all of which are alike. When the metal freezes (crystallizes), it forms into the atom lattice appropriate for nickel, FCC.† Its cooling curve is similar to that shown in Figure 1-28, which is reproduced here as Figure 4-2. When the pure metal crystallizes, it does so at a constant temperature (ignoring the

---

*It is highly unlikely that any large quantity would be 100% pure.
†Nickel's lattice is FCC at all temperatures. Other metals, e.g., iron, might freeze in one lattice and go through other lattice changes before they get to room temperature.

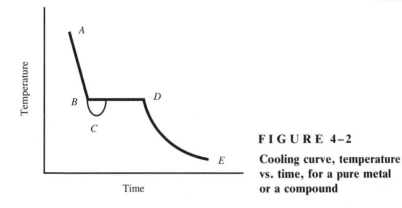

FIGURE 4–2

Cooling curve, temperature vs. time, for a pure metal or a compound

undercooling at point *C*) of 1453°C (2647°F) and is said to change states; i.e., it goes from the liquid state at point *B* to the solid state at point *D*.

Examining one crystal lattice at points *D* or *E* reveals that nickel atoms fill all 14 positions in the lattice. If a specimen of the frozen metal is properly prepared and examined under the microscope, all that is visible are grains and grain boundaries; it appears *homogeneous*.

4.3 SUBSTITUTIONAL SOLID SOLUTIONS

Freezing of a Solid Solution. Now assume that the molten metal being cooled is an alloy of 50% nickel and 50% copper; i.e., the randomly oriented atoms are approximately half nickel and half copper.[‡] When the temperature of the liquid is lowered, it is noticed that at the melting-freezing temperature of nickel, 1455°C (2651°F), nothing happens. At about 1320°C (2410°F) almost all of the metal is liquid, but some tiny dendrite crystals do start to form (point *B* of Figure 4-3). If one of these is lifted out of the melt and the atoms counted in the FCC lattices, it would be found that on the average approximately nine out of every fourteen atoms are nickel and five are copper, and that the copper atoms are dispersed (separated from each other) in the lattice.

If the temperature drops to about 1290°C (2350°F), about half the metal is frozen; if another solid crystal is pulled out of the liquid and analyzed, it would be found that on the average approximately eight out of every fourteen atoms in the atom lattices are nickel and six are copper.

When the temperature drops to 1260°C (2300°F), the metal is all frozen (point *D* on Figure 4-3). This is well above the freezing temperature of copper, which is 1084°C (1981°F). The analysis of a lattice at this temperature shows that, overall,

[‡]The atomic weights of copper and nickel are 63.546 and 58.71, respectively; for this example they are assumed equal. Copper's lattice is also FCC at all temperatures.

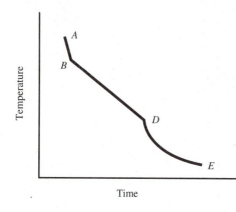

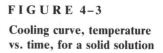

FIGURE 4-3

Cooling curve, temperature vs. time, for a solid solution

approximately half the atoms are nickel and half are copper, and that they are randomly distributed through the lattices.

If a specimen of the frozen metal is prepared metallographically and inspected under the microscope, it will be homogeneous; it will show nothing but grains and grain boundaries. On a macro basis the metal is homogeneous, but internally it is a combination of two kinds of atoms in a solid solution that froze over a range of temperatures rather than at one temperature.

The Effect of Solute Atoms.   When metallic solid solutions form, the metal doing the dissolving is termed the **solvent** and the metal being dissolved is termed the **solute.** That is, the atom that is introduced into the other metal's lattice is termed a *solute atom.*

Solute atoms will not have the same size as the solvent atoms, and when introduced into the lattice they will be either too big or too small to "fit" the lattice exactly. Either way they strain the lattice. Therefore when the solute atom concentration is low, the solute atoms tend to take up locations away from each other so that they have the minimum effect on the energy level of the lattice. As the concentration of solute and solvent atoms nears equality, the atoms tend to be more randomly distributed, but in some cases the atoms become **ordered** and occupy certain positions in the lattice. For example, in a 50-50 atomic percent alloy of copper and zinc, the atoms of the two metals become ordered so that copper and zinc atoms alternate in the lattices. (Ordering is discussed further in Chapter 5.)

Complete Substitutional Solubility.   When one material dissolves in another, it is said to be *soluble* in the other; solubility exists. In the case of copper and nickel, they both have the same type lattice, are next to each other on the periodic table (see Table I-1), and therefore have atom diameters that are about the same size. Because of their closeness in size they can share or fit into the same lattice. The result is that these two metals have 100% or **complete solubility** in each other; that is, any amount of one can be dissolved in the other. Another way to put that is to say that their *solution never becomes saturated; it remains unsaturated.* Copper and nickel can *substitute* into each other's lattice, since it is the same lattice, in any amount and form **substitutional solid solutions.**

In general, those metals with the same lattice that have atom diameters within 8% of each other have complete solubility.[§]

Partial Substitutional Solubility: Size Factor. Copper and silver both have FCC lattices but their atom diameters vary more than do those of copper and nickel. The difference in their diameters is 11.5% to 13%;[||] this is too great a difference in size for the atoms to fit into the lattice in any quantity, but some silver will dissolve in pure copper, and some copper will dissolve in pure silver. The two metals do not form substitutional solid solutions over their full range of compositions, but they are *partially soluble;* that is, *their solutions can become saturated.* Since these ranges of solubility include the pure metals, i.e., the ends of the continuum of Figure 4-1, they are termed **primary** or **terminal, solid solutions.**

In general, this type of **partial solubility** is limited to metals with atom diameters that vary between approximately 8 and 14% of each other.

Partial Substitutional Solubility: Different Lattice. Copper and zinc are immediately beside each other on the periodic table, and their diameters vary only about 4%, but because they do not have the same lattice (copper = FCC, zinc = CPH) they cannot have complete solubility, or their *solutions can become saturated.* Because of their closeness in size, however, they do have a very large terminal partial solubility. This alpha (α) region of terminal solid solubility goes up to almost 40% zinc dissolved in copper and contains common brasses such as cartridge brass (70% Cu-30% Zn) and red brass (85% Cu-15% Zn).

Complete Substitutional Insolubility: Size Factor. Copper and lead both have the FCC lattice, but their atom diameters differ by almost 1 Å, or a minimum of about 25%. This is such a large size difference that virtually no solubility exists between these two metals; they have **complete insolubility.** Metals whose atomic diameters vary by more than 15% generally do not form substitutional solid solutions.

4.4 ■ INTERMEDIATE SUBSTITUTIONAL SOLID SOLUTIONS AND COMPOUNDS

Metals that do not have complete solubility may form solid solutions and/or compounds at compositions other than the terminal compositions. For example, copper and zinc, in addition to the two terminal solid solutions, form four other

---

[§]The relationships for the formation of solid solutions and the so-called electron compounds were identified by W. Hume-Rothery. See W. Hume-Rothery and G. V. Raynor, *The structure of metals and alloys* (3d. ed.), London: Institute of Metals, 1954, and/or W. Hume-Rothery, *Elements of Structural Metallurgy,* Monograph and Report Series No. 26, London: Institute of Metals, 1961, esp. chaps. VIII and IX.

[||](2.889 Å − 2.556 Å) ÷ 2.889 Å = 11.5%; (2.889 Å − 2.556 Å) ÷ 2.556 Å = 13%.

solid solutions, and magnesium and silicon are virtually insoluble in any proportion, except at one composition where they form the compound Mg_2Si.

Intermediate Solid Solutions.   Intermediate solid solutions occur in many substitutional alloy systems. They often have lattices that differ from the terminal solutions; in the copper-zinc system, starting from the 100% copper end, the lattices are FCC, BCC, BCC, cubic, CPH, and CPH. These intermediate solutions form just as those described above, with the atoms of the two metals sharing a single lattice type.

W. Hume-Rothery has pointed out that many of these intermediate solid solutions are compoundlike in that they form at ratios of the number of valence electrons to the number of atoms. For example, the beta (β) brass solid solution forms over a range of compositions that includes 50% copper and 50% zinc. Since their atomic weights are almost equal, if this were a compound its formula would be CuZn; it would have three valence electrons, one from copper and two from zinc, and two atoms, or an electron-atom ratio of $3:2$.

Other electron-atom ratios have been identified as $21:13$ and $7:4$; the gamma (γ) and delta (δ) brass solid solutions occur at these ratios, respectively. We've used the copper-zinc system as an example, but other alloy systems show the same phenomenon.

These solid solutions are compoundlike in their numerical regularity, but their bonding is basically metallic, so they can exist over a range of compositions as opposed to the exact proportions of a compound such as Mg_2Si; also they melt-freeze over a range of temperatures (Figures 4-3) and have properties and internal structures like solid solutions. In this text they are treated as intermediate solid solutions, but they are sometimes called *electron compounds* and *electron phases*.

Intermetallic Compounds.   Intermetallic compounds are formed by elements that are separated on the periodic table so they have electrical attraction for each other. The compounds are simple ratios of the elements and follow the normal valence rules, e.g., Mg_2Si; they have a narrow range of composition, and they melt-freeze at one temperature, as seen in Figure 4-2. Their bonding is ionic or covalent, and their properties tend to be **nonmetallic;** that is, they are hard and brittle. Since intermetallic compounds follow the valence rules, they are also called **valency compounds.**

The crystal structures of intermetallic compounds vary considerably, from simple cubic and body-centered cubic types to unusual variations of cubic and hexagonal configurations.

4.5   INTERSTITIAL SOLID SOLUTIONS

Interstitial Solubility.   In substitutional alloying, if the difference in the atom diameters becomes too great, the metals will not form a solution; however, if the difference in the diameters becomes large enough, the smaller atom is able to fit in the open spaces of the other metal's lattice and form an interstitial (between the spaces) solid solution. This type of alloying does *not* occur between the metals that are in the center of the periodic table; they are too similar in diameter. Instead, interstitial solubility occurs between the metals in the center of the table and the elements in the upper parts of the table, i.e., those with small diameters.

Referring to Table I-1, you can see that the elements *hydrogen, boron, carbon, nitrogen,* and *oxygen*[#] have diameters that are approximately 1.0 to 1.5 Å smaller than those of the common metals; these are the solute atoms that typically form interstitial solid solutions with other metals; more specifically, they are the transition metals. It is believed that this type of alloying is limited to the transition metals because of their unfilled electron subshells. The transition metals are the B group of metals in Table I-1.

Limitations on Interstitial Solubility. The formation of interstitial solid solutions is limited by two conditions. First, the space available is never exactly equal to the diameter of the solute atom that must fit between the spaces of the solvent's lattice. That is, the solvent lattice must be stretched, or put under strain, so it can accommodate the solute atom. The second is that the amount of open space in a lattice is limited; hence the solubility is limited. That is, interstitial solid solutions become easily saturated. Examples of solubilities like those discussed previously (nickel-copper and copper-zinc) are not found in interstitial solid solutions. Typically, the solubility of interstitial solid solutions is very low, amounting to a few percent for terminal solutions and perhaps 10% for intermediate solutions.

Common Interstitial Solid Solutions. Some of the more common interstitial solid solutions are formed between the transition metals—iron, manganese, chromium, vanadium, titanium, molybdenum, and tungsten—and solute atoms of carbon and/or nitrogen. The most important of these are the interstitial solid solutions formed by iron and carbon called *ferrite* and *austenite.*

Characteristics of Interstitial Solid Solutions. Interstitial solid solutions melt-freeze over a range of temperatures (see Figure 4-3) and have metallic properties. A properly prepared specimen viewed through the microscope appears homogeneous, having just grains and grain boundaries.

4.6   INTERSTITIAL COMPOUNDS

When transition metals are combined with small-diameter atoms (B, O, N, C, H) in certain ratios, they form interstitial compounds that are hard and brittle; that have high melting points; that melt-freeze at one temperature; and that have narrow ranges of composition. If the ratio of the solute to solvent atom diameters is low—i.e., the solute atom is small—the solute atom will usually take up inter-

---

[#]As a memory aid, you can make of their chemical symbols a "pronounceable" word—BONCH.

stitial positions in a relatively simple lattice. If the ratio is large, the lattice that results will be very complex. Thus the compound TiC, with a low ratio, has a simple cubic structure, and the compound Fe_3C, with a larger ratio, has a very complex hexagonal-type structure.

Many of the interstitial compounds are extremely hard and wear-resistant and have many industrial applications. Iron carbide or cementite (Fe_3C) and tungsten carbide (W_2C) are examples.

4.7   PHASES

Homogeneous Alloys, or Single Phases.   The metal alloys discussed so far in this chapter are homogeneous alloys. That is, when viewed under a microscope they appear homogeneous; they are uniform and distinct; other structures are not mixed in with them and they are easily distinguishable in the microstructure since the structure is nothing but grains and grain boundaries.** These homogeneous forms are called *phases* or *single phases*. That is, a *phase is completely homogeneous, separate, and distinct when viewed through the microscope.*

There are three ways or forms in which solid metals can exist as single phases—as pure metals, in solid solutions with other metals or nonmetals, and in compounds with other metals or nonmetals.

Solubility and Mixtures.   The only other way metals can exist as solids is in mixtures of two or more of the single-phase forms. Obviously, mixtures are not homogeneous single phases but combinations of phases. *Mixtures come about because of lack of solubility.*

The solubility of metals is determined by their electron configurations, which translates into atomic size differences, lattice differences, etc. The solubility of the metals may vary with the temperature of the metals as liquids and as solids. This will be explored in more detail in the next chapter, but it is important here to understand that when solubility is exceeded, when solutions become saturated, mixtures occur, and that this can occur in different ways. Under the microscope, some mixtures will be as obvious as oil and water; others will be subtle precipitates that are only hinted at when seen with the high magnification of an electron microscope.

There are three possible conditions of saturation that relate to the solubility of solutions:

1. The solution may be **unsaturated;** i.e., the solvent is capable of dissolving more of the solute atoms into its lattice.

**The photomicrographs in Figures A-4, A-6, A-10, A-11, and A-12 all show single-phase solid solutions. The differences in shading and color of Figures A-10, A-11, and A-12 are due to lighting and etching; the parallel stripes are twins. Figures A-13, A-14, and A-15 all show the mixture nature of slow-cooled steels.

2. The solution may be **saturated;** i.e., the solvent has dissolved as much solute as it can. If an attempt is made to add additional solute it will come out of solution; i.e., it will precipitate from the lattice.
3. The solution may be **supersaturated;** i.e., the solvent has dissolved more solute than it would normally be capable of at the particular temperature and pressure. Supersaturated solutions are often created by cooling metals from high temperatures so rapidly that the excess solute cannot escape from the lattice.

Changes of Phase.   When the atoms of a metal go from the random orientation of the liquid state to the crystalline solid state, the metal changes from one distinct homogeneous form to another; it has changed phases.[††] In like fashion, when a metal's atom lattice goes through an allotropic change—a change in lattice due to a change in temperature—it has changed phases. So *changes of state and allotropic changes are changes of phase.*

4.8   SOLID-SOLUTION STRENGTHENING

It has been previously pointed out that in both substitutional and interstitial solid solutions the combining of atoms of different diameters strains the lattice. Atoms have to take up positions that are not at their normal closest-approach distance from neighboring atoms, and the internal energy of the atoms in the lattice goes up. One result of this is that the physical and mechanical properties of the alloy are different from those of the pure metals.

Figure 4-4 shows how the mechanical properties of alloys of two metals with complete substitutional solid solubility—nickel and copper—change with composition. Note that, as small amounts of the weaker copper are added to the stronger nickel, the tensile strength goes up.[‡‡] The properties that result from this alloy system are not an average of the two, but each metal influences the properties of the other. This effect on properties is called **solid-solution strengthening.**

In Chapter 7 we'll find that the mechanical properties of the interstitial solid solution of carbon in iron, called martensite, depend on the amount of the solute carbon atoms present.

As a general rule, the addition of solute atoms, interstitially or substitutionally, to metals will raise strength and hardness and lower plasticity and electrical conductivity.

The effects of solid-solution strengthening, as in the nickel-copper system of Figure 4-4., is not necessarily spectacular; e.g., the difference between the strength of annealed nickel and the maximum strength is only 20%. By contrast, cold-working can more than double the tensile strength of some metals and heat-

[††] As will be discussed later, the liquid may freeze to one solid phase or a mixture of solid phases.
[‡‡] The commercially important alloy monel has an approximate composition of 67% Ni-30% Cu, the composition that has the maximum properties shown in Figure 4-4.

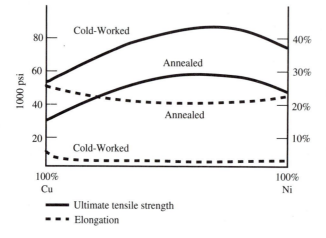

FIGURE 4–4

Mechanical properties of nickel-copper solid-solution alloys

Source: Clyde W. Mason. *Introductory physical metallurgy.* Materials Park, OH: ASM International, 1947, fig. 10, 18. Used by permission.

treating can multiply strengths even more. It is interesting, however, that the change in properties of the nickel-copper system occurs by combining two metals that differed in atom diameter by less than 0.10 Å.

Solid-solution strengthening can be important in improving the strength properties of metals used at high temperatures, where the operating conditions remove the effects of cold-working or heat-treating.

SUMMARY

1. Single phases of metals are homogeneous, distinct, separate forms of the metal.

2. The homogeneous, single-phase forms of metals are pure metals, solid solutions, and compounds.

3. Binary alloys are formed by metals as solid solutions, compounds, and mixtures of the three single phases. The mixture alloy combinations are:
   • pure metal + solid solution
   • pure metal + compounds
   • solid solution + compound
   • two pure metals
   • two solid solutions
   • two compounds

4. Solid solutions are formed when solute atoms reside in the lattice of the solvent atom; the solute atom is "dissolved" by the solvent atom. Their bonding method is metallic.

5. Solid solutions can take two forms:
   a. substitutional, where the solute atoms reside in regular positions in the lattice of the solvent atom.

      b. interstitial, where the solute atoms reside in open spaces within the lattice of the solvent atoms; that is, the solute atoms locate between the solvent atoms, which still occupy their regular positions in the lattice.

6. The two types of solid solutions have different rules for formation:
      a. Substitutional solid solutions have three ranges of solubility that generally depend on the relative diameters of the atoms; those whose atom diameters are
          (1) within 8% of each other and have the same lattice will generally have complete solubility.
          (2) between 8 and 14% of each other will have partial solubility; having the same lattice is not required.
          (3) more than 15% apart will have no solubility, i.e., they are insoluble.
      b. Interstitial solid solutions are formed between transition metals, e.g., iron, chromium, and small-diameter metals such as boron, oxygen, nitrogen, carbon, and hydrogen (BONCH). The extent of solubility in interstitial solutions is usually very low, amounting to only a few percent.

7. Some alloy systems form intermediate solid solutions; most of these are substitutional, but some interstitial types are formed. They melt-freeze over a range of temperatures and have metallic properties.

8. Solutions have three conditions of solubility:
      a. Unsaturated: the solvent can dissolve more solute.
      b. Saturated: the solvent has dissolved all the solute it can at that temperature and pressure.
      c. Supersaturated: the solution contains more solute than it normally could at that temperature and pressure.

9. Introducing solute atoms into a lattice strains the lattice and raises the mechanical properties of the resulting alloy. This is solid-solution strengthening.

10. Compounds are also intermediate alloys, and they take two forms:
      a. intermetallic, which form in simple ratios of the metals and are based on regular valence rules. These melt-freeze at one temperature and are hard and brittle.
      b. interstitial, which form only between the transition elements and the small-diameter atoms. These also melt-freeze at one temperature, have high melting points, and are hard and brittle. Depending on the relative size of solute and solvent atoms, the crystal lattice may be very simple or very complex.

11. Mixtures of single phases occur because the metals lack solubility.

REVIEW QUESTIONS

1. What does it mean for a metal to be homogeneous? What is a single phase?
2. What is a metallic solid solution? Describe what a solid solution of copper and nickel would look like on an atomic level. What would a solid solution of carbon and iron look like?

3. What range of solid solubility would molybdenum and tungsten likely have? Why?

4. What type of alloys would silicon and aluminum likely form? Solid solutions? Compounds? Why?

5. Is it likely that cadmium and lead would have very much solubility for each other? Why?

6. Is it likely that interstitial solid solutions would have large ranges of solubility? Why? Would it be possible to have 100% interstitial solubility?

FOR FURTHER STUDY

Avner, Sidney H. *Introduction to physical metallurgy* (2d ed.). New York: McGraw-Hill, 1974, chap. 5, Constitution of alloys.

Barrett, Charles, and T. B. Massalski, *Structure of metals* (3d rev. ed.). Oxford: Pergamon, 1980, chap. 10.

Hume-Rothery, W. *Elements of structural metallurgy*. Monograph and report series no. 26. London: Institute of Metals, 1961, chaps. VIII and IX.

Reed-Hill, Robert E., and Reza Abbaschian. *Physical metallurgy principles* (3d. ed.). Boston: PWS-Kent, 1992, esp. secs. 9.1–9.3, 9.5, 14.10, 14.11, and 14.13–14.16.

Smallman, R. E. *Modern physical metallurgy* (4th ed.). London: Butterworth, 1985, secs. 5.4–5.6.

Phase Diagrams

OBJECTIVES

After studying this chapter, you should be able to

- describe how the cooling curves of soluble, insoluble, and partially soluble–insoluble alloy systems are used to generate their phase diagrams
- define phase and describe a phase change
- describe liquidus, solidus, and solvus lines on a phase diagram and state what information can be obtained from them
- sketch typical phase diagrams for two metals that are completely soluble, completely insoluble, and partially soluble–insoluble; identify the lines and areas of the diagrams—in other words, relate the solubilities of two metals to the type of phase diagram they form
- describe the formation of the eutectic mixture and state Raoult's law
- identify the three types of metallic single phases—pure metals, compounds, and solid solutions—on phase diagrams
- distinguish single phases from mixture regions on phase diagrams
- differentiate between intermetallic and interstitial compounds and describe their appearance on phase diagrams
- differentiate between terminal and intermediate solid solutions and describe their appearance on phase diagrams
- define ordering and explain how it is shown on phase diagrams
- use a phase diagram to describe how alloys freeze
- explain why the tie-line rule must be used inversely
- use a phase diagram to determine the composition and percent relative amount of phases, given a temperature and alloy composition
- describe, and identify on a phase diagram, the eutectic, eutectoid, peritectic, and peritectoid reactions, and state their consequences

- predict—that is, sketch—the microstructure resulting from slow or fast cooling, given one of the three basic phase diagrams and a composition
- use a phase diagram to explain nonequilibrium cooling and the resulting segregation

DID YOU KNOW?

A few combinations of metals are completely soluble in each other as solids, and most metals have some solid solubility for other metals, although sometimes it is very low. On the other hand, there are some combinations of metals that are not even soluble as liquids. For example, when molten lead is added to molten aluminum, it remains a globule of lead and freezes that way when cooled. The lead particle, being soft, is essentially a void in the aluminum. It is added to aluminum alloys to act as a "chip breaker," to improve machining characteristics.

A PREVIEW

This chapter deals with binary alloys of metals that are soluble in the liquid state; the alloys used for significant engineering applications—structures, vehicles, machine elements—are of this type. Actual alloy systems, the alloys created with two metals, are used for purposes of illustration. In each case the study is intended to illustrate a broader type or class of alloys.

Much of the material in this chapter deals with substitutional alloys, since they offer the widest variety of phase-diagram types. However, the iron-carbon alloy system, the most important of the interstitial alloy systems, is discussed as an example of this other type of alloy.

We'll approach phase diagrams from the standpoint of the solubilities of the metals involved. The phase diagrams for soluble, insoluble, and partially soluble–insoluble systems are developed in turn and the cooling of select alloys is traced. We'll use phase diagrams to determine the compositions and percent relative amounts of phases.

We'll then discuss intermediate phases, i.e., compounds and solid solutions, as they relate to phase diagrams. And we'll briefly touch on ordering and superlattices. In Section 5.8 we'll take up the practical problem of nonequilibrium cooling and explain the formation of nonhomogeneous solid solutions, or cored crystals, by using a phase diagram. This continues the discussion on homogenization treatments begun in Chapter 1.

We'll explain the eutectic, eutectoid, peritectic, and peritectoid reactions and describe their consequences. Lastly we'll discuss the sequence in which phases form during cooling and relate that sequence to the microstructures that result.

5.1   SOLUBILITY

Since only alloys that are soluble in the liquid state are being considered, the solubility of concern is in the solid state that occurs when atoms coexist, either substitutionally or interstitially, in the same lattice. Chapter 4 established the conditions required for this coexistence; ignoring temperature and pressure, these conditions relate to the composition of the alloy and the basic characteristics of the atoms themselves.

However, since alloys are formed by melting different metals together, temperature is very important and cannot be ignored; to the continuum of compositions shown in Figure 4-1 must be added temperature. The study of the resulting diagram of temperature versus composition, the phase diagram, promotes an understanding of the structure and properties of alloys and how they can be modified by heating or cooling treatments.

Pressure can be a variable in the process of creating alloys, but most commercial operations are carried out at atmospheric pressure and that constant pressure is assumed in this text.

5.2   FREEZING OF SOLUBLE ALLOYS

Development of the Phase Diagram. In Chapter 4 the experimental cooling curves (temperature versus time) of pure metals and solid solutions were discussed. Recognizing that soluble alloys can be formed from pure metals that are completely soluble in each other, we can combine these curves to get an idea of how the melt-freeze temperatures vary for all compositions. A third dimension is used in Figure 5-1 to show the cooling curves for five compositions in the copper-nickel alloy system. Notice that the freezing of the two pure metals occurs at single temperatures—i.e., portion *B-D* is flat—while the solid solutions freeze over ranges of temperatures—i.e., *B-D* has a slope.

The time axis of Figure 5-1 goes to the left, into the page. The points where solidification begins and is completed (*B* and *D*) can be projected onto a plane erected along the time axis; a line plotted through the projected points forms a narrow, lemon-peel-shaped figure. This is the phase diagram for the copper-nickel substitutional alloy system, shown in its normal presentation in Figure 5-2. That is, a phase diagram is nothing more than a diagram whose lines indicate the temperatures and compositions where phases change. In Figure 5-1 all the *B* points are where solidification (or freezing, or crystallization) begins, i.e., where the liquid phase begins to change to the solid phase, and all the *D* points are where that process is completed and everything is in the solid phase.

Liquidus and solidus lines. The freezing of a 50-50 alloy of copper and nickel was traced in Chapter 4; now we'll do it again using Figure 5-2.

The cooling metal follows the 50% composition line on the diagram and reaches point *B,* which lies on the **liquidus line.** For any composition of alloy, *the*

FIGURE 5–1 Multiple cooling curves develop the copper-nickel (completely soluble) phase diagram

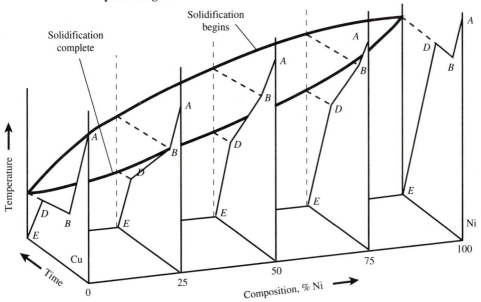

liquidus line represents the lowest temperature where the metal is all liquid, or it is the temperature where the first solid metal starts to form when the liquid metal is cooled. The liquidus divides the single-phase liquid region from the two-phase region below that is a mixture of liquid and the copper-nickel solid solution.

Continued cooling to point *D* completes the solidification process and the metal freezes solid. Point *D* lies on the **solidus line.** For any composition of alloy, *the solidus line is the highest temperature at which the metal is all solid,* or conversely, *it is the lowest temperature at which the liquid phase exists.* The solidus is the dividing line between the liquid-solid mixture region and the single-phase solid solution below.

As solid solutions occur in the continuum of alloys formed by two metals, it is normal practice to identify them in order of the Greek alphabet. Since a completely soluble system such as copper-nickel has only one solid solution, it is called alpha (α). This solid-solution alpha (α) is a single phase and with proper cooling appears under the microscope as a separate, distinct, homogeneous alloy.

Composition of Phases During Cooling of an Alloy of 50% Ni. Tracing again the cooling of the 50% Ni alloy in Figure 5-2, we see that at point *B* (1320°C, 2408°F) the liquid has the composition of approximately 50% nickel. Some crystals have just begun to form, but their composition can't be 50% Ni or they would be liquid at that temperature. Since a temperature on the *solidus* is the highest temperature at which a solid of the corresponding composition can exist, the first solid must be being formed at point *M*, and it must have a composition of about 63% nickel (approximately nine of the fourteen atoms per lattice are nickel).

FIGURE 5–2   Phase diagram for soluble metals copper and nickel

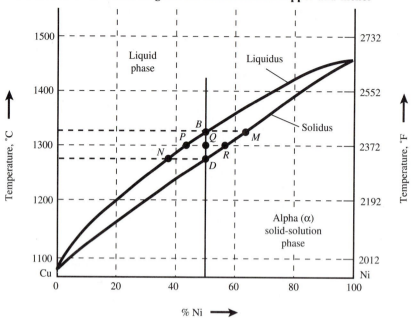

Source: *Metals handbook* (9th ed.). Materials Park, OH: ASM International, 1973, *8,* 294. Annotations by author.

At point *D* (approximately 1280°C, 2336°F), the solid has the composition of approximately 50% nickel, but the last liquid that was available to freeze couldn't have been of the composition of *D* or it would have been frozen already. Since a temperature on the *liquidus* is the lowest temperature at which liquid of the corresponding composition can exist, the last liquid must be of the composition at point *N,* or about 37% nickel.

In this alloy system, the solidification process prefers to freeze nickel atoms first (since they freeze at higher temperatures) and then copper atoms. So, although the liquid starts out with about equal numbers of each atom, it becomes more concentrated in copper as the temperature drops, so that the final liquid contains much more copper than nickel. Conversely, the first solid to form has much more nickel in it than copper. In Figure 5-2 the composition of the *liquid* initially corresponds to point *B* and progresses to point *N* during cooling; the *solid* starts with a composition corresponding to point *M* and moves to point *D*

For purposes of illustration, if cooling stops at point *Q* (1300°C, 2372°F), the metal will be a *mixture* of the liquid and solid-solution phases. The liquidus and solidus lines are the boundaries of this region, and at any temperature within this mixture region, the two lines represent the composition of the liquid and solid phases, respectively, that can exist at that temperature. Thus at the temperature represented by point Q the composition of the liquid corresponds to point P (about 43% nickel), and the composition of the solid corresponds to point R (about 57% nickel).

Two general rules develop from this discussion: In a mixture region of a phase diagram, e.g., inside the envelope of liquidus and solidus lines, *the compositions of the phases that makes up the mixture at that temperature are read from the appropriate line, i.e., liquid from liquidus, solid from solidus.* And in single-phase regions of a phase diagram, *the composition can be read directly from the diagram, since the metal is not a mixture but all one phase.*

5.3   FREEZING OF INSOLUBLE (MIXTURE) ALLOYS

Development of the Phase Diagram. When metals are insoluble in the solid state, the freezing process completely segregates the two metals and must result in a mixture of the two pure-metal phases. That is, an alloy of metals insoluble as solids must freeze as the pure metals themselves; since they are insoluble they can't freeze and exist in each other's lattices.

Figure 5-3 is a three-dimensional view of the cooling curves at different compositions that develop the phase diagram for bismuth and cadmium, which are insoluble as solids. Note that the cooling curves of the two pure metals have the flat *B-D* portion where melting-freezing takes place at one temperature, and that

FIGURE 5-3 Multiple cooling curves develop the phase diagram for insoluble metals bismuth and cadmium

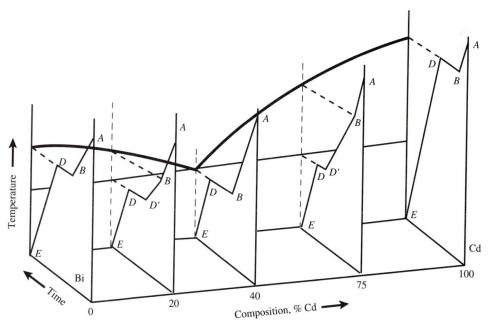

the cooling curve of 40% Cd also has a flat *B-D* portion. However, the intermediate compositions, 20 and 75% Cd have a different shape; they have a slope from *B* to *D'* and then a flat portion from *D'* to *D*. Projecting these points of change onto a plane erected on the time axis results in the phase diagrams shown in Figure 5-4.

Comparing Figures 5-3 and 5-4, we find that the pure metals bismuth and cadmium freeze at temperatures *BD*. The alloys with intermediate compositions of 20 and 75% Cd, however, freeze over ranges of temperature that begin with *B* and continue down to *D'D*, where they become solid.

Composition of Phases During Cooling of an Alloy of 20% Cd (Figure 5-4).

When the temperature reaches *B*, which is on the liquidus line, crystals of solid metal begin to form with the composition of point *M*. This happens because bismuth and cadmium are insoluble as solids, and the only metal that can freeze is one of the pure metals. Also note that in Figure 5-4 the solidus line includes parts of the vertical lines representing the pure metals. Remembering the definition of the solidus, at the temperature of point *B*, we note that the composition of the first metal to freeze is 100% bismuth (which is 0% Cd), or point *M*.

At the temperature of point *Q* the metal that is freezing still has the composition of 0% Cd, or point *R*. However as the pure bismuth freezes, there is less of it available in the liquid; the remaining liquid must therefore contain proportionally more pure cadmium. At the temperature of point *Q* the composition of the liquid corresponds to point *P*, or approximately 30% Cd.

As the temperature continues to fall, pure bismuth (of a composition of 0% Cd) continues to freeze and the proportion of pure cadmium in the liquid continues to increase. At a temperature just above point *D'D*, the only metal that has

F I G U R E 5–4 Phase diagram for insoluble metals bismuth and cadmium

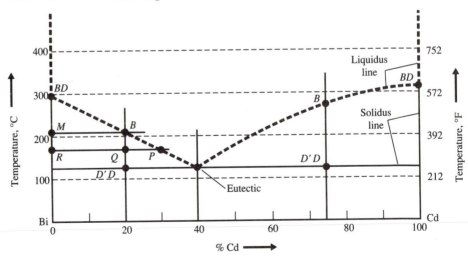

Source: *Metals handbook* (8th ed.). Materials Park, OH: ASM International, 1973, *8*, 272.
Annotations by author.

frozen so far is pure bismuth, but the composition of the liquid has followed the liquidus line down to the point labeled *BD* at 40% Cd. That is, the liquid remaining to freeze has the same composition as the alloy of 40% Cd shown in Figure 5-3.

Figure 5-3 shows that this alloy of 40% Cd has a cooling curve with a flat portion. In alloy systems of *insoluble metals* the composition of the last liquid to freeze always will be at a point similar to that labeled "eutectic" in Figure 5-4; that is, for any alloy, as freezing proceeds, the liquid will move toward a composition that has the lowest melting-freezing temperature; this point is called the *eutectic*. This explains why the cooling curves for the 20 and 75% Cd alloys of Figure 5-3 have a flat *D'D* portion.

The formation of the **eutectic** is described by **Raoult's law,** which essentially says that *alloys of two metals that are soluble as liquids but insoluble as solids will melt-freeze at a temperature lower than the melt-freeze temperature of either pure metal*. Note that this is not the result for soluble metals, as discussed in Section 5.2; there the melt-freeze temperature for all alloys was somewhere *between* that of the two pure metals. (See the solidus line of Figure 5-2.)

The end result of the freezing of the alloy of 20% Cd is a microstructure containing pure bismuth and the eutectic mixture of bismuth and cadmium.

Freezing the Liquid of Eutectic Composition. In the liquid state the atoms of alloys move about in random fashion; the atoms of metals that are soluble as liquids will be reasonably uniformly distributed throughout the melt. However, if the metals are insoluble in the solid state, it means that at the freezing temperature this uniform distribution must cease; i.e., the metals must become segregated because below this temperature they are no longer soluble, and their atoms can no longer exist in this random distribution.

From the cooling curves of Figure 5-3 note that *this segregation takes place at one temperature, but it does not occur instantaneously*. The cooling curves all include a time lag during which the metals segregate and the latent heat is removed; the closer the alloy composition is to that of the eutectic, the greater the time required.

Whether the two metals are of the same lattice type or not, if Raoult's law operates they can't freeze in the same lattice. *They must lose their latent heat of fusion and, while at the same temperature, relocate so that they don't freeze together.* However, as solidification occurs, the atoms lose their ability to move, which means that the atoms can't move very far. All the atoms of one metal won't be completely separated from the other; instead the atoms will congregate in smaller groups.

Since metals freeze in preferred lattice directions ($\langle 100 \rangle$ for cubic, $\langle 10\bar{1}0 \rangle$ for hexagonal) and since freezing relates to temperature gradients, *metals of the eutectic composition freeze in a laminated structure of thin, alternating layers or plates*. In this way, the atoms have to move a minimum distance to segregate. The slower the cooling rate, the more time there is for the atoms to segregate, so more atoms collect together, and the plates are thicker; faster cooling rates generate thinner laminations. The laminated nature of a eutectic mixture can be seen in Figures A-27 and A-28.

Tracing the Cooling of the Alloy of 75% Cd (Figure 5-4). When the temperature is lowered to point *B*, metal with the composition of pure cadmium begins to freeze. As cooling continues, only pure cadmium freezes, while the composition of the liquid follows the liquidus line; at the eutectic temperature (*D'D*) the liquid reaches the eutectic composition of 40% Cd. Cooling below the eutectic temperature results in a microstructure containing pure cadmium and the eutectic mixture of bismuth and cadmium.

Note that the process of the freezing of this alloy is similar to that of the 20% Cd alloy, except that the first metal to freeze is bismuth. This difference occurs because the two alloys are on opposite sides of the eutectic. So any alloy that is to the left of the eutectic point (composition *less than* the eutectic, or **hypoeutectic**) will first freeze the pure metal on the left side of the diagram; any alloy that is to the right of the eutectic (composition *greater than* the eutectic, or **hypereutectic**) will first freeze the pure metal on the right side of the diagram. Regardless of which side of the eutectic, *for all alloys in completely insoluble systems, the last liquid to freeze is of the eutectic composition.*

5.4   FREEZING OF PARTIALLY SOLUBLE–INSOLUBLE ALLOYS

Complete Insolubility? The previous section dealt with metals that were completely insoluble in the solid state and used bismuth and cadmium as examples. In reality, no metals are *completely* insoluble; there is always some amount of solubility, although with some alloy systems it is very small.

The aluminum-silicon alloy system is insoluble by Hume-Rothery's substitutional criteria. Aluminum is FCC with an atom diameter of 2.863 Å; silicon is cubic with an atom diameter of 2.351 Å. The difference in their diameters is 18% compared with aluminum and 22% compared with silicon, well beyond the 15% guideline. Figure 5-5 shows the phase diagram for the aluminum-silicon system. Since these metals are insoluble, there is a eutectic, but note that there is an area of primary or terminal solid solubility in the extreme lower-left corner labeled alpha (α). This small region indicates that over a very small range of composition silicon atoms will dissolve in aluminum's lattice. The data on the diagram indicate that up to 1.65% Si will dissolve in aluminum at the eutectic temperature of 577°C; the number 99.83 at the right end of the eutectic temperature line indicates that 0.17% Al will dissolve in silicon. These are not very large solubilities; for most practical purposes these two metals are insoluble.

Development of the Phase Diagram. Phase diagrams for alloy systems of metals that are partially soluble–insoluble can be developed by using cooling curves in exactly the same way as we've done with the soluble and insoluble cases. The difference is that, as the name suggests, over part of the composition range the metals are soluble, and over another part they are insoluble; the resulting phase diagram combines the two types.

FIGURE 5–5 Phase diagram for aluminum-silicon, virtually insoluble

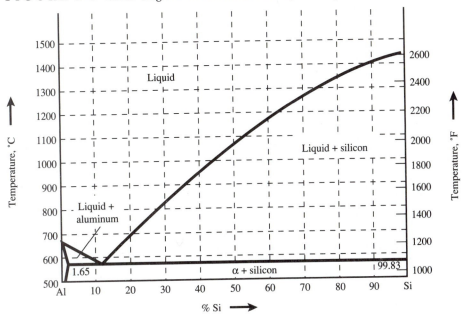

Source: *Metals handbook* (8th ed.). Materials Park, OH: ASM International, 1973, *8*, 263.

Silver and copper form a substitutional alloy system typical of the partially soluble–insoluble type. Both metals are FCC; copper has an atom diameter of 2.556 Å; silver has an atom diameter of 2.889 Å. The percent difference in their diameters is 13% compared with copper and 11.5% compared with silver, which fit the criteria for a partially soluble–insoluble alloy system.

Figure 5-6 shows cooling curves for four compositions that could be used in developing part of the silver-copper phase diagram. For 100% copper the cooling curve is that of a pure metal,* and at 28% copper the cooling curve is for the eutectic composition; both of these types have been discussed previously.

The cooling curve for the 75% copper alloy is influenced by both the solubility and the insolubility in the alloy system; its cooling curve is a combination of the two. From *A* to *B* the metal cools as a liquid. From *B* to *D'* the terminal solid-solution beta (β) freezes. But at *D'*, the eutectic temperature, there is still liquid available. This liquid, of eutectic composition, freezes as eutectic mixture from *D'* to *D*, the level portion of the curve. This alloy freezes as a combination of crystals of solid-solution beta (β), plus the eutectic mixture of solid solutions alpha (α) and beta (β).

The cooling curve for the 5% copper alloy is that of a solid solution. From *A* to *B* the metal cools as a liquid; from *B* to *D* it solidifies as the terminal solid-solution

*The cooling curve for pure silver (0% Cu) has been left off so that the curve for 5% Cu can be seen.

F I G U R E 5–6 **Multiple cooling curves develop the silver-copper (partially soluble–insoluble) phase diagram**

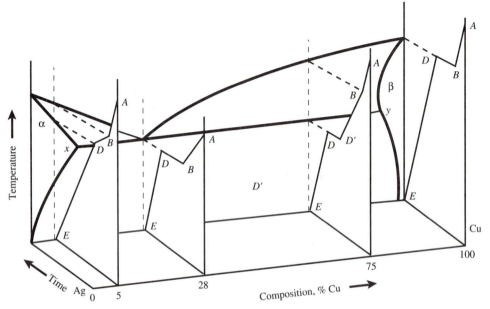

alpha (α); from D to E it cools as a solid. Note that although the overall alloy system has a eutectic, any liquid with a composition up to point x freezes as the solid-solution alpha (α) before it ever reaches the eutectic temperature.

Composition of Phases During Freezing of Silver-Copper Alloys. Figure 5-7 shows the phase diagram for the silver-copper system in its normal presentation. The freezing of the 75 and 5% copper alloys will be traced; the letter identifications can be related to the cooling curves of Figure 5-6.

Cooling the 75% copper alloy to point B, about 970°C (1778°F), initiates the formation of crystals of the terminal solid-solution beta (β) at point M; these initial crystals have a composition of about 94% Cu. At point B the liquid would still be essentially 75% copper and 25% silver.

Cooling to point Q at 900°C (1652°F), the metal is a mixture of liquid (point P, composition 55% Cu), and solid-solution beta (β) (point R, composition 93% Cu). As this mixture is cooled to just above point $D'D$, the composition of the liquid follows the liquidus to point BD, 28% Cu, and solid-solution beta (β) follows the solidus to point y, 92% Cu. Thus the last remaining liquid, being of eutectic composition, freezes in the same manner as the 28% Cu alloy of Figure 5-6, i.e., at one temperature, the eutectic temperature.

In order to more easily study the freezing of the 5% copper alloy, let's expand the alpha portion of the phase diagram, as shown in Figure 5-8. This alloy freezes in a much more straightforward manner than the 75% alloy. When the liquid cools to point B, crystals of the terminal solid-solution alpha (c·) begin to form at point M

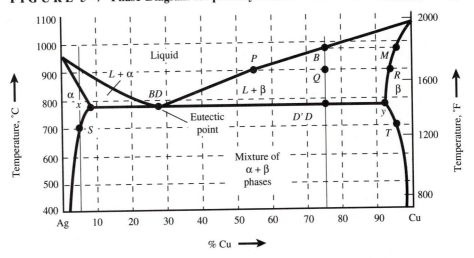

Source: *Metals handbook* (8th ed.). Materials Park, OH: ASM International, 1973, *8*, 253. Annotations by author.

FIGURE 5–8 Alpha (α) region of the silver-copper phase diagram

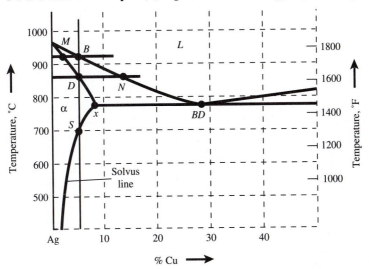

Source: *Metals handbook* (8th ed.). Materials Park, OH: ASM International, 1973, *8*, 253. Annotations by author.

103

with a composition of about 2% Cu. When the metal cools to point *D*, on the solidus line, it becomes completely solid alpha (α). Note that the last liquid available, point *N*, has a composition of about 14% Cu, not the eutectic composition; if cooled slowly, this alloy should never have eutectic in its microstructure.

At temperatures between points *B* and *D*, the alloy is a mixture of liquid and solid-solution alpha (α). As this cooling occurs, the composition of the liquid changes from *B* to *N* along the liquidus line, and the composition of the solid changes from *M* to *D* following the solidus line.

Continued Cooling of the 5% Copper Alloy. In Figure 5-8 note that if the 5% alloy continues to cool it arrives at a point *S* on a line not described before. This line is a **solvus line** and represents the limits of the solubility of copper in silver in the solid state at the corresponding temperature. To the left of the solvus line the alpha solid solution is unsaturated, so the solvus line is the limit of saturation or solubility in the solid state.

While this alloy cools between *D* and *S* it is a solid solution; solute copper atoms, amounting to 5% of the total weight, are residing with silver in the solvent silver lattice. However, when the temperature gets below point *S*, some of the copper atoms must leave the silver lattice and form their own copper lattices. That is, *the copper comes out of solution, or precipitates.*

At the temperature of point *S*, about 700°C (1292°F), the other end of the phase diagram (see Figure 5-7) shows that at point *T* the copper lattice can dissolve about 5% silver as beta (β) solid solution. So, when the copper leaves the alpha (α) lattice at point *S*, it does so as solid-solution beta (β), taking a small amount (about 5%) of silver with it. When the 5% Cu alloy cools below point *S* it becomes a mixture of the two solid solutions alpha (α) and beta (β).

If cooling is at a reasonably slow rate, the beta (β) precipitates at the boundaries of the alpha (α) grains and may also show up as globules within the grains.

At room temperature the solubility of the two metals for each other is so low (about 1%) that the two solid solutions are just slightly impure silver and copper; the resulting alloy is essentially just a mixture of silver and copper. In this alloy system the solubility is limited because of the relative size of the atoms. It is somewhat logical that with an increase in temperature the lattice will expand, making it easier to fit atoms of a different diameter. Note that the shape of the solvus lines is such that as the temperature approaches the eutectic temperature, the solubility of alpha (α) and beta (β) approach their maximums, points *x* and *y*, respectively.

Alloy systems that have solid-solution areas where the solubility *increases* as the temperature increases enable the creation of supersaturated solutions. That is, the 5% Cu alloy of Figures 5-7 and 5-8, if heated to 750°C (1382°F), can dissolve all 5% of the copper. If cooled rapidly enough, that copper may stay trapped in the silver lattice, in spite of the decreasing solubility shown on the phase diagram. The creation of supersaturated solutions is termed *solution heat-treating* and is discussed briefly later in this chapter; it will be covered in detail in Chapter 10.

Continued Cooling of the 75% Cu Alloy. The freezing of the 75% Cu alloy began with the development of crystals of the terminal solid-solution beta (β); then, at the eutectic temperature, the remaining liquid froze as the eutectic mixture. The microstructure consists of grains of the solid-solution beta (β) surrounded by the laminations of alpha (α) and beta (β) making up the eutectic mixture.

Since a solvus line is a limit of solubility, the composition of the alpha (α) and beta (β) phases below the eutectic temperature can be read from that line at any particular temperature. Earlier it was determined that both alpha (α) and beta (β) could dissolve about 5% of the other metal at 700°C (1292°F), and as the temperature dropped to room temperature both solubilities declined significantly. This means as the 75% Cu alloy cools below the eutectic temperature each phase must precipitate the other. The beta (β) phase that solidified above the eutectic temperature precipitates some alpha (α), and both the alpha (α) and the beta (β) phases in the eutectic precipitate some of the other phase.

Interstitial Terminal Solid Solutions.   The examples used so far in this chapter have been for substitutional alloys. Figure 5-9 is the simplified phase diagram for the iron–iron carbide system and shows two interstitial solid solutions: alpha (α) ferrite and gamma (γ) austenite.[†] On the phase diagram they have the same appearance as terminal substitutional solid solutions, but they are *interstitial*.

Ferrite, with a BCC lattice, does exist down to room temperature, but has a very low solubility even at elevated temperatures. Austenite, with a FCC lattice,

[†]A third solid solution, delta (δ), is not considered here; it is discussed briefly in Chapter 6.

FIGURE 5–9   Simplified iron–iron carbide phase diagram

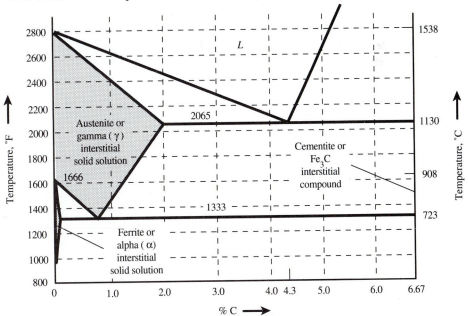

Source: *Metals handbook* (8th ed.). Materials Park, OH: ASM International, 1973, *8*, 275. Annotations by author.

has somewhat greater solubility and can dissolve up to 2.0% carbon at 2065°F (1130°C), but it exists only at temperatures above 1333°F (723°C).

5.5   INTERMEDIATE PHASES

In addition to pure metals and terminal solid solutions, intermediate solid solutions and compounds are single-phase regions that are also shown on phase diagrams.

Intermediate Solid Solutions. Figure 5-10 is a portion of the phase diagram for the substitutional copper-zinc alloy system. Copper (FCC) and zinc (CPH) form the alpha (α) terminal solid solution and the intermediate solid solutions beta (β) and gamma (γ), which have the BCC lattice, different from either of the alloying metals. A number of other metals form alloy systems that have similar diagrams;[‡] note that in Figure 5-10 the solubility of the terminal solid solution alpha (α) *decreases* as temperature goes up, unlike the silver-copper system, but that the solubility ranges of solid solutions beta (β) and gamma (γ) increase as temperature increases.[§]

The beta (β) and gamma (γ) solid solutions include the compositions that give electron-atom ratios of 3 : 2 and 21 : 13 and are sometimes referred to as *electron compounds*. (See Section 4.4.) Intermediate solid solutions appear as open areas in the central portion of a phase diagram and are usually identified with a Greek letter.

Intermediate interstitial solid solutions are formed, for example, between carbon and some transition metals, but usually at relatively low levels of solubility. Interstitial solutions rely on the ability of the solute atom to fit open spaces in the solvent lattice, and because the amount of this space is limited, intermediate solutions of this type are uncommon.

Ordering. In the Cu-Zn system, Figure 5-10, as the temperature goes up the BCC lattice of the beta (β) phase expands and thus its solubility goes up; i.e., as the temperature increases atoms of a different size can more easily fit in the BCC lattice. Many of the alloy systems involving copper and the metals nearby in the periodic table have V-shaped intermediate solid-solution regions like the beta (β) region of Figure 5-10.

With the wide range of solubility that exists at elevated temperatures, the copper and zinc atoms fit into the beta (β) BCC lattice in a random fashion; there is no order as to how they are arranged. However, as the temperature decreases, the ability of the lattice to accept atoms of different size in any sequence or order disappears, and at temperatures between 456 and 468°C (853 and 875°F), the beta (β) solid solution changes to beta-prime (β').

[‡]Copper-silicon, copper-tin, silver-cadmium, and silver-zinc are among these systems.
[§]The copper-zinc system also has the intermediate solid solutions delta (δ) and epsilon (ε) that are of higher zinc content than shown on Figure 5-10.

FIGURE 5-10 Portion of the copper-zinc phase diagram, showing intermediate phases

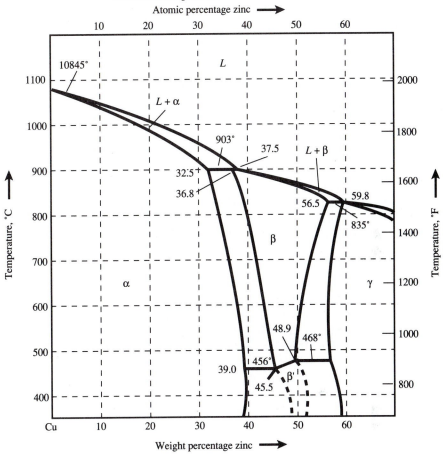

Source: *Metals handbook* (8th ed.). Materials Park, OH: ASM International, 1973, *8*, 301. Annotations by author.

The difference between these two is that the atoms in the beta (β) phase are in random orientation, and in beta-prime (β') phase they are in order, or "ordered." Since this ordering is occurring at a composition of approximately equal numbers and weights of atoms, the order is "every other one." Visualize the BCC lattice as a repeating series of cubes, with the corner of each cube located at the center of another cube. Ordering merely means that starting with a copper atom, every other atom, in all three directions, is a copper atom, and starting with a zinc atom, every other atom is a zinc atom. Thus *ordering is an overall arrangement of the atoms superimposed on the regular lattice known as a* **superlattice.**

Since all compositions of the beta-prime (β') phase are not exactly 50%, there are not always exactly equal numbers of atoms, so the ordering (superlattice)

continues for varying distances through the lattice, corrects for the variation in composition, and continues on. ‖

The practical significance to ordering is that it changes the properties of the alloy. For example, alloys of the beta (β) phase of Figure 5-10 are known as "beta brasses." These alloys normally must be hot-worked because at room temperature, which is the beta-prime (β') phase, they lack the plasticity to be cold-worked. This is discussed further in Chapter 10.

The example of the copper-zinc beta (β) phase involves ordering of approximately equal numbers of atoms. Ordering involving other ratios of atoms is also common.

Compounds.   *An intermetallic compound has a simple ratio of elements;* i.e., it has an exact composition. Thus on a phase diagram it appears as a *vertical line.* Figure 5-11 is the phase diagram for magnesium and silicon. As discussed in Section 4.4, these metals are largely insoluble, but they do form the compound Mg_2Si. This compound is represented on the phase diagram by the vertical line, identified as beta (β), which occurs at 37% Si by weight. Note that the atomic percentage is 33⅓% Si and 66⅔% Mg, or two atoms of magnesium for each atom of silicon.

Gold-lead is an example of an alloy system containing two compounds; beta (β) and gamma (γ) are indicated by the vertical lines in the central portion of Figure 5-12. From their atomic percentages, the metals are in a 2:1 ratio, but in both directions. That is beta (β) is $AuPb_2$ and gamma (γ) is Au_2Pb.

The interstitial compound cementite, or iron carbide, Fe_3C, is indicated on

‖The terms used to describe this are "short-range" and "long-range" ordering.

F I G U R E 5–11   Phase diagram for magnesium and silicon, which form an intermetallic compound

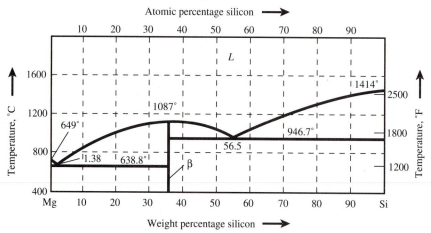

Source: *Metals handbook* (8th ed.). Materials Park, OH: ASM International, 1973, *8*, 315.

FIGURE 5-12 **Phase diagram for gold and lead, which form two intermetallic compounds**

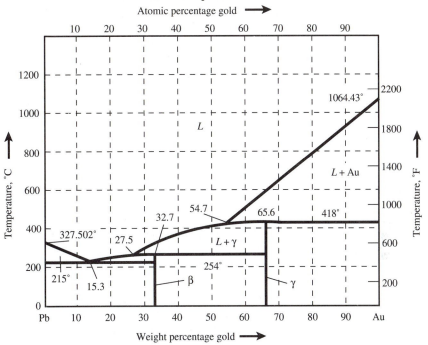

Atomic percentage gold ➡

Weight percentage gold ➡

Source: *Metals handbook* (8th ed.). Materials Park, OH: ASM International, 1973, *8,* 268.

Figure 5-9. Other transition elements form compounds with small-diameter solute atoms, but cementite is the best known. Since cementite has a single composition of 6.67% C, it appears on the phase diagram as a vertical line.

5.6 EQUILIBRIUM PHASE DIAGRAMS

Equilibrium. Phase diagrams are formally known as *equilibrium phase diagrams,* inferring that the cooling or heating used in developing the diagram is done under "equilibrium conditions." Transferring heat to or from metal with temperature in equilibrium is theoretically impossible. A more practical concept of equilibrium is that time must be recognized as a factor. The cooling curves used in Figures 5-1, 5-3, and 5-6 show that phases do require time to change. *Equilibrium then can be viewed as cooling or heating at a rate such that whatever phase changes will occur have time to do so;* the equilibrium phase diagram indicates the temperatures *and compositions* where the phase changes take place.

In the analyses done previously for the copper-nickel, bismuth-cadmium, and silver-copper alloy systems (Figures 5-2, 5-4, and 5-7), an important assumption was made in reading the compositions from the phase diagrams. Not only must time be allowed for the temperature to equalize, but also for the composition to equalize. For example, in Figure 5-2, the alloy of 50% nickel is solid at point D and should be half nickel and half copper. However, it was previously stated that the last liquid to solidify is of the composition at point N, or 37% Ni; the overall composition may be equal amounts of nickel and copper, but the last metal to freeze will have a different composition. *For true equilibrium, time is required for the atoms of this last metal to diffuse through the already frozen metal to equalize the distribution of atoms.*

It is important to realize in reading compositions on phase diagrams that in the real world of nonequilibrium cooling, diffusion will not be complete, and the compositions will probably not be as read from the diagram. The diagrams should be used as theoretical guides to the approximate composition. The effect of cooling rates on the phase diagram will be discussed in a later section.

Changes of Phase. Although equilibrium conditions are difficult to achieve during heating or cooling, they can be approached by holding or annealing the metal for long periods of time at a specific temperature. Phase changes that might not be revealed by cooling a specimen may become evident in this manner. In some cases, fast cooling from the holding temperature retains the elevated-temperature structure at room temperature so it can be analyzed metallographically.

As illustrated earlier in this chapter, the temperature at which changes of phase occur can be determined by plotting cooling-rate data and finding where the rate changes. Phase changes due to allotropy, a change in lattice due to a change of temperature, can be found through X-ray diffraction methods. (See Section 1.4.)

Diagram of Phases. The phase diagram is simply a diagram of phases; that is, the composition of the metallic *single* phases can be read directly from the diagram. Mixtures cannot. In a liquid region whatever the composition of the alloy is will be the composition of the liquid.# Within a solid-solution region whatever the composition of the alloy is will be the composition of the solid solution. For compounds whatever composition is read from the diagram is the composition of the compound. However, as we'll see below, the composition of a mixture region cannot be read directly from the diagram.

Determining Compositions of Phases in Mixture Regions. For any particular temperature the composition of the liquid portion of a mixture must be read from the liquidus line, the solid portion from the solidus line, and if the mixture is of solid solutions their compositions are read from solvus lines.

For example, when the 75% Cu alloy of Figure 5-7 (the silver-copper system) cools to point Q, 900°C (1652°F), it is a mixture of liquid and solid-solution beta (β). The compositions of these two phases can be read from the liquidus and solidus lines at that temperature, i.e., points P and R. In a similar fashion, when

#This is true for metals soluble as liquids.

this alloy cools to 700°C (1292°F), it is a mixture of alpha (α) and beta (β) solid solutions. The compositions of the two solid solutions are read at their respective solvus lines, points *S* and *T*.

Note that in both these examples of mixture regions the composition of the alloy itself, 75% Cu, does not determine the composition of the *phases*. At 900°C (1652°F) all alloys of 55 to 93% Cu would have the liquid and beta (β) phases present and their compositions would be read at points *P* and *R*. Similarly, at 700°C (1292°F) all alloys from 5 to 95% Cu would have the alpha (α) and beta (β) phases present and their compositions would correspond to points *S* and *T*.

5.7   RELATIVE AMOUNTS OF PHASES IN MIXTURES

Composition vs. Relative Amounts.   The variables of phase diagrams are temperature and the relative amounts of the two pure metals; the amount of each metal is not given in absolute units, e.g., pounds or grams, but as a percentage of the total amount of the two metals. We have already discussed the method of determining the *composition* of phases, but a question that has not been answered is *how much* of a phase exists of that composition. To better explain how this is determined, let's enlarge the central part of the copper-nickel phase diagram, Figure 5-2, as Figure 5-13. The original identification letters have been retained, but the compositions at the liquidus and solidus lines have been added, as well as data for 1290°C (2354°F) and 1310°C (2390°F).

Figure 5-13 shows that for the alloy of 50% Ni at 1300°C (2372°F) the liquid and alpha (α) phases exist together in a mixture region, and their compositions correspond to points *P* and *R,* or about 43% Ni and 57% Ni, respectively. This tells what the two phases are made of, but it does not tell how much of each phase there is. Above point *B*, say at 1400°C (2552°F), the metal is all liquid, i.e., its

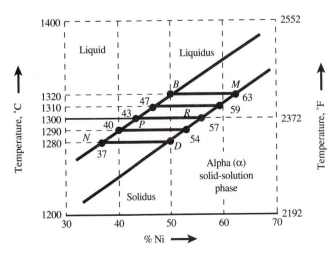

FIGURE 5–13

Copper-nickel phase diagram, enlarged view, for a 50% alloy

Source: *Metals handbook* (8th ed.). Materials Park, OH: ASM International, 1973, *8,* 294. Annotations by author.

relative amount is 100% liquid, and its composition is 50% Ni. Below point *D*, say at 1200°C (2192°F), the metal is all solid alpha (α); i.e., its relative amount is 100% alpha (α), and its composition is 50% Ni. Between points *B* and *D* the relative amount of the two phases changes but can be calculated by use of the tie lines (heavy, solid lines) that run between the liquidus and solidus lines.

The length of the tie line can be used to represent all of the phases present in a mixture region. For example, at 1320°C (2408°F) the alloy is exactly on the liquidus line, and the line *BM*, 63 − 50 = 13 units long, represents 100% of the *liquid phase*.

When the metal cools to 1310°C (2390°F) the line is 59 − 47 = 12 units long, and those 12 units represent the *total* of the two phases, liquid and solid alpha (α), that are present at that temperature. The alloy composition, 50%, divides the 12-unit line into two pieces or segments; *the relative length of these two segments, i.e., their length relative to the total length of the line, gives the relative amounts of the two phases.* At 1310°C (2390°F) the lengths of the two segments of the line are 59 − 50 = 9 and 50 − 47 = 3. Since the total of the two phases is represented by 12 units, the *relative amounts* of the two phases must be 3 ÷ 12, or 25%, and 9 ÷ 12, or 75%.

The question now is which one is which? At 1310°C (2390°F) the metal has cooled to just below the liquidus; just 10°C ago it was all liquid and it has to cool another 30°C before it will be all solid. So, at 1310°C (2390°F) most of the metal is liquid (75%) and the lesser amount is solid alpha (α) (25%).

Considering the metal at 1290°C (2354°F), the total tie line is 54 − 40 = 14 units long, and the two segments are 54 − 50 = 4 and 50 − 40 = 10. At 1290°C (2354°F) the metal is very close to the solidus, very close to being all solid, so the larger of the relative amounts, 10 ÷ 14, or 71%, is solid alpha (α), and the smaller amount, 4 ÷ 14, or 29%, is liquid.

Note that in these calculations *a phase is represented by the segment of the tie line away from the phase.* At 1290°C (2354°F) the 10 units that represent the *solid* constitute the distance from the alloy at 50% to the *liquidus* at 40%; that segment is on the side of the alloy line *away from* the solid alpha (α) phase.

Inverse Tie-Line Rule. This illustrates the *inverse tie-line rule: The relative amount of a phase is determined by dividing the length of the line segment away from the phase by the length of the tie line going between the two phases.***

In the examples given above, the temperatures were chosen so that the metal was almost all one phase or the other, and so the concept could be developed by reason. The inverse rule is needed to analyze conditions where the answer is less

---

**The inverse tie-line rule can be shown to be valid by accounting for the alloy metal in the two phases. For example, for the 50% Ni alloy at 1290°C, some of the nickel is in solid form and some in liquid; if x is the decimal amount that is solid, then $(1 - x)$ is the liquid. The amount of nickel in the solid is x times the composition of the solid, which is 54% Ni, or 0.54 x. The amount of nickel in the liquid is $(1 - x)$ times the composition of the liquid, which is 40% Ni, or $0.40(1 - x)$. The sum of the nickel in the solid and liquid must equal the total of the nickel in the alloy, or 50%, so the following equality exists: $0.54x + 0.40(1 - x) = 0.50$, or $0.54x - 0.40x = 0.50 - 0.40$; $0.14x = 0.10$; $x = 10 ÷ 14$. That is, the relative amount that is solid is 10 ÷ 14, which agrees with the answer determined above.

obvious, that is, where the two segments of the tie line are equal or almost equal in length. For example, at 1300°C (2372°F), point *Q*, the tie line is 57 − 43 = 14 units, and the two segments are each seven units. In this case which end of the line is which is academic since the relative amounts are 50-50, but by the inverse tie-line rule segment *QR* (57 − 50 = 7) represents the liquid, and *PQ* (50 − 43 = 7) represents the solid alpha (α).

In Section 5.3 the cooling of an alloy of 75% cadmium, 25% bismuth was traced using the phase diagram, Figure 5-4. A portion of that diagram is reproduced in Figure 5-14. As the alloy cools to just above the eutectic temperature, the two phases are the liquid, composition 40% Cd, and the solid, 100% pure cadmium.

In Figure 5-14 the tie line between these two phases is 60 units long (100 − 40 = 60). The segments representing the phases are indicated in the figure. The length of the segment representing the liquid is 100 − 75 = 25 units, or the relative amount of liquid is 25 ÷ 60, that is, 42%; the segment representing the solid cadmium is 75 − 40 = 35 units, so the relative amount of solid cadmium is 35 ÷ 60, or 58%.

Use of Composition and Percent Relative Amount. The final stages of solidification of this alloy can now be discussed more fully. When the alloy cools to just above the eutectic temperature, about 58% of it has already frozen as pure cadmium; the remaining 42% is liquid of the eutectic composition, 40% cadmium-60% bismuth, which will freeze as the eutectic mixture. If the total metal available is 75% cadmium and 25% bismuth, and 58% is frozen cadmium, then 75 − 58 = 17% of the total cadmium is left to join with the 25% bismuth to form the 42% eutectic (17 + 25 = 42), which is the amount of liquid from the initial calculation. This process is illustrated in Figure 5-15.

Note that the same numbers can be determined on the basis of the composition and percent relative amount of the eutectic—if 42% of the metal is to be

F I G U R E 5–14 **Portion of the bismuth-cadmium phase diagram for an alloy of 75% cadmium**

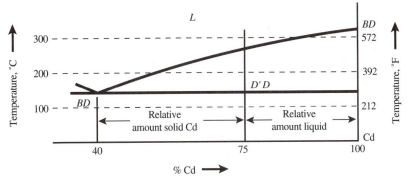

Source: *Metals handbook* (8th ed.). Materials Park, OH: ASM International, 1973, *8*, 272. Annotations by author.

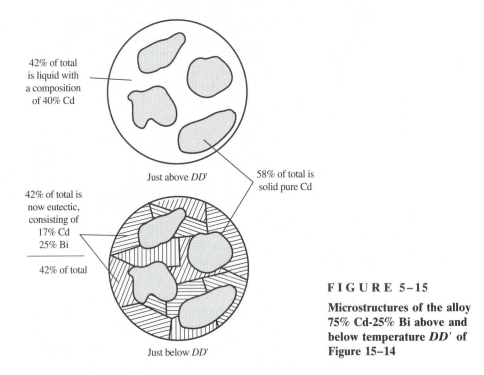

42% of total
is liquid with
a composition
of 40% Cd

Just above *DD'*

58% of total is
solid pure Cd

42% of total is
now eutectic,
consisting of
17% Cd
25% Bi
──────────
42% of total

Just below *DD'*

FIGURE 5–15

Microstructures of the alloy 75% Cd-25% Bi above and below temperature *DD'* of Figure 15–14

eutectic mixture, and if the composition of the eutectic is 40% cadmium-60% bismuth, then

42 × 0.40 = 16.8, or 17% of the total is cadmium in the eutectic

42 × 0.60 = 25.2, or 25% (all) of the total is bismuth in the eutectic

Therefore the total relative amount that is eutectic equals 42% (17% + 25%).

Percent Relative Amount in the Solid State. The inverse tie-line rule can also be used to calculate the relative amounts of phases in the solid state. In Section 5.4 the cooling of an alloy of 75% copper was described using the silver-copper phase diagram, Figure 5-7. Referring to that diagram, at 700°C (1292°F) there are two phases present, alpha (α) and beta (β), and their respective compositions correspond to points S and T. The tie line that extends between these two points represents all of the alloy and is approximately 95 − 5 = 90 units long. The segment that represents the alpha (α) phase is the segment away from alpha (α), or 95 − 75 = 20 units; the relative amount of alpha (α) is then 20 ÷ 90, or 22%. The segment that represents the beta (β) phase is 75 − 5 = 70 units; the relative amount of beta (β) is then 70 ÷ 90, or 78%. This is a logical result since the alloy composition is closer to that of beta (β) and would therefore have more of that phase in the mixture.

The compositions of the alpha (α) and beta (β) phases in the 5% alloy of copper (Figures 5-7 and 5-8) were determined earlier. It would be useful to be able to calculate how much beta (β) is available to precipitate from a 5% copper alloy at room temperature. If the solubilities of these two phases in each other are 1% at room temperature, then the tie line between them is $99 - 1 = 98$ units long. The segment that represents the beta (β) phase is the segment away from beta (β), or $5 - 1 = 4$; the relative amount of beta (β) is $4 \div 98$, or just a little more than 4%.

This says that if a 5% Cu alloy was cooled sufficiently fast that a supersaturated solution was created, it would be possible that 4% of the microstructure could later precipitate as beta (β).

5.8   EQUILIBRIUM VS. NONEQUILIBRIUM COOLING AND CORING

Nonequilibrium Effects.   The calculations done in the preceding section are theoretically valid only if the equilibrium conditions required by the phase diagram are followed. Although commercial processes operate at nonequilibrium heating and cooling rates, many have rates slow enough that an analysis using the phase diagram is still correct qualitatively, if not also quantitatively.

Nonequilibrium heating and cooling affect the temperatures and compositions at which phases change, i.e., the location of the lines on the phase diagram. In general, the faster the rate of temperature change, the more the lines move, and they move in the direction the temperature is changing. That is, a rapid *heating* rate tends to raise the lines; a rapid *cooling* rate tends to lower the lines.

Cored Crystals.   In the *equilibrium* phase diagram the assumption is made that the newly frozen and previously frozen atoms diffuse into each other so that the composition of the solid equalizes to that corresponding to the solidus at any particular temperature. *A more realistic view is that the center or core of the first crystal freezes with a composition that changes very little as cooling proceeds* because there is not enough time for **diffusion.** As the temperature drops, the composition of the *newly* frozen metal follows the solidus line, but also without much diffusion; this results in layers of differing compositions freezing around the original crystal. The lack of diffusion causes **cored crystals** to be created, that is, crystals that have a variation in composition.

The effect of nonequilibrium cooling, the process for creating cored crystals, is diagrammed in Figure 5-16. This figure is the copper-nickel phase diagram of Figure 5-2, with lines and points added for the rapid cooling of a 50% alloy. Point *M* is where the first crystal forms with a composition of about 63% Ni. As cooling continues, its composition follows line *MX;* not enough diffusion occurs for the composition to follow the solidus, *MD.*

As additional metal freezes, it does so with a composition corresponding to the solidus, *MDV,* but little diffusion occurs. So, if the inside of the crystal has a composition varying along the line *MX,* and if the outside of the crystal has a composition varying along the line *MDV,* the *average* composition must be some-

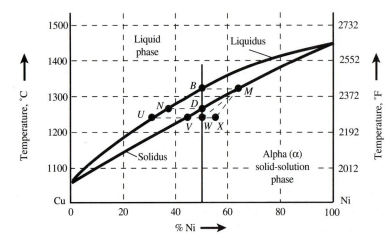

FIGURE 5–16

Copper-nickel phase diagram, showing effects of nonequilibrium cooling

Source: *Metals handbook* (8th ed.). Materials Park, OH: ASM International, 1973, *8*, 294. Annotations by author.

where between the two lines, e.g., line *MW*. In effect, *MW* becomes the new solidus line for the *average* composition of the alloy.

When the temperature drops to where this new solidus line, *MW*, crosses the average composition, which is known to be 50% Ni, the metal will be completely solid; so *the freezing temperature of this alloy has been lowered* from point *D* to point *W*. (Note that the last liquid available to solidify has a composition corresponding to point *U*, not point *N*.) Thus an alloy that is supposed to be a homogeneous solid solution of one composition might have a microstructure that varies from the composition at point *M* to that at point *V*.

Figure 5-17 is a schematic representation of the microstructure resulting from the freezing process just described. The nickel and copper atoms are represented by solid and open circles, respectively. The frozen metal is to the lower left of the drawing, with atoms in a geometric pattern, and the liquid is to the upper right, with atoms randomly oriented. The nickel atoms (solid circles) are concentrated in the core of the dendrite arms and in the liquid that is away from the dendrites, i.e., in the upper-right corner. The copper atoms are concentrated between the arms and in the liquid adjacent to the dendrites. Where dendritic crystals have spread to meet each other, the resulting grain boundaries are shown. Note that some of the atoms are not in perfect alignment to represent dislocations.

The photomicrograph in Figure A-1 shows the cored structure in an as-cast piece of another copper-nickel alloy—33% copper-67% nickel, known as monel metal. The nickel-rich dendrites have been starved for liquid metal and can be seen sticking up into the cavity. The nonuniform composition caused the attack of the etchant used in preparation of the specimen to vary across the microstructure so that the dendrites can be seen.

Cored crystals are primarily of concern in those alloy systems that are supposed to form a homogeneous solid solution over all or part of their composition range. In insoluble systems, or the insoluble portion of the systems, the two metals are going to segregate anyway, so the effect of changing of the lines on the phase diagram is of lesser consequence.

F I G U R E   5-17   Schematic representation of cored dendrites of nickel atoms (solid circles) and copper atoms (open circles) freezing from liquid

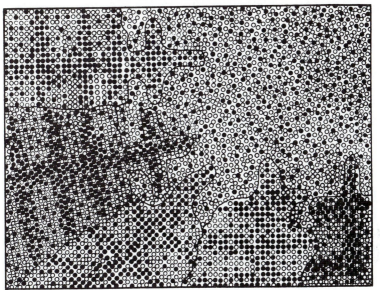

Source: Clyde W. Mason. *Introductory physical metallurgy*. Materials Park, OH: ASM International, 1947, fig. 14, 24. Used by permission.

Homogenization.   The effects of the segregation caused by coring during casting can be reduced, if not eliminated, by holding the as-cast metal at a high temperature, usually for a number of hours, to homogenize the structure. The high temperature provides the energy the solid atoms need to relocate and become uniformly distributed in the lattices. The higher the temperature that can be used, the shorter the time required to effect the homogenization. However, care must be used not to exceed the reduced freezing temperature caused by the rapid cooling; for the alloy described in Figure 5-16 the homogenizing temperature must be below point W; heating between points W and D would cause a partial melting of the cast structure.

The photomicrographs in Figures A-2 through A-6 show the cored structure of yet another copper-nickel alloy—85% copper-15% nickel. The first three views show the effect of a rapid casting rate (Figure A-2), a rapid casting rate followed by homogenizing for 3 h at 750°C (1382°F) (Figure A-3), and a rapid casting rate followed by homogenizing for 9 h at 950°C (1742°F) (Figure A-4). Figure A-2 shows that the fast cooling of the casting process has had the effect of coring the structure. Homogenizing for the shorter time and lower temperature (Figure A-3) makes the microstructure somewhat more uniform, but a little longer time and higher temperature (Figure A-4) are required to complete the process.

It would seem logical that if fast cooling caused coring it could be eliminated by slower cooling. Figure A-5 shows that *casting the same alloy in a very hot mold, so that it cooled very slowly, still did not give a uniform structure.* Note that the large raised portions are parallel and form a regular pattern; they don't look like dendrite arms, but they are. In order to make this microstructure uniform it has to be homogenized for *16 h* at 950°C (1742°F) (Figure A-6).

The slower freezing rate did not prevent the variation in composition; instead it just made the dendrite arms farther apart. When the metal was homogenized, it had to be held at the higher temperature for a longer time because the atoms had to move farther to equalize the composition. In some commercial metal-casting processes the metal is purposely cast rapidly to get small dendrite-arm spacing so it can be homogenized at lower temperatures and shorter times than would otherwise be possible.

5.9   EUTECTOID, PERITECTIC, AND PERITECTOID REACTIONS

Eutectoid Reaction. The solubilities of a few terminal solid solutions and many intermediate solid solutions, upon cooling, terminate at a temperature above room temperature and go through a eutectoid reaction. The similarity of the words *eutectoid* and *eutectic* is intentional. *The eutectic reaction involves the loss of solubility of the liquid phase; the eutectoid involves the loss of solubility of a solid phase.*

The general appearance of eutectics and eutectoids in phase diagrams is shown in Figure 5-18. The eutectic reaction involves the change of one liquid phase to a mixture of two solid phases, here identified as alpha (α) and beta (β). The eutectoid reaction involves the change of one solid phase, gamma (γ), to a mixture of two solid phases, alpha (α) and beta (β).

The freezing of a liquid of eutectic composition was described in Section 5.3 as a process of segregating the two insoluble metals into a laminated microstructure. A very similar process occurs in the eutectoid reaction, except all the phases are now solid.

In the eutectoid reaction of Figure 5-18, at some temperature and composition, the metals are soluble as solids and form a solid-solution gamma (γ). However, at the solvus lines, identified as *S*, this solubility begins to change. If the

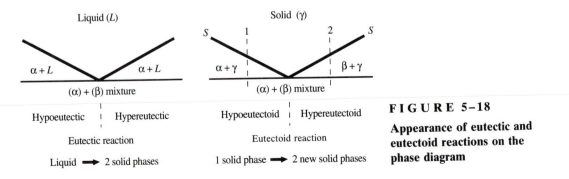

FIGURE 5–18

Appearance of eutectic and eutectoid reactions on the phase diagram

alloy is *hypoeutectoid* (less than the eutectoid composition), the gamma (γ) first precipitates alpha (α) at the solvus line, changes composition following the solvus line to the eutectoid point *e*, and then changes (transforms) to the eutectoid mixture of alpha (α) and beta (β) when the temperature goes below the eutectoid. If the alloy is *hypereutectoid* (greater than the eutectoid composition), the gamma (γ) precipitates beta (β) before forming the alpha (α)-beta (β) eutectoid mixture at the eutectoid temperature.

The most commercially important eutectoid is that formed in the steel region of the interstitial iron-carbon system; it will be discussed in detail in Chapter 6. The laminated nature of the eutectoid structure can be seen in Figure A-13. Many substitutional alloy systems have eutectoids, e.g., many of the alloy systems formed with copper, but few of them are of importance from a practical standpoint.

Peritectic Reaction. *When the melting temperatures of two insoluble metals are fairly close to being the same, they follow Raoult's law, and the alloy system contains the eutectic reaction.* The bismuth-cadmium (Figure 5-4) and copper-silver (Figure 5-7) systems are examples of this generality. On the other hand, if *melting temperatures of the insoluble metals are very different, then the alloy systems tend to contain the peritectic reaction;* they often contain the eutectic also. Peritectics appear on phase diagrams in the forms seen in Figure 5-19.

The gold-lead system (Figure 5-12) is an example. In this system two intermetallic compounds (metallic single phases), beta (β) and gamma (γ), are formed by cooling from two-phase regions that do not contain either compound. Using Figures 5-12, let's consider the freezing of an alloy of 67% gold, the composition of compound gamma (γ). At the liquidus line, at about 600°C (1112°F) pure gold (a metallic single phase) begins to freeze; pure gold continues to form and as the temperature approaches 418°C (784°F), the phases present are pure gold and liquid. According to the phase diagram, at 418°C *all* the metal is supposed to become the intermetallic compound gamma (γ). This is the peritectic reaction: From a liquid and one solid phase a new solid phase is formed, or in equation form:

$$\text{Liquid} + 1 \text{ solid} \Rightarrow \text{New solid}$$

or in this case,

$$\text{Liquid} + \text{Gold (Au)} \Rightarrow \text{Gamma } (\gamma)$$

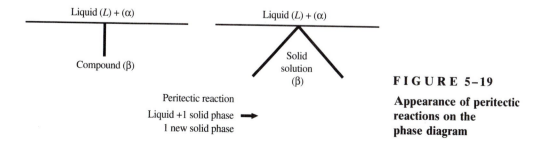

FIGURE 5–19

Appearance of peritectic reactions on the phase diagram

Now gamma (γ) is a compound with a 2 : 1 ratio of gold to lead, but at 418°C (784°F) a significant amount of gold is already frozen as an FCC lattice of gold atoms only, no lead. In order for the gamma (γ) compound to form, the gold atoms must physically relocate and combine with the lead atoms that are just in the process of becoming solid, and they must do this as the temperature goes down. This is very difficult to accomplish under realistic cooling rates, and the microstructure that typically results is a mixture of the compound gamma (γ) and pure gold rather than the single-phase compound gamma (γ). Such a mixture may completely transform to the compound gamma (γ) if held for a long period of time at a temperature below 418°C (784°F).

The formation of the beta (β) compound, 33% Au, a 1 : 2 ratio of gold to lead, occurs in the same way, except the phases involved are different. An alloy of 33% Au first freezes the compound gamma (γ) as it cools. At 254°C (489°F) all the remaining liquid, and the solid already frozen, are supposed to become beta (β); the same problems as described above occur with this alloy.

In the gold-lead system, any alloy of more than 27% Au first freezes as a phase that, per the equilibrium diagram, should change to another phase. Those alloys with less than 27% gold form an ordinary insoluble system and have the mixture-type microstructure predicted by the phase diagram.

Figure 5-19 shows that the new phase created by the peritectic reaction can be a compound or a solid solution. The gold-lead system just discussed is an example of the first type; the silver-platinum system, Figure 5-20, is an example of the second.

Any alloy in the silver-platinum system that contains between 42 and 66%

FIGURE 5–20 Silver-platinum phase diagram

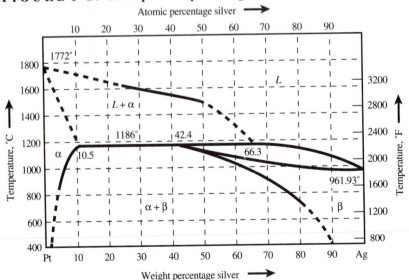

Source: *Metals handbook* (8th ed.). Materials Park, OH: ASM International, 1973, *8*, 255.

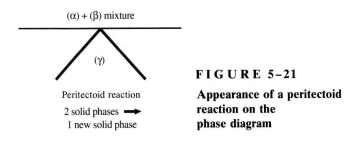

FIGURE 5-21

Appearance of a peritectoid reaction on the phase diagram

silver first freezes the alpha (α) terminal solid solution before reaching the peritectic temperature of 1186°C (2167°F). At that temperature, and between the compositions given, the solid-solution beta (β) forms, and to follow the phase diagram must do so by making use of the atoms already frozen as alpha (α). This peritectic reaction is

$$\text{Liquid} + \text{Alpha } (\alpha) \Rightarrow \text{Beta } (\beta)$$

The result is similar to the gold-lead system discussed above; at temperatures and compositions where the microstructure is supposed to be a solid solution, it will be a mixture.

As a generality, *alloys containing peritectics result in segregated microstructures; lengthy homogenization treatments may achieve the results predicted by the phase diagram.*

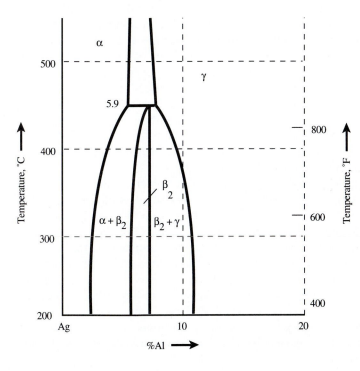

FIGURE 5-22

A portion of the silver-aluminum phase diagram

Source: *Metals handbook* (8th ed.). Materials Park, OH: ASM International, 1973, *8*, 252. Alterations by author.

Peritectoid Reaction.   Just as the eutectic and eutectoid reactions are related, so are peritectic and peritectoid. The peritectoid reaction is merely a peritectic where everything is in the solid state. The peritectoid reaction is

2 solid phases ⇒ 1 new solid phase

The typical appearance of a peritectoid is shown in Figure 5-21; the mixture of solid-solutions alpha (α) and beta (β) forms a new solid-solution gamma (γ) when the temperature goes below a certain point.

Figure 5-22 shows an area of the silver-aluminum system that contains a peritectoid. As we can see, an alloy containing about 7% Al, if heated to 500°C (932°F), would be in a mixture region between the alpha (α)-gamma (γ) solid solutions. Cooling below 450°C (842°F) puts the alloy into a new phase, beta (β); two solid phases have cooled to form a new solid phase, the peritectoid reaction.

The consequences of the peritectoid reaction are similar to those of the peritectic—the reaction rarely goes to completion unless lengthy homogenization is performed. In the case of the 7% Al alloy, cooling of the alpha (α)-gamma (γ) mixture does not produce beta (β) unless extensive heating at under 450°C (842°F) is done.

5.10   EFFECT OF FREEZING SEQUENCE ON THE MICROSTRUCTURE

When alloys that are soluble as liquids are cooled, they have three possible conditions of solid solubility; they can form alloy systems that are soluble, insoluble, or partially soluble–insoluble. Alloy systems with solidus lines at angles (soluble and partially soluble–insoluble types) vary the composition of the freezing metal. Alloy systems with vertical solidus lines (insoluble type) freeze metals of one composition but segregate the metals because of a eutectic. Slanted solvus lines in alloy systems indicate that the solid solubilities of phases vary with temperature, which means a phase or phases can precipitate. How the solubility of systems changes with temperature determines which phases form first, or which will precipitate, and these things have much to do with the final microstructures, which in turn have much to do with the properties of the alloys.

To illustrate the above, the alpha (α) region of the silver-copper system (Figure 5-8) is shown in Figure 5-23 and is used to trace the cooling of alloys 1 and 2. The microstructures at three temperatures, resulting from this cooling process, are shown in Figure 5-23(a) through (e).

Following the slow cooling of alloy 1 from the liquid, alpha (α) first begins to freeze. Figure 5-23(a), at about 900°C (1652°F), shows three grains being formed, with some liquid remaining. In Figure 5-23(b) the cooling has gone below the solidus, the three grains have grown, and the liquid has disappeared. As cooling continues, the solvus line is crossed; at 500°C (932°F), Figure 5-23(c) shows the beta (β) phase precipitating along the alpha (α) grain boundaries and within the grains. Because of the slanted solvus line, the alloy, which initially froze as a solid solution, becomes a mixture.

FIGURE 5-23 Alpha (α) region of the silver-copper phase diagram, with sketches of microstructures for two alloys

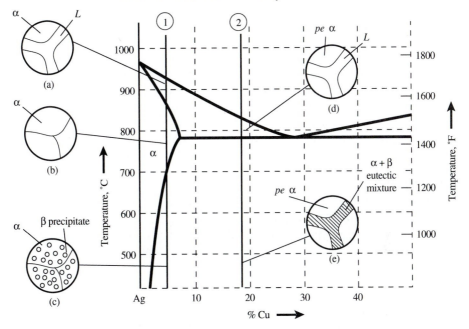

Source: *Metals handbook* (8th ed.). Materials Park, OH: ASM International, 1973, *8*, 253. Alterations and annotations by author.

In some metal systems with phase diagrams like that of Figure 5-23, rapid cooling (solution heat-treating) from the alpha (α) region forms a supersaturated solid solution of beta (β) in alpha (α). Depending on the metals involved, the beta (β) might stay in solution indefinitely, slowly precipitate at room temperature, or require some heating to cause it to precipitate. The is called **precipitation-hardening,** or **aging.** In such treatments care is taken to have the precipitate distributed throughout the grains rather than concentrated at the grain boundaries, which increases the strength and hardness of the metal.

If the alloy composition were to the *left* of the solvus line at room temperature, the beta (β) phase would not precipitate and the alloy would remain the alpha (α) solid solution.

At 900°C (1652°F) alloy 2 is all liquid and no sketch is shown. At about 800°C (1472°F), just above the eutectic temperature, Figure 5-23(d) shows that some alpha (α) has frozen but liquid still remains. This liquid is of the eutectic composition and with further cooling forms the eutectic mixture of Figure 5-23(e).

In Figure 5-23(d) and (e) the alpha (α) is preceded by the letters *pe,* which stand for **proeutectic;** that is *proeutectic* alpha (α) formed before the *eutectic* temperature was reached. Cooling the alloy below the eutectic temperature freezes the alpha (α)-beta (β) eutectic mixture. Now, the alpha (α) in the eutectic is the same solid solution as the proeutectic alpha (α), but it formed at a lower temperature and a later time than the proeutectic alpha (α)—thus the need to distinguish between the two.

The term *proeutectic* may be used to designate any phase that forms during cooling before the eutectic temperature is reached. In alloy systems with *eutectoids* a similar terminology is used; a **proeutectoid** phase is one that forms, or transforms, before the *eutectoid* temperature is reached.

The term **primary** is often used to identify the first crystal or phase that separates from the molten liquid. Such an identification is correct in any alloy system, but it would be especially appropriate in systems that do not have eutectics, or where the eutectic is not involved in that alloy. The alpha (α) solid solution of alloy 1 in Figure 5-23 could be termed *primary alpha* (α); it was the first crystal or phase to separate from the liquid.

SUMMARY

1. When metals melt or freeze, energy must be added or removed to change states. This change in energy can be detected and related to temperature and composition (temperature-versus-time cooling curves) and provides one way to establish phase diagrams for metal alloy systems.

2. Phase diagrams of binary alloy systems are plots of temperature versus the relative amounts of the two metals, in percent (usually in weight); the lines on the diagram show where changes of phase occur.

3. The lines on the diagram also show the limits of the solubility of the phases that make up the alloy system.

4. The uppermost lines on the diagram are liquidus lines. The liquidus separates the single-phase liquid region from the two-phase liquid-solid mixture region.

5. The next-lower lines are solidus lines. The solidus separates the two-phase liquid-solid mixture region from the solid region.

6. For any composition the liquidus line shows the lowest temperature at which the metal is all liquid, and the solidus is the highest temperature at which the metal is all solid.

7. When the combination of temperature and percent composition place an alloy in a liquid-solid mixture region of a phase diagram, the composition of the liquid part of that mixture is read at the liquidus line, and the composition of the solid part from the solidus line, both at that temperature.

8. The relative amount of phases present in mixtures is determined by using the *inverse tie-line rule:* The length of the line between the two phases represents the total of the metal at that temperature; the length of the line segment from the alloy composition to a phase, relative to the length of the total tie line, represents the percentage of the *other* phase.

9. Metals that are soluble as liquids but insoluble as solids follow Raoult's law—each metal lowers the melting-freezing temperature of the other until a minimum temperature, the eutectic temperature, is reached; this temperature is lower than the melting-freezing temperature of either metal. The composition of the metal that freezes at the eutectic temperature is the eutectic mixture.

10. The insoluble metals in this eutectic mixture cannot freeze in each other's lattices so they segregate and freeze in laminations, or layers, of the two metals called the *eutectic structure*.

11. The atoms of some intermediate phases will randomly distribute themselves throughout their joint lattice above a certain temperature, but below that temperature they will become ordered. That is, the atoms of each metal will only occupy certain sites in the lattice; they form a superlattice. This tends to reduce the plasticity of the metal and, in the case of the beta (β) brasses, requires that they be hot-worked.

12. When analyzing mixture regions for specific compositions and temperatures, it is best to proceed in an orderly sequence—first identify the phases present and their composition and then calculate their relative amounts. The following examples demonstrate the use of a table to make such analyses.

   a. The phases of the aluminum-silicon system (Figure 5-5) for an alloy of 60% silicon at 600°C:

Alloy	Temp.	Phases	Composition, % Si	% Relative Amount
60% Si	600°C	Liquid	14	$\dfrac{100-60}{100-14} = \dfrac{40}{86} = 47\%$
		Si	100	$\dfrac{60-14}{100-14} = \dfrac{46}{86} = 53\%$

   b. The phases of the gold-lead system (Figure 5-12) for an alloy of 70% gold at 419°C, just *above* the peritectic temperature:

Alloy	Temp.	Phases	Composition, % Au	% Relative Amount
70% Au	419°C	Liquid	55	$\dfrac{100-70}{100-55} = \dfrac{30}{45} = 67\%$
		Au	100	$\dfrac{70-55}{100-55} = \dfrac{15}{45} = 33\%$

   c. The phases of the gold-lead system (Figure 5-12) for an alloy of 70% gold at 417°C, just *below* the peritectic temperature:

Alloy	Temp.	Phases	Composition, % Au	% Relative Amount
70% Au	417°C	γ	66	$\dfrac{100-70}{100-66} = \dfrac{30}{34} = 88\%$
		Au	100	$\dfrac{70-66}{100-66} = \dfrac{4}{34} = 12\%$

13. The compositions of phases making up solid mixtures are read from solvus lines, or from lines representing a pure metal or a compound.

14. The solubilities of the metals determine the basic type of phase diagram they form.
    a. Metals that are completely soluble form a diagram with a very thin mixture region, somewhat like lemon-peel shape, and have just one solid solution.

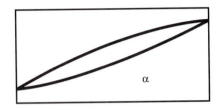

    b. Metals that are completely insoluble form a diagram containing a eutectic. At room temperature the alloys are mixtures.

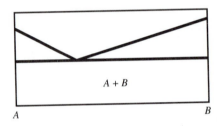

    c. Metals that are partially soluble–insoluble have regions of limited solubility where alloys exist as solid solutions, but they also have regions of insolubility where eutectics are formed and the alloys exist as mixtures.

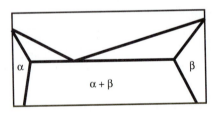

    d. Insoluble metals that have greatly different melting temperatures tend to form peritectics (a liquid and solid freeze to form a new solid phase, often a compound).

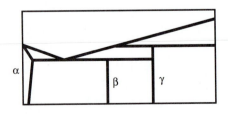

15. Some metals in certain compositions go through changes of phase termed *reactions;* these reactions, upon cooling, are as follows:

> Eutectic: Liquid phase ⇒ A mixture of two solid phases
> Eutectoid: 1 solid solution ⇒ 2 new solid phases
> Peritectic: 1 liquid and 1 solid phase ⇒ 1 new solid phase
> Peritectoid: 2 solid solutions ⇒ 1 new solid solution

16. *Equilibrium,* as applied to phase diagrams, means that whatever changes should happen will have enough time to do so. Some phase changes will not go to completion by slow cooling alone, but will require reheating and holding the alloy at an elevated temperature, i.e., giving it a *homogenizing treatment.*

17. When metals go from the random orientation of the liquid state to the crystalline solid state, time is required for the atoms to get into their proper location. Also, if the composition of the solid varies as it is freezing, time is required for the atoms to diffuse through the lattice and achieve the proper distribution, i.e., create a *homogeneous* structure.

18. When metals are heated or cooled faster than equilibrium, atoms do not have time to diffuse and the liquidus and solidus lines tend to move in the direction of the temperature change—heating raises the lines, cooling lowers the lines.

19. Nonuniform cooling of solid solutions will create cored crystals. That is, the composition of the dendrites formed will not be uniform, but will have atoms of the higher-freezing-temperature metal at their center or core, and atoms of the lower freezing temperature at their extremities.

20. Upon cooling, phases that solidify or precipitate before reaching the eutectic temperature are referred to as *proeutectic phases.* Similarly, solid phases that transform before reaching the eutectoid temperature are referred to as *proeutectoid phases.*

21. When cooling an alloy causes it to cross a solvus line and go from a solid-solution region to a mixture region, the phase in solution will tend to precipitate. Rapid cooling may create a supersaturated solid solution and the phase may stay in solution.

REVIEW QUESTIONS

1. How do the melting-freezing curves of pure metals, compounds, and solid solutions differ?

2. Why do insoluble metals freeze in the platelike eutectic structure? Why don't they freeze as concentric spheres?

3. Discuss the role the solubility of two metals plays in the form or type of phase diagram they develop.

4. An alloy of 80% nickel-20% copper (Figure 5-2) freezes while being cooled at a fairly rapid rate. What is the likely composition of the center of the dendrite that is formed?

What is the likely composition of the outer portion of the dendrite? Explain your reasoning.

5. In the aluminum-silicon system (Figure 5-5), for all alloys with more than 1.65% Si what is the composition of the last liquid to freeze under slow cooling?

6. For an alloy of 25% Zn-75% Cu (Figure 5-10) at 600°C (1112°F), identify and calculate the composition and percent relative amount(s) of the phase(s) present.

7. For an alloy of 40% Zn-60% Cu (Figure 5-10) at 600°C (1112°F), identify and calculate the composition and percent relative amount(s) of the phase(s) present.

8. For an alloy of 49% Zn-51% Cu (Figure 5-10), discuss the difference between the atom lattice at 700°C (1292°F) and the lattice at 400°C (752°F).

9. In the iron-iron carbide system (Figure 5-9), for an alloy containing 1.2% carbon at 1400°F (760°C), identify and calculate the composition(s) and percent relative amount(s) of the phase(s) present.

10. Discuss the similarities and differences of eutectics, eutectoids, peritectics, and peritectoids.

FOR FURTHER STUDY

Avner, Sidney H. *Introduction to physical metallurgy* (2d ed.). New York: McGraw-Hill, 1974, chap. 6, Phase diagrams.

DeGarmo, E. Paul, J. Temple Black, and Ronald A. Kohser. *Materials and processes in manufacturing* (7th ed.). New York: Macmillan, 1988, chap. 4, Equilibrium diagrams.

Hume-Rothery, W. *Elements of structural metallurgy*, Monograph and report series no. 26. London: Institute of Metals, 1961, chaps. VIII and IX.

Reed-Hill, Robert E., and Reza Abbaschian. *Physical metallurgy principles* (3d. ed.). Boston: PWS-Kent, 1992, chap. 11, Binary phase diagrams.

Smallman, R. E. *Modern physical metallurgy* (4th ed.). London: Butterworth, 1985, secs. 3.1–3.2.

FERROUS AND NONFERROUS METALS

PART 2 USES THE METALLURGICAL BASICS FROM PART 1 AND APPLIES them to specific alloy systems. Chapter 6 emphasizes the iron–iron carbide alloy system, its phase diagram, and the results of the slow-cooling of carbon steels. Chapter 7 investigates the results of nonequilibrium cooling of steels and the generation of martensite, giving more emphasis to the time-temperature transformation diagram than to the phase diagram. We'll look at the distortion problem caused by the formation of martensite in the course of heat-treating steels.

Chapter 8 examines the influence on steels of alloying elements in general. Although steels can be categorized according to their specific applications—alloy steel and tool steel, for instance—the alloying elements used in each category will often be used for the same effect. So our emphasis will be on similarities rather than differences. Chapter 9 reveals that cast iron is steel with so much carbon that the carbon is insoluble.

Chapter 10 will concentrate on copper and aluminum alloys and also look at other alloy systems. But because there are so many possible nonferrous alloy systems we'll focus on the more general alloying and heat-treating principles rather than on individual alloys.

Although compositions and properties of specific steels, cast irons, and nonferrous metals are discussed in Part 2, the specific purpose of these five chapters is to illustrate the metallurgy of the alloys and their elements and suggest applications for the metals.

You can supplement the material covered here by heat-treating ferrous and nonferrous metals in the laboratory. Use of Appendix E, along with the performance of a Jominy end-quench test, is particularly helpful in understanding the principles of hardenability. The appendix also contains practical applications of the Jominy test.

C H A P T E R   6

Iron and Steel

OBJECTIVES

After studying this chapter, you should be able to

- duplicate the simplified iron–iron carbide phase diagram, identifying the phases, mixture regions, and transformation lines
- define the ferrite, austenite, and cementite phases
- explain the principal difference between slow-cooled hypoeutectoid and hyper-eutectoid steels
- trace the transformation of hypoeutectoid and hypereutectoid steels as they are slow-cooled from the austenite region, sketch their microstructures at any temperature, and explain how their mechanical properties vary with microstructure

DID YOU KNOW?

Iron is an ancient metal that has given its name to one of our prehistoric ages. The biblical book of Daniel, written about the sixth century B.C., describes a beast of iron. The book of Genesis, covering a period more than a thousand years before that, mentions the forging of iron. But despite the great age of the technology, despite all the aerospace technologies developed since World War II, iron and steel are still our most commonly used metals. Our most modern buildings, vehicles, machines, and tools all make extensive use of iron and steel. People working in the manufacturing sector of any of the world's economies are dependent on these metals, in one or more of their many forms, either in their products or the tools and equipment used to produce them.

A PREVIEW

The metallurgical advantage that steels and cast irons have over almost all other metals—their ability to go through an allotropic transformation—is our first topic. After examining the phase diagram, we'll consider what happens when the metal is slow-cooled from a solid solution called austenite.* From this study we'll then develop the microstructures of slowly cooled steels and go on to examine how mechanical properties vary with carbon content. Lastly, we'll discuss the other elements present in plain carbon steels.

---

*This material was named for W. S. Roberts-Austen, metallographer.

6.1   ALLOTROPIC TRANSFORMATIONS

Allotropy. A few metals are capable of polymorphic[†] transformation—that is, capable of existing in more than one atom lattice in the solid state. When the change in lattice is reversible with temperature, the transformation is called *allotropic*.

At room temperature pure iron has a body-centered cubic (BCC) lattice, is magnetic, and is designated alpha-iron (α), or **ferrite.** At 1414°F (768°C) α-ferrite becomes nonmagnetic, but there is no change in the crystal structure.[‡] At 1666°F (908°C), the lattice structure of pure iron changes to face-centered cubic (FCC) and is designated gamma-iron (γ), or **austenite.**

As the melting point of iron, 2800°F (1538°C), is approached, the metal undergoes another transformation to the BCC lattice and is now termed delta-iron; in most applications this region is of little interest and is mentioned here only because you'll often find the term in metallurgical texts and journals.[§]

In Chapter 3 we discussed the ability of metals to recrystallize through cold-working and annealing. Steel is capable of that same type of *recrystallization,* but because of its allotropic properties it can also easily recrystallize when heated above the ferrite-to-austenite ($\alpha \rightarrow \gamma$) transition temperature where a new grain structure is formed. If the temperature goes high enough into the γ-austenite range, grain growth will occur, as explained previously in our discussion of annealing. As the steel cools through the austenite-to-ferrite ($\gamma \rightarrow \alpha$) transformation, a second recrystallization takes place with grain refinement. This process can be repeated, making it important in improving the properties of cast and forged ferrous products.

Alloying Elements. Being allotropic, all these transformations are reversible with temperature, but alloying elements will alter the temperatures at which the transformations take place. Carbon, the element that changes iron to steel, has the marked effects seen in the iron-carbon phase diagram, Figure 6-1. Also, substitutional alloying of the iron can slow the *rate* at which these transformations take place on cooling; thus alloying is used to change the heat-treating characteristics of steels and cast irons.

The ability of iron to undergo the allotropic transformation is what makes possible the heat treatment of steels and cast irons. Steels are considered to be alloys of iron containing up to 2.0% carbon; cast irons are alloys of iron containing 2.0 to 6.67% carbon. Most commercial steels contain less than 1.5% carbon, and most commercial cast irons contain less than 4.0% carbon.

The different allotropic forms (α, γ) have differing abilities to dissolve alloying elements. Carbon is only slightly soluble in α-ferrite—0.008% at room tempera-

---

[†]The word comes from *poly* meaning "many" and *morph* meaning "form," hence "many forms."

[‡]At one time this metal was called beta-iron (β) and its transformation temperature was called A_2. But this is not considered a phase change, and these designations are no longer used.

[§]This peritectic region is at such a high temperature that possible segregation is of little importance by the time the steel has cooled to normal working temperatures.

FIGURE 6-1 **Iron–iron carbide phase diagram, with phases, transformation lines, and reactions**

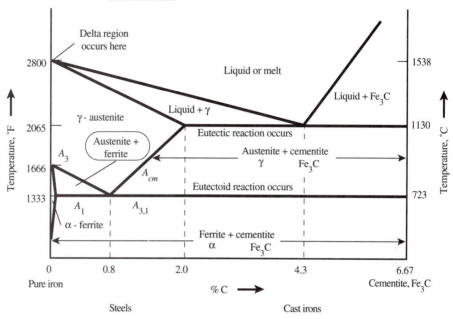

ture, 0.025% at 1333°F (723°C)—but in γ-austenite the solubility of carbon can be as high as 2.0% at 2065°F (1130°C). These are *interstitial solid solutions* of carbon and iron. Both α-ferrite and γ-austenite also form *substitutional solid solutions* with many other alloying elements.

At 6.67% carbon, well above the limits for solid solubility in α-ferrite or γ-austenite, the interstitial compound iron carbide (Fe₃C), or **cementite,** is formed, which can precipitate from the γ-austenite solid solution on cooling.

6.2   IRON–IRON CARBIDE PHASE DIAGRAM

Transformation Lines. The iron–iron carbide phase diagram (Figure 6-1) shows how the temperature for these transformations changes with carbon content. The temperatures for the polymorphic transformations of pure iron are shown on the left axis at 0% carbon. As carbon is added, the temperatures of these **transformations** change and begin to occur over ranges of temperatures.‖ Pure iron, for example, changes from alpha-ferrite to gamma-austenite ($\alpha \rightarrow \gamma$) at 1666°F (908°C). As carbon is added, that transformation begins at lower and lower temperatures; for carbon contents of 0.025% C and greater it begins at 1333°F (723°C).

‖This is similar to the influence of alloying elements on the melt-freeze temperatures of pure metals.

For ease of identification, the transformation lines of the diagram, also known as **critical lines,** are given letter-number designations. These designations change according to whether the carbon content is less (hypo-) than or greater (hyper-) than the eutectoid composition. To the left, on the hypoeutectoid side, the horizontal line at 1333°F (723°C) is called A_1; the next line above that is termed the A_3 line.[#] To the right on the hypereutectoid side, the first line is a merger of the two lines on the left and is therefore called $A_{3,1}$; the next line above it is termed A_{cm}.

Phases Present. The A_1 line, 1333°F (723°C), is the minimum temperature at which α-iron begins transforming to γ-iron as the temperature is raised; as carbon content is raised above 0.025% this temperature doesn't change. The second line, A_3, *represents the conclusion of the allotropic transformation of the α-iron to γ-iron.* To the left of the eutectoid, in the region between A_1 and A_3, there is a mixture of the two solid-solution phases, α and γ. However, to the right of the eutectoid, the region between the two lines is a mixture of only γ and cementite; no α phase is present. Since A_1 and A_3 represent the start and finish of the allotropic transformation of α to γ, to the right of the eutectoid the designations are merged into $A_{3,1}$, indicating α changes at the line; there is no mixture region of α and γ above $A_{3,1}$.

The right side of this mixture region is bounded by the interstitial compound Fe_3C, or cementite, and the left side by γ-austenite. Since the cementite composition is a constant 6.67% carbon, the composition represented by the right end of the tie line through this region is always Fe_3C. The other side of the region is a different matter; the composition of the γ-austenite changes with the temperature, from 2.0% carbon at 2065°F (1130°C) to 0.8% carbon at 1333°F (723°C). The carbon in this range of temperatures and compositions must either be in solution in the γ-austenite or with the iron as the compound Fe_3C-cementite. Thus the line to the left is referred to as A_{cm}, the line along which cementite, "cm," is soluble in austenite or precipitates from the austenite, depending on whether the metal is heated or cooled through the range.

The A_{cm} line should be recognized as a *solvus* line defining solubility in the solid state.

Summary. The *simplified* iron–iron carbide system has four solid phases present: (1) pure iron; (2) the ferrite, or alpha (α), phase; (3) the austenite, or gamma (γ), phase; and (4) the cementite, or Fe_3C, phase. The system has four significant temperatures: (1) 1333°F (723°C), the eutectoid temperature or A_1 and $A_{3,1}$; (2) 1666°F (908°C), where pure iron changes from BCC to FCC; (3) 2065°F (1130°C), the eutectic temperature; and (4) 2800°F (1538°C), the melting point of pure iron. It also has four significant compositions of carbon: (1) 0.8%, the eutectoid composition; (2) 2.0%, the maximum solubility of austenite and the nominal maximum carbon content for steels; (3) 4.3%, the eutectic composition; and (4) 6.67%, the composition of cementite.

[#]The letter A stands for the French word *arret* for "stop" or "arrest." The cooling curves will arrest or pause when these temperatures are reached.

6.3   TIME

Normally phase diagrams deal with *equilibrium* conditions, but the data are generated by either heating or cooling specimens. These *nonequilibrium* conditions cause the transformation lines to move up with heating and down with cooling. In our discussions in this chapter we used diagrams that assume that approximate equilibrium conditions exist.

In spite of these approximate equilibrium conditions, the iron–iron carbide phase diagram can still be used to study commercial processes such as annealing, normalizing, spheroidizing, hot-working, and malleabilizing. The microstructures formed by these processes can still be understood from the phase diagram. However, the quenching treatments of steels and cast irons are too far from equilibrium to be understood from the phase diagram.

Figure 6-1 shows that above 1333°F (723°C), but below the solidus, all compositions contain γ-austenite, both steels and cast irons. In the alloying and heat-treating of steels and cast irons, the presence of this γ-austenite—with its ability to dissolve carbon interstitially, and other elements substitutionally—is critical.

It is important to realize that the dissolving of other metals in the γ-austenite takes time and that, conversely, it also takes time for them to come out of solution when their solubility limit is exceeded upon cooling. Under the conditions assumed in the equilibrium diagram, time is of no concern; but as commercial conditions are approached, time does become a factor. In fact, that is how steel and cast-iron hardening is accomplished; the γ-austenite is cooled faster than it can precipitate, or reject out of solution, the alloying elements, principally carbon, that are dissolved in it.

The great usefulness and versatility of the iron-carbon system is primarily dependent upon its ability to go through the *allotropic* transformation from γ-austenite to α-ferrite.

6.4   EUTECTOIDS

We discussed eutectoids in Chapter 5. The solid solutions of some alloy systems undergo an "unsolutioning" when cooled, in a manner very similar to the formation of a eutectic from a liquid solution. That is, what was a single solid solution forms two new solid solutions that separate into laminated layers of the two metals that had been in solution. The appearance of the eutectoid on the phase diagram, like that of a eutectic, is a "V," and the reaction appears to take place like that of a eutectic does, except that the solid-solution phase has a crystal structure, whereas a liquid would not. This preexisting crystal structure will have an influence on the nucleation of the new precipitating phases and the growth of the eutectoid.

The iron-carbon system of steels and cast irons contains the best-known example of a eutectoid.** The single solid-solution phase γ-austenite transforms to a mixture of two phases, α-ferrite and Fe_3C-cementite, with an overall composition of 0.8% carbon; using the tie line, we can calculate that this eutectoid structure will be 88% α and 12% Fe_3C:

$$\frac{6.67 - 0.8}{6.67} = 88\% \ \alpha \qquad \frac{0.8 - 0}{6.67} = 12\% \ Fe_3C$$

In appearance (see Figures A-13 and A-16) the laminations resemble mother-of-pearl, and the structure is called **pearlite.** (Note that this is a recognizable *structure,* not a phase; it is a *mixture* of the two phases α-ferrite and γ-cementite.)

6.5   STRUCTURES OF SLOW-COOLED IRON AND CARBON

Figure 6-2 is similar to Figure 6-1 except that it identifies the structures and mixtures, as well as the phases, that exist in certain portions of the diagram. We'll discuss the total diagram first and then look at the steel portion in detail; the cast-iron portion is covered in Chapter 9.

Freezing Liquid to Solid.   The slow-freezing of all compositions above 2.0% and up to 6.67% guarantees the presence of the eutectic structure; it is present in greater amounts the closer the composition is to the 4.3% eutectic (the inverse tie-line rule). The slow-freezing of all steel compositions (carbon < 2.0%) results in no eutectic structure being present because under equilibrium conditions liquid is completely solidified before its composition can reach the eutectic composition. In Figure 6-2, the first region of solid metal is all γ-austenite up to 2.0%, γ-austenite plus eutectic from 2.0 up to 4.3%, and eutectic plus Fe_3C-cementite from 4.3 to 6.67%. If cooling proceeds from these solids, different structures result as the transformation lines are crossed.

Cooling as a Solid.   For hypoeutectoid steels (less than 0.8% carbon) the next transformation starts at the A_3 line, where the γ-austenite starts to transform to α-ferrite; when A_1 is reached *the remaining γ-austenite is at the eutectoid composition and transforms to pearlite.* Thus, at room temperature, the structure present is α-ferrite with pearlite.

For hypereutectoid steels (steels with greater than 0.8% carbon) the next transformation starts at the A_{cm} line; here the γ-austenite starts to give up its carbon, which joins with the iron to form Fe_3C-cementite; at $A_{3,1}$ this transformation stops and *any remaining γ-austenite transforms to pearlite.* Thus the room temperature structure is Fe_3C-cementite and pearlite.

**Copper-aluminum and copper-tin are examples of nonferrous alloy systems that contain eutectoids; these are discussed in Chapter 10.

FIGURE 6–2   Iron–iron carbide phase diagram, with structures resulting from slow cooling

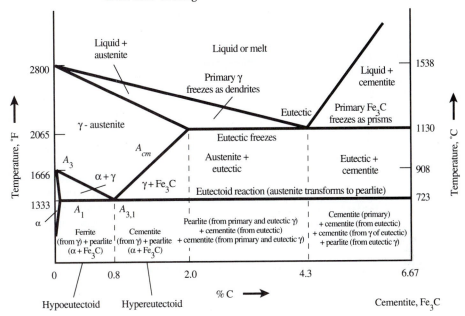

Source: Clyde W. Mason. *Introductory physical metallurgy*. Materials Park, OH: ASM International, 1947, fig. 50, 86. Used by permission.

For compositions of more than 2.0% carbon, the structures at room temperature are all combinations of pearlite and Fe_3C-cementite, but as shown in Figure 6-2, the origins of each are complex.

The FCC γ-austenite phase can dissolve carbon interstitially over a large temperature range, and up to 2.0% maximum at the eutectic temperature. As the temperature falls from this maximum, γ-austenite precipitates carbon as Fe_3C-cementite along the A_{cm} solvus line; as the temperature drops from 2065°F (1130°C) to 1333°F (723°C), the solubility of carbon in γ-austenite decreases from 2.0 to 0.8%.

The BCC α-ferrite phase, different from γ-austenite, can dissolve little carbon at any temperature, and the A_3 line has a completely different significance than the A_{cm} line. *When hypoeutectoid γ-austenite cools, the ability of γ-austenite to dissolve carbon actually goes up* as the temperature drops below 1666°F (908°C).

Cooling Hypoeutectoid and Hypereutectoid Steels.   We'll now follow Figures 6-3, 6-4, and 6-5 in tracing the slow-cooling of two steels from the γ-austenite region. Steel number 1 is a hypoeutectoid steel and steel number 2 is a hypereutectoid steel. Although both steels will contain pearlite, the major difference between them is in the phase that forms before (*proeutectoid*) the formation of the pearlite.

As the hypoeutectoid steel, steel number 1, cools from the γ-austenite region, it undergoes the γ-austenite $\rightarrow$ α-ferrite polymorphic transformation before any

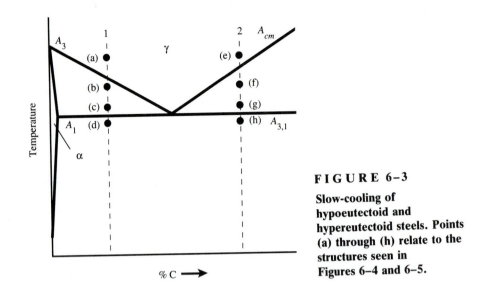

FIGURE 6-3

Slow-cooling of hypoeutectoid and hypereutectoid steels. Points (a) through (h) relate to the structures seen in Figures 6-4 and 6-5.

FIGURE 6-4 Microstructures of slow-cooled hypoeutectoid steel no. 1. Parts (a) through (d) refer to the points in Figure 6-3.

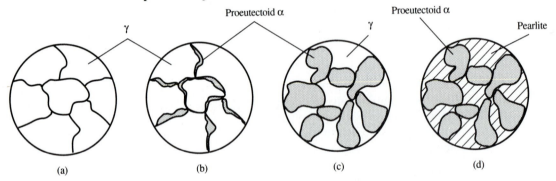

FIGURE 6-5 Microstructures of slow-cooled hypereutectoid steel no. 2. Parts (e) through (h) refer to the points in Figure 6-3.

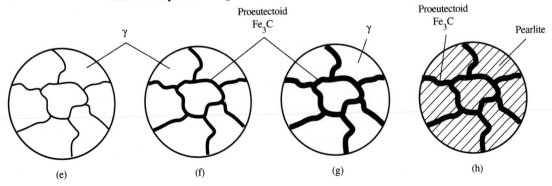

Fe_3C-cementite is precipitated. This occurs at a temperature below that for pure iron (1666°F, 908°C), because the carbon dissolved in the γ-austenite lowers the A_3 transition temperature (similar to Raoult's law, where one metal lowers the melting point of another in a eutectic system). As the proeutectoid α-ferrite, which is almost carbon-free, forms from the γ-austenite, the carbon content of the γ-austenite increases from its original value and continues to increase as the temperature falls, following the A_3 line. This process continues until A_1 (1333°F) is reached, where the γ-austenite composition is 0.8% carbon, the eutectoid composition, and pearlite forms. Under equilibrium cooling, this is the lowest transformation temperature, analogous to the minimum freezing temperature of mixtures encountered in eutectics.

A hypereutectoid steel, number 2 in Figure 6-3, on cooling through the γ-austenite region begins to reject excess carbon (the carbon it can no longer keep in solution) as proeutectoid Fe_3C-cementite when the saturation line, A_{cm}, is reached. The Fe_3C-cementite precipitates at the γ-austenite grain boundaries as the temperature drops until, at 1333°F (723°C) or $A_{3,1}$, the remaining γ-austenite goes through its allotropic transformation to α-ferrite. Because α-ferrite can dissolve little carbon, however, the rejected carbon forms additional Fe_3C-cementite along with the α-ferrite and then forms the fine-grained, laminated eutectoid, pearlite.

Whether the steel is hypoeutectoid or hypereutectoid, by reason of the inverse tie-line rule, the closer the composition is to the eutectoid composition of 0.8% carbon the more pearlite will be present in the final structure. If the steel is of a composition of 0.8%, or close to it, no *proeutectoid* α-ferrite or γ-cementite will separate from γ-austenite, and its structure will be all pearlite.

Formation of Microstructures.   Figure 6-4 shows the microstructures during the cooling process just described for steel number 1. At point (a) the structure is γ-austenite grains and grain boundaries, i.e., a solid solution. At point (b) the A_3 line is crossed and some α-ferrite is formed at the γ-austenite grain boundaries. This process continues as the temperature drops to point (c). The γ-austenite increases in carbon content and its composition approaches 0.8% carbon, although the volume of γ-austenite grains remaining has gone down. At point (d) the A_1 line is crossed; all the γ-austenite of point (c) transforms to α-ferrite and Fe_3C-cementite in the form of pearlite. The proeutectoid α-ferrite remains as it was at point (c). Figure A-14 is a good example of what the structure of a slow-cooled hypoeutectoid steel looks like through the microscope.

Figure 6-5 shows the progression of microstructures for a slow-cooled hypereutectoid steel such as steel number 2 of Figure 6-3. At point (e) there is nothing but γ-austenite grains and grain boundaries, as at point (a) in Figure 6-4. At (f) the A_{cm} line is crossed and proeutectoid Fe_3C-cementite has started to precipitate at the γ-austenite grain boundaries. As the temperature drops to point (g), this precipitate or network of Fe_3C-cementite gets wider. When the $A_{3,1}$ line is crossed, the remaining γ-austenite transforms to pearlite, but the original proeutectoid Fe_3C-cementite network remains. Figure A-13 shows this Fe_3C-cementite network in a high-carbon steel.

Cooling Rate.   For both hypoeutectoid and hypereutectoid steels, if the cooling rate is speeded up, the amount of the proeutectoid phase network precipitated will be reduced; it does not have time to travel to the γ-austenite grain boundaries. Therefore some of the proeutectoid phase may come out of solution within the γ-austenite grain itself and some of it may become part of the pearlite when it forms at a lower temperature. We can see the first phenomenon in Figure A-15; proeutectoid α-alpha formed within the old γ-austenite grain in a hypoeutectoid steel. In Figure A-19 the white slashes within the old γ-austenite grain are a portion of the proeutectoid Fe_3C-cementite that would have formed at the grain boundaries of a slower-cooled hypereutectoid steel. Also notice that the extent of the Fe_3C-cementite network around the old γ-austenite grain is very thin. This is especially noticeable if Figures A-19 and A-13 are compared; they have the same carbon content but A-19 was cooled faster than A-13.

Pearlite.   The transformation of γ-austenite to pearlite occurs after the proeutectoid phases have precipitated, that is, when the cooling γ-austenite reaches the A_1 or $A_{3,1}$ line. The pearlite starts to form as small islands at the juncture of the proeutectoid phase, if present, and the γ-austenite grain, and then progresses through the remaining γ-austenite. This can be seen in Figure A-20. The slower the cooling rate, the coarser the pearlite laminations of α-ferrite and Fe_3C-cementite will be; that is, at slower cooling rates the laminations will be thicker and farther apart. Conversely, faster cooling rates will produce finer laminations. The fine pearlite will be stronger than the coarse pearlite, but the coarse pearlite will have higher plasticity.[††]

These slowly cooled steels have only a proeutectoid phase, α-ferrite or Fe_3C-cementite, and pearlite. So if, as just explained, increasing the cooling rate suppresses the formation of the proeutectoid phase, then the percent pearlite will go up. Thus, although the nominal composition of pearlite is 0.8% carbon under equilibrium cooling, faster cooling can produce steels of essentially all pearlite, with carbon contents from 0.4 to 1.4%.

6.6   MECHANICAL PROPERTIES OF SLOW-COOLED STEELS

The major difference between the two steels of Figure 6-3 is in the makeup of the proeutectoid phase that first precipitates. For the hypoeutectoid steel 1, the proeutectoid phase is soft α-ferrite, which has high plasticity. For the hypereutectoid steel 2, however, it is hard and brittle Fe_3C-cementite. For the hypoeutectoid steel, all the Fe_3C-cementite is within the pearlite formation. The α-ferrite laminations help the pearlite overcome the brittle nature of Fe_3C-cementite, and as the carbon content of a *hypoeutectoid* steel increases, the strength and hardness increase almost in direct proportion to the amount of pearlite formed.

[††]Plasticity indicates both tensile ductility and compressive malleability.

Because of this relationship, the tensile strength of hypoeutectoid steels can be estimated from the carbon content. For example, the tensile strength of annealed iron is approximately 40,000 psi; the tensile strength of annealed pearlite (0.80% carbon) is approximately 120,000 psi, or a difference of 80,000 psi; each 0.10% carbon contributes about 10,000 psi. Thus an annealed 0.30% carbon steel would have an estimated tensile strength of 70,000 psi (40,000 + 30,000).

Table VI-1 shows tensile property data typical of low-to-medium carbon steels in the hot-rolled and cold-drawn conditions. Note that the hot-worked tensile strengths are close to those that would be predicted by the carbon-content formula discussed above. The cold-drawn data illustrates the point made in Chapter 2 that plastic deformation has a large influence on the yield strength, bringing it closer to the tensile strength.

When the eutectoid composition (0.80%) is exceeded, the brittle network of proeutectoid Fe_3C at the γ-austenite grain boundaries (see Figures 6-5 and A-13) weakens the steel; the tensile strength of annealed (furnace-cooled) carbon steel peaks at about 120,000 psi and then levels off or goes down as carbon goes above 0.80%. If the cooling rate is speeded up to that of normalizing (air cooling), rather than annealing, the Fe_3C-cementite network will be reduced, appearing like that seen in Figure A-19. The strength will continue to increase as carbon goes above 0.80%, but eventually even normalizing will not reduce the brittle Fe_3C network enough, and when the carbon gets to about 1.3%, the tensile strength will start to decline.

Although the data for Figure 6-6 are based on hot-worked steels, which would approximate normalized properties, the graph still illustrates the effects discussed above. Initially, as the carbon increases, the strength goes up, but at about 1.0% C

T A B L E VI–1 Typical mechanical properties for AISI/SAE* carbon steels

UNS Alloy No.	AISI/ SAE Alloy No.[†]	Hot-Rolled				Cold-Drawn			
		Tensile Strength, ksi[‡]	Yield Strength, ksi	Elon-gation, %	Hard-ness, HB[§]	Tensile Strength, ksi	Yield Strength, ksi	Elon-gation, %	Hard-ness, HB
G10080	1008	44	24.5	30	86	49	41.5	20	95
G10150	1015	50	27.5	28	101	56	47	18	111
G10200	1020	55	30.0	25	111	61	51	15	121
G10300	1030	68	37.5	20	137	76	64	12	149
G10400	1040	76	42.0	18	149	85	71	12	170
G10500	1050	90	49.5	15	179	100	84	10	197

*A standard designation system established by the American Iron and Steel Institute/Society of Automotive Engineers.

[†]In this system the last two digits of the number are the carbon content in points; e.g., 1020 = 0.20% C.

[‡]ksi = thousands of psi.

[§]HB = Brinell hardness.

Source: *Metals handbook* (8th ed.). Materials Park, OH: ASM International, 1961, *1*, 188.

F I G U R E 6–6 Mechanical properties vs. percent carbon for hot-worked steels

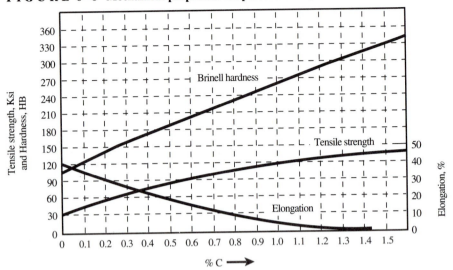

it levels off. Also notice that the ductility, as measured by the elongation, drops rather rapidly with increased carbon and then levels off.

Hardness behaves somewhat differently; at low carbon levels it parallels the tensile strength, but it retains its slope and does not level off at higher carbon contents. This is because the hardness test is a compressive or bearing-strength test, and the hard, brittle Fe_3C-cementite is additive to the hardness reading; it does not pull apart as it would in a tension test.

6.7   COMPOSITIONS OF COMMERCIAL CARBON STEELS

For the purposes of commercial production the compositions of steels (and other metals) are identified by various designation systems.[‡‡] In commercial production the 10XX series, or **plain carbon steels,**[§§] which presumably would be only iron and carbon, contain small amounts of sulfur, phosphorus, manganese, and silicon, but they are still called carbon steels. Sulfur and phosphorus are impurities that come from the ore and/or coke, and manganese (not magnesium) and silicon are elements that are typically added.

[‡‡]A unified numbering system (UNS) has been established to avoid confusion among the numerous identification systems established by societies and trade associations. Where possible, the identification numbers of previously existing systems are incorporated into the UNS designation.
[§§]Steel carbon contents are referred to in "points"; each 0.01% carbon is 1 point, each 0.10% carbon 10 points. The Society of Automotive Engineers (SAE) and the American Iron and Steel Institute (AISI) steel-numbering systems use four digits, the last two of which are the carbon content in points. The first two digits refer to the alloy content; plain carbon steel is designated as 10XX. That is, a steel with 0.15% carbon would be 1015. In UNS it would be G10150.

Sulfur. Sulfur comes from the original ore and carries over into the steel, where it forms iron sulfide (FeS). This compound forms a eutectic with iron that has a lower melting point than the steel itself; the eutectic tends to concentrate at the grain boundaries. In fabricating operations that require high temperatures (forging and rolling, for instance) this eutectic melts and reduces the bond between the steel grains. Thus the steel becomes weak and brittle at high temperatures and is termed **hot-short;** usually the sulfur content is controlled by the use of manganese to 0.05% or below to prevent this condition.

For free-machining steels, or resulfurized steels, the sulfur content is purposely raised to between 0.08 and 0.35%. These steels are designated by a system similar to that used for plain carbon steels, except for the first two digits, 11XX. The numerous sulfide particles act as chip breakers, reducing tool wear.

Phosphorus. A phosphorus content above about 0.12% reduces the ductility of steel, causing brittleness or weakness at room temperature; the word *short* is used to describe the brittle condition, and since it occurs at reasonably cold temperatures the term **cold-short** is used.

To prevent cold-shortness the phosphorus content is usually kept below 0.04%; at this level the phosphorus dissolves in α-ferrite, increasing strength and hardness slightly.

Some steels are rephosphorized *and* resulfurized to improve machinability. In such steels the phosphorus content is 0.04 to 0.12% and the sulfur content is 0.16 to 0.35%. These steels carry a 12XX designation.

Manganese. A manganese content of 0.03 to 1.00% is used in commercial plain carbon steels. Some part of this *reduces the effects of sulfur by forming MnS, which floats out with the slag* during melting or becomes a fine inclusion distributed throughout the structure. For this purpose, the amount of manganese is kept at 2 to 8 times the amount of the sulfur.

The amount of manganese present, over that needed to react with the sulfur, will form Mn_3C, which associates itself with the iron carbide, Fe_3C, in the cementite. Manganese can also react with oxygen during the melting process and act as a **deoxidizer,** thus improving the soundness and cleanliness of steels.

Manganese greater than 1.0% is an alloying element and is discussed further in the next two chapters.

Silicon. A silicon content of 0.05 to 0.3% is used in most commercial steels for two purposes: it serves as a *deoxidizer* because of its strong affinity for oxygen (forming SiO_2, or "sand"), and it strengthens the steel with little reduction of plasticity by *substitutionally dissolving in the ferrite.*

Silicon in larger quantities can control the form carbon takes in cast irons and will be discussed further in Chapter 9.

Impurities. Impurities usually come from either the original ore used to make the iron or from the scrap used in the BOF (basic oxygen furnace), electric furnace, or open-hearth steel-making processes. Thus many other **tramp,** or unwanted, elements and compounds can be present in steels.

SUMMARY

1. Iron has the property of allotropy and can thus change atom lattice, or recrystallize, through change of temperature. It is this property that enables steels and cast irons to be heat-treated so readily.

2. The solid single phases present in the iron–iron carbide system include the following:
    a. *Pure iron.* Body-centered cubic and magnetic at room temperature, it becomes face-centered cubic and nonmagnetic above 1666°F (908°C).
    b. *Ferrite or α-iron.* A magnetic, soft, ductile, malleable, interstitial solid solution of carbon in body-centered cubic iron, it has a very low solubility for carbon at any temperature.
    c. *Austenite or γ-iron.* A nonmagnetic, soft, ductile, malleable, interstitial solid solution of carbon in face-centered cubic iron, it has a maximum solubility for carbon of 2.0% at 2065°F (1130°C), and 0.8% at 1333°F (723°C). In unalloyed carbon steels and cast irons, it exists only at higher temperatures.
    d. *Cementite or iron-carbide or Fe_3C.* An interstitial compound of 6.67% carbon in iron, it is very hard and brittle and has a high melting point.

3. Steels are considered to have carbon contents up to 2.0% C, although commercially they seldom have more than about 1.2% carbon. Cast irons have a carbon content of over 2.0%, but seldom exceed 4.0%.

4. In the slow-cooling of steels, the γ-austenite that reaches 1333°F (723°C), or $A_1/A_{3,1}$, will have a carbon content of 0.8% and on further cooling will form a eutectoid mixture of α-ferrite and Fe_3C-cementite called *pearlite*. The eutectoid reaction occurs because, although iron and carbon are soluble in the austenite phase, they are almost completely insoluble below $A_1/A_{3,1}$.

5. Under equilibrium cooling, pearlite consists of 88% α-ferrite, 12% Fe_3C-cementite.

6. The difference in microstructures and properties between hypoeutectoid and hypereutectoid steels is the proeutectoid phase that precipitates before reaching $A_1/A_{3,1}$. This phase is soft, plastic α-ferrite for hypoeutectoid steels and brittle Fe_3C-cementite for hypereutectoid steels.

7. Speeding up the cooling rate will suppress the formation of the proeutectoid phases and increase the percent pearlite formed. Rapid cooling can produce steels with almost complete pearlite microstructures over a wide range of carbon contents.

8. The precipitation of the proeutectoid phases and pearlite tends to begin at the grain boundaries of the austenite.

9. A number of other elements are present in steels called "carbon steels." Sulfur and phosphorus are usually unwanted elements that come into the steel from the original iron ore. Manganese is added to help control the sulfur.

Manganese and silicon are added as deoxidizers. Silicon also dissolves in the ferrite and can help strengthen the steel.

REVIEW QUESTIONS

1. From memory, sketch the iron–iron carbide phase diagram and show the four important temperatures, compositions, and phases.
2. Define the three solid single phases present on the iron–iron carbide phase diagram. Include in your definition their type of solution or compound, the lattice structures of the solutions, and important physical and mechanical properties.
3. Explain why pearlite is always 88% ferrite and 12% cementite at room temperature under equilibrium cooling conditions.
4. Explain why the A_{cm} line is a solvus line.
5. Trace the slow-cooling of a hypoeutectoid carbon steel from the austenite region to room temperature and explain what changes in structure occur as the critical lines are crossed.
6. Reconsider Question 5 for a hypereutectoid carbon steel.
7. A hypoeutectoid carbon steel of 0.38% C is slow-cooled from the austenite region. What phases are present at room temperature, what composition do they have, and what is their percent relative distribution?
8. Reconsider Question 7 for a hypereutectoid carbon steel of 1.05% C.
9. A hypoeutectoid carbon steel of 0.38% C is slow-cooled from the austenite region to room temperature. What percent of the microstructure would be proeutectoid ferrite and what percent would be pearlite? Sketch and label the microstructure.
10. Reconsider Question 9 for a hypereutectoid carbon steel of 1.05% C.
11. Explain why the tensile strength of slow-cooled carbon steel tends to level off when the carbon content gets above about 0.80% C.
12. Explain what the term *short* means when used in describing steels.
13. Carbon steels are basically alloys of iron and carbon. Why are the elements manganese and silicon added?

FOR FURTHER STUDY

Alexander, W. O., G. J. Davies, K. A. Reynolds, and E. J. Bradbury. *Essential metallurgy for engineers*. Wokingham, U.K.: Van Nostrand Reinhold, 1985, 61–63.

Avner, Sidney H. *Introduction to physical metallurgy* (2d ed.). New York: McGraw-Hill, 1974, chap. 7.

DeGarmo, E. Paul, J. Temple Black, and Ronald A. Kohser. *Materials and processes in manufacturing* (7th ed.). New York: Macmillan, 1988, 91–99.

Flinn, Richard A., and Paul K. Trojan. *Engineering materials and their applications* (3d ed.). Boston: Houghton Mifflin, 1986, secs. 6.1–6.5.

Mason, Clyde W. *Introductory physical metallurgy*. Materials Park, OH: ASM International, 1947, chap. VI.

Reed-Hill, Robert E., and Reza Abbaschian. *Physical metallurgy principles* (3d ed.). Boston: PWS-Kent, 1992, chap. 18.

CHAPTER 7

Steel Heat-Treating

After studying this chapter, you should be able to

- describe the annealing processes of steels using the phase diagram, state the purpose of each process, and describe their influence on the microstructures of hypoeutectoid and hypereutectoid steels
- define and describe martensite and describe what influences its transformation and properties
- use the TTT diagram to explain the transformation of austenite to pearlite, bainite, and martensite
- explain the role of alloying and grain size on the transformation of austenite to martensite
- differentiate between hardness and hardenability and state what determines each and how they are related on the Jominy test
- describe the quenching process and the factors that affect each stage
- predict whether surface and internal stresses will be compressive or tensile, given a metal and type of quench
- name and describe methods to reduce the residual stresses of quenching
- describe the tempering process and its effects on properties for carbon and alloy steels
- using the TTT diagram, describe and compare the tempering, austempering, and martempering processes and explain why each process is used
- describe the two approaches used to harden the surface of steels and identify which of the common surface-hardening methods are used in each approach

DID YOU KNOW?

The slow-cooling of steel can produce tensile strengths of about 150,000 psi maximum. However, by rapid cooling, quenching, and other further treatments it is possible to produce steels that are very hard, 60 HRC* and over, with tensile strengths of over 300,000 psi. Without steels of these properties, such common things as engines, aircraft, and cutting and forming tools would be impossible.

A PREVIEW

Steel can be heat-treated by two different kinds of processes—those that can be described and understood from the equilibrium phase diagram and those in which the cooling rate is so rapid that the time-temperature transformation diagram must be used. In Chapter 6 the results of equilibrium and near-equilibrium cooling were investigated. This chapter now considers first the slow-cool commercial processes and then the faster cooling or quenching processes. We'll investigate the quenching process, its consequences (the formation of martensite and residual stresses), and the processes used to offset these consequences. Your study of the material in this chapter will be greatly aided by laboratory investigation of the properties of steels cooled at various rates. Use of the Jominy test and other materials discussed in Appendix E will also prove very helpful.

*HRC = Rockwell hardness number, C scale, as defined in Appendix C.

7.1   SLOW-COOLED STEELS

The slow-cooling processes used with steels are shown in Figure 7-1.

Process Annealing.   Steels are often annealed to remove or modify the effects of cold work. Cold-worked steels, heated below the A_1 transformation temperature (1000 to 1250°F or 540 to 680°C), will recrystallize if there is a proper combination of cold work and temperature. This **subcritical annealing** is usually carried out as part of a cold-working and annealing sequence and is often referred to as a process anneal or intermediate anneal. In low-carbon steels, α-ferrite will recrystallize because of the cold work, but since the A_1 temperature is not reached, pearlite will not transform to γ-austenite; thus the pearlite will retain whatever directional orientation the cold-working process has given it. Also, since no γ-austenite is present, the cooling rate from the annealing temperature does not affect hardening; obviously, it should not be so rapid that it introduces distortion or residual stresses.

Stress-Relief Annealing.   The same temperature range (1000 to 1200°F or 540 to 680°C) used in process annealing is also used to reduce or remove residual stresses

F I G U R E   7–1 **Schematic diagram of the typical temperature ranges used in slow-cooling heat treatments of steel**

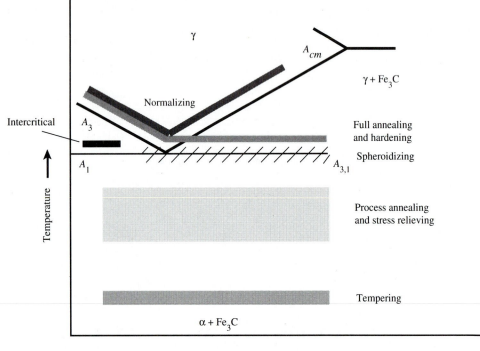

due to welding, machining, or forming operations. This process is a recovery anneal (see Chapter 3) and is called *stress-relief annealing*.

Full Annealing. Full annealing involves slow cooling from temperatures that are high enough that the steel undergoes phase changes, yielding microstructures very close to those predicted by the phase diagram. In general, full annealing is used to return the steel to a very soft, workable condition, although it can also refine the grain size and improve magnetic, electric, and machining properties. During the slow-cooling process, the steel is left in the furnace and the rate of cooling controlled with the furnace. This is an expensive operation because the furnace cycle is very long, and the furnace itself is subjected to the thermal shock of cycling to high temperatures and then cooling to room temperature. The resulting expansion and contraction of furnace metal and refractory increases the rate of deterioration of a very expensive piece of equipment.

The results of full annealing of a hypoeutectoid steel were described in Chapter 6 for steel number 1 in Figure 6-3 and in the microstructures shown in Figure 6-4. These results occur when the steel is slow-cooled from the γ-austenite range (i.e., above the A_3 critical line), which is the normal full-annealing treatment given low-carbon steels.

Since the proeutectoid phase for a slow-cooled hypoeutectoid steel is soft α-ferrite, this treatment does not cause the problem it does with a hypereutectoid steel. The complete transformation of the steel to γ-austenite reduces any directional orientation of grains imparted by prior hot- or cold-working and may cause the loss of desirable mechanical properties. In recent years **intercritical annealing** of hypoeutectoid steels has been introduced to avoid this problem. That is, heating just above A_1 transforms only the pearlite to γ-austenite; on slow-cooling the metal is softened for further cold work, but the untransformed α-ferrite retains the desired directional orientation. The consequences of quenching from this intercritical temperature will be discussed later.

For hypereutectoid steels, full annealing is done intercritically, primarily to avoid the formation of a brittle cementite network, usually an undesirable condition, especially if machining is to be done. Recall that above the $A_{3,1}$ line there is no α-ferrite phase; to transform all the α-ferrite of a hypereutectoid steel to γ-austenite therefore requires heating only above the lower critical line. Thus full annealing of hypereutectoid steels is done by heating *between* the critical lines $A_{3,1}$ and A_{cm}, or intercritically. This process also reduces the loss of the preferred orientation of grains imparted by prior hot- or cold-working.

Spheroidizing. To overcome the machining problems caused by the hard Fe_3C-cementite network and/or large amounts of pearlite, **spheroidize annealing** is used on steels with carbon contents above approximately 0.5%. The result of this anneal is shown in Figure A-24. The proeutectoid cementite network is formed into globs, or spheroids, and the Fe_3C-cementite from the pearlite is formed into smaller spheroids. These spheroids act as chip breakers and materially improve the machinability of the steel. Spheroidize annealing can be accomplished by any one of three processes:

1. heating and holding the steel just below 1333°F (723°C)
2. cycling the steel above and below 1333°F (723°C)
3. heating the steel above 1333°F (723°C) and either cooling very slowly or holding just below 1333°F (723°C), and then cooling slowly

All these processes involve long furnace times, e.g., 20 h or more.

Normalizing. Another way to change the influence of proeutectoid networks is to increase the cooling rate from the γ-austenite region. As seen in Figures 6-4 and 6-5, slow cooling gives proeutectoid phases an opportunity to form at the γ-austenite grain boundaries. Compare Figures A-13 and A-19 to see the effect of increased cooling rate. **Normalizing** is a faster cooling process; it involves heating the steel into the γ-austenite range (above A_3 or A_{cm}) and air cooling. With more rapid cooling, the proportions of the eventual mixture structure cannot be predicted from the phase diagram, and some noneutectoid steels can become almost entirely pearlitic. The process of normalizing produces a steel with higher strengths and hardness than does annealing, so normalizing may be a final heat treatment for some applications.

Because many steel-fabricating operations are done in the γ-austenite range, air cooling from the final temperature, if above the upper critical line, will give normalized properties. Thus hot-rolled and normalized properties have similar values.

Tempering. The temperature range of tempering is shown in Figure 7-1 to indicate its approximate location. Tempering is not truly an annealing process; its role in reducing the brittleness of martensite will be discussed later in this chapter.

7.2   QUENCHED STEELS: TIME-TEMPERATURE TRANSFORMATIONS

Art vs. Science in the Hardening of Steels. For many years it was known that quenching steels made them harder and, especially for higher-carbon steels, very brittle. The blacksmith knew how to forge steel, reheat it cherry red, quench it in water or oil, and draw or temper it at a lower temperature. He knew that he had to do the quenching rather quickly; if he waited too long or quenched it too slowly, he would not get the strength and hardness he wanted. The metallurgists knew that quenched steel had a different microstructure; it was needlelike rather than laminated pearlite and was called **martensite.**[†] Though the art was known, the actual mechanism was not well understood.

In 1930 E. C. Bain and E. S. Davenport published the first report of studies of the transformation of austenite with time at constant temperatures below A_1. The diagrams resulting from these studies are called *time-temperature transformation diagrams,* or **TTT diagrams.** The diagrams are also referred to as *isothermal*

[†]This hard form of steel was named for Adolf Martens, 1850–1914.

transformation diagrams, or **IT diagrams** (*iso* meaning "equal" or "uniform") because they are done at a constant temperature for any specific specimen. Because of their general shape (see Figure 7-3) they are also known as *C-curves* or *S-curves*.

Developing the TTT Diagram.   Bain and Davenport took numerous small specimens of a known steel, heated them to the γ-austenite region, quenched them all to the same temperature—say, 1200°F (650°C)—held them for various lengths of time, and then quenched them quickly to room temperature. (To simplify the explanation, assume the steel to be 0.80% C, a eutectoid steel.) The specimens were then examined metallographically.

The process Bain and Davenport used for their study is illustrated in Figure 7-2. Assume three specimens (a, b, c) are heated to the γ-austenite range, cooled quickly to 1200°F (650°C), held at that temperature for different lengths of time, and then quenched rapidly to room temperature. Specimen a, held at temperature only a short time, might have very little pearlite present, essentially transforming to all martensite, a very hard form of steel discussed shortly. Specimen b, held a little longer, might be half pearlite and half martensite; and specimen c, held still longer, might have very little martensite, essentially being all pearlite. These data gave the researchers three points for the temperature 1200°F (650°C). This same process was then repeated at temperatures from A_1 down to near room temperature.

Possible Transformation Products.   At temperatures above about 1000°F (540°C) the austenite transformed into either martensite, part martensite and part pearlite, or all pearlite.

At temperatures below approximately 1000°F (540°C), instead of transforming into pearlite, the γ-austenite transformed into a different structure called **bainite.**

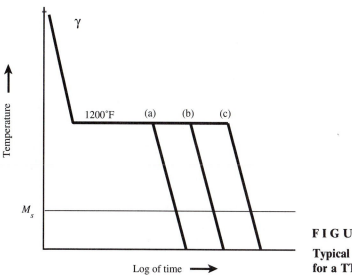

FIGURE 7–2

Typical cooling paths for a TTT diagram

F I G U R E 7–3   Generic TTT diagram for a eutectoid steel

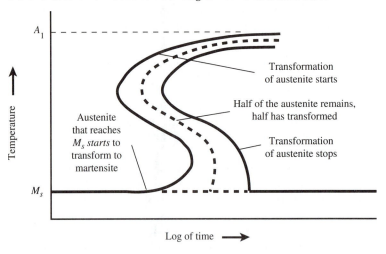

That is, in this temperature range the specimens for the three points contained all martensite, half martensite and half bainite, and all bainite, respectively.

Thus for a given steel Bain and Davenport ended up with a series of points of temperature and time that indicated when the austenite started, finished, and was half-and-half transformed into something other than martensite. These points were connected together to form the typical curves shown in Figure 7-3.

Formation of Pearlite and Bainite with Time.   From the above discussion it is clear that the formation of pearlite and bainite requires time. In Chapter 6 it was shown that under slow-cooling, γ-austenite transforms to pearlite and a proeutectoid phase. However, if the time-transformation of γ-austenite can be delayed until a lower temperature is reached, it forms bainite rather than pearlite. *Bainite, like pearlite, is a mixture of α-ferrite and Fe₃C-cementite;* it is *not* a single phase.

On the other hand, if the α-austenite is cooled rapidly (i.e., quenched) and does not have the opportunity to transform to either of these softer products, it will change to martensite.

7.3   AUSTENITE-TO-MARTENSITE TRANSFORMATION

Allotropic Change of Austenite.   *The γ-austenite phase is an interstitial solid solution* containing up to 2.0% carbon in FCC iron; α-ferrite (BCC) can dissolve a very small amount of carbon compared with γ-austenite. Therefore the transformation of γ-austenite to pearlite requires time for the carbon to leave the γ-austenite and form Fe₃C-cementite, the only other phase that can consume the carbon escaping from the γ-austenite. If that time is not available, i.e., if the steel is quenched rapidly, martensite is formed.

Martensite.   When γ-austenite is cooled, it must undergo the allotropic change of its lattice from FCC to BCC, and the carbon that has not had time to diffuse, i.e., escape, is trapped in the BCC lattice. Because there is not enough room for a large amount of carbon in the BCC lattice, the lattice is strained; one side becomes elongated and its volume increases about 4%. It is no longer cubic, but becomes an elongated square solid, or tetragon. It is still body-centered; i.e., it still has eight corner atoms and one in the center. *Martensite is a supersaturated solid solution of interstitial carbon trapped in a body-centered tetragonal, or BCT, lattice.* (See Figure 7-4.)

The Transformation Process. Austenite does not begin to transform to martensite until it reaches the M_s, or martensite start, temperature, and only continues to transform as temperature decreases. That is, unlike the change of γ-austenite to pearlite or bainite, time is not important—but the reduction of temperature below M_s is. Holding the austenite steel just below M_s even for an extended period of time will *not* cause it to transform to martensite; this can only occur by continuing to reduce temperature.

Because the M_s temperature is well below $A_1/A_{3,1}$ for most steels, martensite is formed at temperatures where the steel is not very plastic and where atomic movement or diffusion is difficult. If time were available and the temperature high enough, the excess carbon could migrate and form Fe_3C-cementite, as it does when it makes pearlite and bainite; instead the carbon remains trapped in the iron lattice. As the temperature decreases, the martensite is formed almost instantaneously as needles oriented to planes of the original γ-austenite lattice. (See Figure A-21.)

The M_s Temperature. The M_s decreases with the amount of elements added to the steel, carbon having the greatest effect. Carbon is 9 times more powerful than the next most powerful elements, manganese and chromium. Other elements that significantly lower the M_s are molybdenum and nickel, in that order. The M_s temperature for low-carbon, unalloyed steels can be as high as 900°F (480°C); for high-carbon, high-alloy steels it can be as low as 300°F (150°C).

M_f and Retained Austenite. Since martensite is a nonequilibrium product, it can be considered unstable. That is, it can be considered a transition between γ-austenite and the equilibrium products α-ferrite and Fe_3C-cementite. At low-car-

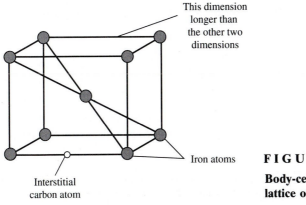

This dimension longer than the other two dimensions

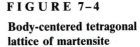

Iron atoms

Interstitial carbon atom

FIGURE 7-4

Body-centered tetragonal lattice of martensite

bon contents there is an M_f (martensite finish) temperature where most of the γ-austenite has completed its transformation to martensite. At higher carbon contents, above about 0.6%, all the γ-austenite may not transform, and the maximum hardness may not be obtained. This austenite is termed **retained austenite**; i.e., it is "retained" below the temperature where it should have transformed to martensite.

As a quenched steel moves through the M_s-to-M_f region, the generation of martensite is not uniform with temperature. When M_s is crossed, a few needles are formed; the rate of formation increases to a peak in the approximate middle of the M_s-M_f range and then decreases as M_f is approached.

Martensite Hardness. The higher the martensite carbon content, the greater the lattice strain and the higher the hardness. Martensite hardness is about 35 HRC at 0.2% C and increases rapidly to about 63 HRC at 0.6% C; at 0.8% C it reaches its maximum of 65 HRC. The maximum varies depending on the amount of untransformed, or retained, austenite.

It is important to realize that carbon is what makes the big contribution to the hardness of martensite. Other alloying elements contribute to lowering the M_s, and later we'll discuss their role in quenching a steel to martensite. But here we want to emphasize that it is carbon that makes martensite hard and strong.

Summary.

1. There is no diffusion during the austenite-to-martensite transformation; small amounts of γ-austenite change suddenly to the BCT lattice, whose volume expands by approximately 4%.
2. The transformation only continues if the temperature continues to fall below M_s; if the cooling is interrupted, the transformation stops and won't start again until the temperature drops. Thus the transformation is independent of time.
3. The transformation process is not linear with temperature, but starts and finishes slowly, with the highest rate in the middle of the M_s-M_f range.
4. The M_s temperature decreases with increasing carbon content, and to a lesser extent, with the addition of other alloying elements.
5. Martensite is a nonequilibrium product; it is theoretically unstable, even though the structure is known to have lasted for centuries.
6. Martensite's significant property is hardness, which is determined almost entirely by the carbon content.

7.4   **USING THE TTT DIAGRAM**

Now that we have an idea of the importance of time[‡] and have examined all the structures and phases involved in the cooling of steel, we'll explore the uses of the TTT diagram.

[‡]Note that the austenite-to-martensite transformation is shown on a graph of temperature vs. time, in spite of the fact that that transformation does not proceed with time.

The More General Case.   This study of the TTT diagram began by considering the special case of a eutectoid steel. The unusual thing about a eutectoid steel is that with slow-cooling all the austenite transforms to pearlite; there is no proeutectoid phase. The more general case of noneutectoid steels must now be investigated. Figure 7-5 has a side-by-side comparison of the TTT and phase diagrams. To simplify the discussion, compositions 1 and 2, shown on the phase diagram, are chosen so that they both intersect their upper critical lines at the same temperature. The TTT diagram has three cooling paths identified as *A*, *B*, and *C*.

Proeutectoid Phases and Pearlite.   Chapter 6 explained that the γ-austenite of a slow-cooled noneutectoid steel first transforms to either α-ferrite or Fe_3C-cementite, depending on whether the composition is hypoeutectoid or hypereutectoid. In Figure 7-5 the γs above and to the left of the TTT curve indicate that the steel is still austenitic in these areas. Thus the steel of cooling path *A* is γ-austenite until it reaches point *U*, where it crosses a line and enters a region marked $\gamma + pe$ (proeutectoid) α or Fe_3C.

   These are the same phases that the steel would cool through, as shown on the phase diagram to the right. The uppermost line, *U-V*, takes the place of the A_3 and A_{cm} lines on the phase diagram, and as it curves up to the right it approaches (i.e., is asymptotic to) the equilibrium temperature predicted by the phase diagram; any steel cooling through *U-V* will first transform to a proeutectoid phase.

   If the cooling continues toward the right and crosses the line *W-X*, the γ-

F I G U R E   7–5   **Comparison of the TTT diagram for noneutectoid steels and the steel portion of the iron–iron carbide phase diagram. (*pe* labels a proeutectoid.)**

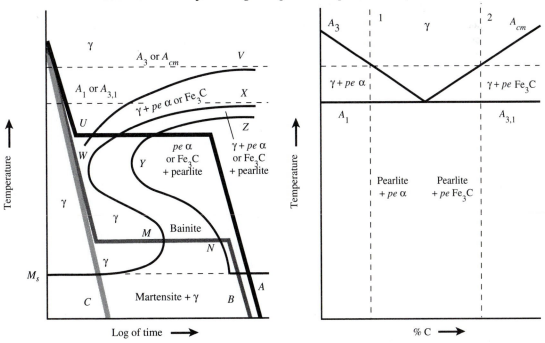

austenite remaining will begin to transform to pearlite; this is also predicted by the phase diagram. The *W-X* line also approaches the $A_1/A_{3,1}$ line, the eutectoid temperature predicted by the phase diagram. If cooling continues, and crosses line *Y-Z*, all the γ-austenite will transform to pearlite, which is the case for cooling path *A*. *Thus cooling paths that take the steel through the upper portion of the TTT diagram result in the same structures as those predicted by the phase diagram*, but they differ in the amounts calculated from the phase diagram.

Notice that, for the two steels of Figure 7-5, cooling horizontally through the nose of the curve is below the line *U-V* and no proeutectoid phase will form. The initial cooling rate is so great, and carried to a low enough temperature, that proeutectoid α-ferrite or Fe_3C-cementite does not have an opportunity to form; the structure will be all pearlite.

Bainite. Cooling path *B* misses the nose of the curve, delays the transformation of the γ-austenite, and keeps it from going below the M_s temperature. By staying above M_s, the γ-austenite has time to transform to bainite. This transformation begins when the steel crosses the TTT curve at *M* and is completed at *N*. (Cooling pattern *B*, used in a process called *austempering*, is discussed later.) If, during the time between *M* and *N*, the steel were suddenly quenched, the austenite remaining would form martensite, and the microstructure would be a combination of bainite and martensite.

Photomicrographs. An analysis of the photomicrographs in Figures A-20 and A-22 may be helpful in understanding the TTT diagram and the slower cooling rates represented by cooling paths *A* and *B*. Both photomicrographs are of hypereutectoid steels. Figure A-20 is of steel that was quenched through the upper part of the curve. Thin traces of white Fe_3C-cementite show at the original γ-austenite grain boundaries when the upper line (*U-V*) was crossed. Then the second line (*W-X*) was crossed and the large, dark, rounded areas of pearlite were formed. From that temperature the quenching was fast enough that the remaining γ-austenite avoided crossing *Y-Z* and some of it reached M_s; thus the centers of the old γ-austenite grains transformed to martensite and show as needles.

Figure A-22 shows pearlite, bainite, and martensite. The dark, rounded areas of pearlite are at the old γ-austenite grain boundaries and formed first because of slow-cooling at a high temperature, such as crossing line *W-X*. The γ-austenite was then cooled to a lower temperature (e.g., a little above point *M*) where the bainite was formed, the feathery structure in the lower-right quadrant of the photo. All the γ-austenite was not transformed, however, so when the temperature fell below M_s, the remaining γ-austenite transformed to martensite, the needlelike structure that makes up most of the photo.

The feathery bainite seen in Figure A-22 is typical of bainite formed at higher temperatures and resembles pearlite in that it is alternating plates of ferrite and cementite. Bainite formed at lower temperatures, closer to the M_s, appears much like martensite and is termed *acicular* (needlelike) *bainite*.

7.5   CRITICAL COOLING RATE

In Figure 7-5, cooling path C completely misses the upper part of the TTT curve; this means the γ-austenite arrives at M_s without transforming. As it crosses M_s and continues cooling, the austenite transforms as discussed in Section 7.3, and the steel arrives at room temperature with a completely martensitic structure. The hypereutectoid steel shown in Figure A-21 was quenched rapidly; all the γ-austenite transformed to martensite.

Since cooling rate C was fast enough that the steel became all martensite, the γ-austenite cooled faster than the critical cooling rate. That is, *the **critical cooling rate** is the minimum cooling rate that just prevents austenite from transforming to anything but martensite.* Any rate slower than the critical cooling rate allows some or all of the austenite to transform to ferrite, cementite, pearlite, and/or bainite. In many applications, e.g., in cutting tools, a hard, strong steel is required, so the cooling objective becomes "beating the nose of the curve" to get all martensite.

CT Diagram.   When a steel is quenched, like C in Figure 7-5, and not held for isothermal transformation, the Bain-Davenport curves are technically not valid. Continuous transformation, or quenching, is different from cooling isothermally, and the resulting *CT* diagram[§] lies slightly to the right and below the TTT diagram.

Normal quenching produces cooling rates that are not uniform and are difficult to determine. This makes the generation of CT diagrams difficult, and TTT diagrams are the more commonly used; also, since the CT diagram is to the *right* of the TTT, the *critical* cooling rate determined on a TTT diagram is more conservative.

Figure 7-6 compares the CT and TTT diagrams for a hypoeutectoid steel. The CT diagram, in heavy lines, is superimposed on the TTT diagram. The uppermost heavy line corresponds to the temperatures and times where γ-austenite begins to transform to α-ferrite, the second heavy line is where γ-austenite begins to transform to pearlite, and the third is where the transformation to pearlite stops. Lines Q, R, S, and T represent cooling paths.

Following T, the path with the slowest of the four cooling rates, the γ-austenite begins to transform to α-ferrite at the first line, starts to form pearlite at the second, and by the third line all the γ-austenite is transformed. The microstructure consists of pearlite and α-ferrite. All cooling paths to the *right* of S would have this result.

For cooling paths between R and S the situation is somewhat different; the third line is not solid but dotted. At the upper lines α-ferrite and pearlite form as before, but the dotted third line means that the austenite-to-pearlite transformation is not completed, and there is some γ-austenite left to form martensite at the M_s temperature. Thus cooling paths to the left of S have some amount of martensite. For cooling paths between R and Q the upper line is gone so they no longer

[§]This graph is also termed the *continuous cooling transformation diagram*, or *CCT*.

FIGURE 7-6 Continuous transformation (CT) diagram superimposed on a TTT diagram for a hypoeutectoid steel

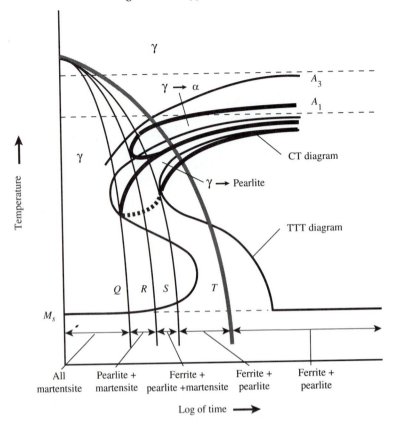

form α-ferrite; instead they are pearlite and martensite. *Cooling path Q represents the critical cooling rate;* all cooling paths equal to or greater than (to the *left* of) Q will produce all martensite.

Achieving Martensite. To achieve a completely martensitic steel, the "nose of the curve" must be avoided. This can be done two ways—*cool faster than the critical cooling rate,* or do something to the steel that *moves the TTT or CT diagram to the right.* If the curve moves to the right, then a slower cooling rate can avoid the nose of the curve; the critical cooling rate will have been reduced. If this happens then the **hardenability** of the steel has been improved.

7.6   HARDENABILITY

The *hardenability* of a metal or alloy is its ability to achieve the hard transformation product of γ-austenite (martensite) at slower cooling rates. Since the hardness of martensite depends on the carbon content, hardenability is not the same as hardness but is rather the *ability to harden*, the ability to become martensite. Perhaps "martensiteability" would be a better name.

Factors That Control Hardenability. To achieve the austenite-to-martensite transformation at slower cooling rates, the γ-austenite must be prevented from starting to transform to pearlite or bainite; it must be prevented from recrystallizing from the FCC austenite lattice to the BCC lattice of α-ferrite and the Fe_3C-cementite compound. In Chapter 3, Heating and Hot-Working of Metals, the general conditions that influence recrystallization were identified. Although locked-in strain energy is the chief factor in initiating recrystallization, large initial grain size and dissolved foreign atoms or solute atoms are adverse factors; that is, they tend to *retard* recrystallization. In the context of annealing, these adverse factors can be offset by additional prior cold work or by raising the temperature. In heat-treating, the temperatures are so high that strain energy is no longer a factor, but substitutional solute atoms and grain size can slow the recrystallization process to help achieve fully martensitic steels. That is, both solute atoms and larger austenite grain size are capable of moving the TTT diagram to the right and improving steel hardenability.

Austenite grain size. Grain boundaries are considered irregularities, or discontinuities, that are areas of higher energy; atoms are not at their lowest energy level since they are not at the closest-approach distance from each other. The smaller the grain size, the more grain boundaries there are and therefore the more energy available to help initiate the transformation of the γ-austenite to pearlite. The larger the grain size, the less of this energy there is available and the more the transformation is slowed, thus contributing to hardenability.

Whereas alloying to improve hardenability is a common practice, purposely using large grain sizes is not common because the improvement is achieved at the expense of ductility, and stress cracking of the final product is often encountered. For this reason austenite grain size must be controlled, and since grain size is a factor in hardenability, it must be considered in any comparison of steels. For example, if specimens of the same composition but different grain sizes are compared, the TTT diagram for the steel with the largest grains will be farthest to the right.

Alloying elements. The effect of alloying elements on hardenability is similar in some ways to the effects of grain boundaries. The dissolved foreign atoms are thought to migrate to grain boundaries, where they seem to act as pins, preventing the boundaries from shifting; the recrystallization or transformation of the austenite is thus impeded. This same explanation was given in Chapter 3 to describe how solute atoms retard recrystallization during annealing.

Alloying elements have varying effects on steel hardenability; among the more common ones, vanadium, tungsten, molybdenum, and manganese have strong influence, while chromium, silicon, and nickel have a weaker effect. Cobalt is the only element that is known to reduce hardenability. The atom diameters of these elements are all within 15% of iron and thus would have at least some substitutional solubility with iron. All of the elements just listed, except silicon, should be recognized as transition elements.

Although interstitial carbon is primarily responsible for the hardness of the martensite, it does make a small contribution to hardenability. Raising the carbon content results in a slightly slower critical cooling rate; however, other alloying elements are far more effective in achieving this; for example, manganese has over 10 times the influence on hardenability as an equal amount of carbon. The purpose and relative influence of the commonly used alloying elements is discussed at greater length in Chapter 8, Alloy Steels.

7.7   AUSTENITIZING

The hardening-strengthening of steel requires the conversion of γ-austenite to martensite. Therefore converting the steel to austenite is an obvious requirement. It is also a requirement that the alloying elements be in substitutional solid solution with the γ-austenite; this is critical to slowing the γ-austenite-to-pearlite transformation so that the γ-austenite-to-martensite transformation can take place. Thus part of the heat treatment must be to ensure that the alloying elements are dissolved in the γ-austenite.

Depending on carbon content, the steel is heated above A_3 or $A_{3,1}$, usually 25 to 75°F (15 to 40°C) above the temperature predicted by the phase diagram, to facilitate the solution process.

In addition to being heated to a high enough temperature, the steel must be held long enough that all areas have an opportunity to transform. This is dependent on the thickness of the steel. A good rule of thumb is that the part should be held at temperature for *one hour per inch of thickness*.

Excessive temperature, and excessive time at temperature, deteriorates the steel by oxidation and decarburizes the surface (see Figure A-18). Such excess should thus be avoided. Tool steels and other high-quality products are usually heated in controlled-atmosphere furnaces to minimize this problem. Overheating also enlarges the γ-austenite grain size, which was previously identified as a quality problem.

7.8   INTERCRITICAL HEAT-TREATING

Figure 7-7 uses a TTT diagram to show the quenching of a hypoeutectoid steel from an intercritical temperature. The steel was reheated from room temperature but was not completely austenitized. The steel was held between A_1 and A_3, permitting the pearlite-to-γ-austenite transformation to be close to that predicted by the phase diagram, but the temperature was not high enough to change the α-ferrite to γ-austenite. The carbon from the pearlite raised the carbon content of the γ-austenite to approximately 0.80% C, so that when quenched, it forms martensite of a high carbon content. The resulting steel has hard martensite interlaced with soft α-ferrite, as seen in Figure A-23.

F I G U R E 7–7 Intercritical heat-treating of a hypoeutectoid steel

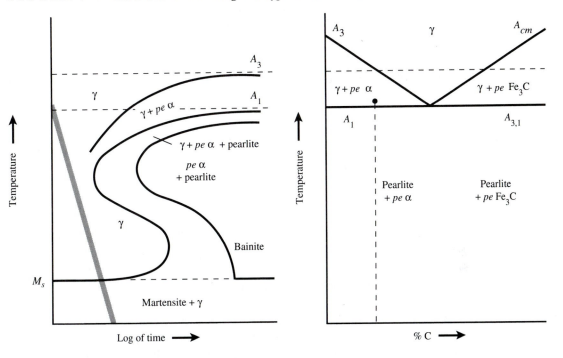

The advantages of intercritical heat-treating are that the directional characteristics of prior working are retained, the presence of the α-ferrite gives the steel some ductility, and the martensite provides strength and hardness. Intercritical heat-treating practice is usually applied to hypoeutectoid steels, and especially to steel sheet that will later be formed.

7.9   JOMINY HARDENABILITY TEST

The TTT diagram is very helpful in understanding the mechanism of austenite transformation, but it is a very difficult diagram to develop and it does not give a quantitative measure of hardenability. The Jominy test, discussed in Appendix E, solves both those problems.

Basis of the Jominy Test. The Jominy test is performed by austenitizing a standard 1-in.-diameter specimen of steel, quenching it at one end under standard conditions (see Appendix E) and measuring the hardness at various distances from the quenched end. Virtually all the heat travels out through one end (see Figure 7-8), so the quenched end cools the fastest. Remembering the relationship between the critical cooling rate and the formation of martensite, you'll see that the quenched end will be the hardest. The hardness will decrease from the

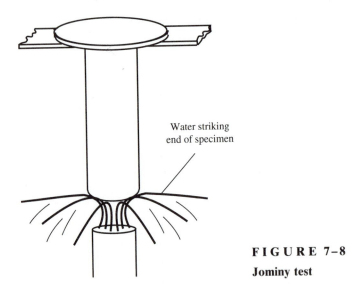

Water striking
end of specimen

F I G U R E 7–8

Jominy test

quenched end to the end in the fixture, which is losing most of its heat to the air. Since the thermal conductivities of various steels can, with very little error, be assumed constant, direct comparisons of hardenability can be made among steels.

Figure 7-9 shows that, as the measurements are taken farther and farther from the quenched end, the hardness decreases. Thus each position on the Jominy bar can be thought of as a cooling curve on a TTT diagram, or more properly, a continuous-cooling, or CT, diagram. For any particular steel, some positions on the Jominy bar may cool faster than the critical cooling rate and martensite may be achieved. Other positions cool more slowly than the critical rate, but still cool fast enough that some, but not all, of the γ-austenite transforms to martensite. Of course, some positions cool so slowly that all the γ-austenite transforms to the

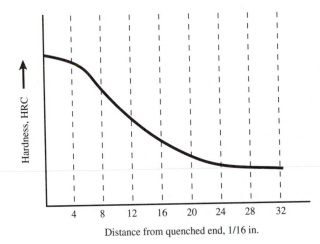

Distance from quenched end, 1/16 in.

F I G U R E 7–9

Typical Jominy curve

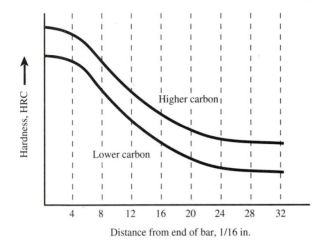

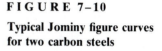

FIGURE 7–10

Typical Jominy figure curves for two carbon steels

slower products; it may become part pearlite and part bainite and may also contain a proeutectoid phase, α-ferrite or Fe₃C-cementite.

Influence of Carbon and Alloy Content. Figure 7-10 shows Jominy curves for two carbon steels with different carbon contents. The steel that has more carbon develops a slightly harder martensite at the quenched end. Not too far from the quenched end the hardness drops off for both steels; at the slower cooling rates it cannot develop the harder structures—i.e., it has low hardenability. The hardness of both steels drops off, but the one with the higher carbon is still the harder steel.

Figure 7-11 shows a similar comparison, but with carbon and alloy steels. Away from the quenched end, the hardness of the carbon steel (1) falls quickly. The alloy steels (2 and 3) do not; their curves are flatter. Notice that the alloy and carbon steels that have the same carbon content (1 and 2) develop the same

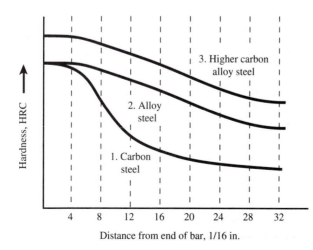

FIGURE 7–11

Typical Jominy curves for carbon and alloy steels

hardness at the quenched end; that is, hardness is dependent on carbon content, not alloy content. The alloy content is responsible for the ''flatness'' of the curves; for any given carbon content, the flatter the curve, the more hardenability the steel has.

The composition of the two alloy steels of Figure 7-11 varies only as to carbon content. Thus they have about the same hardenability, but the one with the higher carbon content will always have a higher hardness value than the other.

H Alloy Steels.   When metals are purchased, their compositions are usually specified. In steels intended for use where hardness is important, variation in composition, even though within the commercial specification, may be enough that the desired mechanical properties are not achieved when the steel is heat-treated. Therefore it is possible to purchase H steels, which are produced to meet composition limits but which also develop Jominy curves that fall within a specified standard range. These steels are identified by an ''H'' after their four-digit number, e.g., 4130H.

Figure 7-12 shows typical curves for an H steel. The upper and lower curves are for compositions that are in the higher and lower extremes, respectively, of the composition specification. In the standard alloy series, the variation in the alloying elements, e.g., manganese or chromium, may be as much as 2:1; the carbon usually varies a maximum of 10 points. Thus the variation in hardness from high to low values is very small at the quenched (martensite) end where carbon is the determining factor. This variation in hardness increases, however, at the slower cooling rates (i.e., farther from the quenched end) where the variation in alloying elements changes hardenability.

Applications of the Jominy Test.   The curves for H steels can be used to ensure that a part will develop the specified hardness when quenched. Methods are available to relate the cooling rates of various shapes to a position on the Jominy bar; once that position is known, a steel meeting the specification, i.e., a steel that

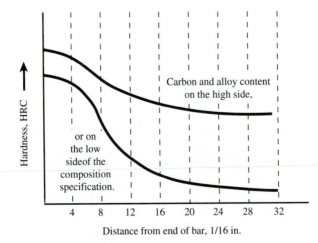

F I G U R E 7–12

Jominy diagram for a typical H alloy steel

has a hardness at that cooling position that falls within the range specified, can be selected from handbooks of Jominy curves. By careful analysis of Jominy curves, it is also possible to ensure that one area of a part has a minimum hardness (be below the H band) while another area has a maximum value (always be above the H band). For example, a steel and a quenching method can be chosen to ensure that the surface is hard in order to resist wear, and the interior is softer but tougher. This design procedure is described and illustrated in Appendix E.

7.10   QUENCHING, OR HEAT REMOVAL

Quenching involves the principles of heat transfer, not of metallurgy, but since the eventual structure, hardness, and strength of steels and nonferrous metals are determined by the cooling rate in the heat-treating process, it is dealt with here.

When a metal at a high temperature is submerged in a liquid, i.e., when it is quenched, the temperature of the metal piece will generate a temperature-versus-time plot similar to that shown in Figure 7-13. In this graph the rate at which heat is removed is the rate of change of temperature with time, dT/dt, or the slope of the curve at any point. Note that this rate goes through three distinct stages as the temperature of the metal part is lowered to room temperature.

The Quenching Process. For purposes of explanation, assume that a piece of metal at 1600°F (870°C) (1) is held in a **quenching medium,** or **quenchant,** (2) of water (3) that is at room temperature. The water in contact with the piece heats rapidly to its boiling point and converts to steam; it "sizzles." Initially, steam forms a film or blanket around the metal and insulates it from the liquid water. This film impedes the transfer of heat from the metal to the liquid water and prevents a fast quench. This is the **film stage** of quenching and it has only a moderate rate of heat removal.

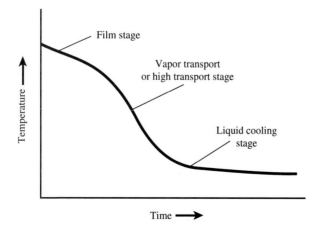

FIGURE 7–13

Typical curve showing the three stages of quenching

As the metal cools somewhat, the violent generation of steam subsides and some of the liquid comes in contact with the metal surface; boiling is still going on, but the steam film no longer completely encloses the metal. During this **high-transfer stage,** or **vapor-transport stage,** the heat-removal rate is the highest of the three stages.

When enough heat is removed, the metal loses the ability to convert the liquid to steam and the **liquid-cooling stage** of quenching begins. The water near the metal removes heat and becomes warmer, but without any boiling action, the warm water becomes stratified; now cooler water is supplied only by normal convection currents. Also, as the metal cools, there is a reduction in the temperature difference between the metal and the adjacent water; there is, therefore, a decreasing rate of heat removal. The liquid-cooling stage of quenching has the slowest cooling rate of the three.

Quenching Variables.   This example assumed the metal was simply held in room-temperature water. There are a number of factors in that process that can be changed to improve the quench. An improved quench reduces the time required to lower the temperature of the metal to some critical transformation temperature. That is, an improved quench increases the dT/dt of a stage, or reduces the time in the slower stages, or both.

Since most transformation temperatures, especially in steels, are fairly high, the cooling rates through the vapor-transport stage, and especially the film stage, are important. However, the liquid-cooling stage occurs at such a low temperature that reducing the time in that stage will probably do little to beat the nose of the TTT or CT curve.

Agitation of the quenchant is one of the most common and effective ways to both increase the dT/dt and reduce the time spent in the film stage; agitation brings the metal into the high-transfer stage sooner. This is shown in Figure 7-14. The function of agitation is to wipe away, or disrupt, the steam film and enable the contact of liquid and metal to occur earlier in the quenching process. In hand quenching this is done by simply "swishing" the metal in the quenchant. Com-

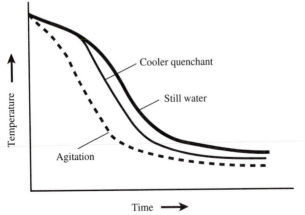

F I G U R E   7–14

Quenching curves showing factors affecting the rate of cooling

mercial equipment often sprays or circulates the quenchant. In some facilities the metal is dropped into the quenchant, which accomplishes agitation.

Another change that can be made to influence the process is in the quenching medium itself. Water was used in our example, and it is a popular quenchant, but other media are also used. Various natural and synthetic oils can be used to achieve specific quench rates. Oils present an unusual situation because their characteristics change with temperature. They do not transfer heat rapidly, and so boiling tends to persist; however, as oils become warmer they become more fluid and their cooling rate increases. The end result depends on which of these trends is the more significant for a particular oil and quenching temperature.

In recent years the use of aqueous solutions containing organic polymers has become popular. Varying the concentration and type of polymer can alter the characteristics of the quenchant to meet specific requirements.

For very rapid quenching, chilled brine is used. Many tool-and-die steels can achieve martensite with much slower air cooling; the low dT/dt develops very little distortion and residual stresses. Isothermal transformation (e.g., austempering, which will be discussed later in this chapter) is done in heated media such as molten salt or molten lead.

The temperature of the quenchant is another variable. A lower quenchant temperature requires more heat extraction before the steam film is formed; more cooling is accomplished initially and the effect shown in Figure 7-14 results. Note that the initial slope of the curve is the same as that in the original quench; transition to the high-transfer stage just occurs sooner. Controlling quenchant temperature is often accomplished by using cold makeup water and/or circulating the liquid through a heat exchanger system. In some cases the medium is actually chilled, as in "chilled brine."

Variables Due to the Part Itself. The surface-to-volume ratio (S/V) of the metal part is very important in determining how fast it quenches. Since heat is removed through the quenchant in contact with the metal surface, the more surface contacted for a given volume (the higher the S/V), the faster the heat is removed. Notice that the S/V ratio is not a constant; a surface (length squared) divided by a volume (length cubed) always leaves one unit of measure in the denominator. For rectangular parts, the S/V can be thought of as the surface divided by the surface times the thickness; that is, the S/V is inversely related to the thickness or minimum dimension of parts. For a sphere, the S/V is inversely related to the diameter. These relationships jibe with our experience—smaller or thinner things cool faster than bigger or thicker things.

So far the concern has been to remove the heat from the surface of the part, but the heat must first get to the surface. As in the Jominy bar, the thermal conductivity rate of the metal and the distance the heat must travel determine the cooling rate of the interior of the part. Since conductivity rates are not infinite, the center of metal parts will not cool as fast as the surface. Thus there will be a limit, regardless of how fast the surface is quenched, to the size of part that can be heat-treated to the hardest structure throughout.

The Jominy test itself is interesting from a quenching standpoint. Although water impinges on the end of the specimen, the agitation is rather moderate compared with the velocities used in some commercial heat-treating facilities.[||] Also, the S/V ratio is low since most of the heat exits through the quenched end of the bar. Thus the highest hardness achieved in a Jominy test is not necessarily the highest possible for that steel.

7.11   RESIDUAL STRESSES INDUCED BY QUENCHING

When metals are heated and cooled, they go through a cycle of physical expansion and contraction. If heating and cooling took place slowly, or if rates of thermal conductivity were infinite, there would be no residual stresses induced by the changes in temperature. Of course, that is not what happens in most commercial processes or with real metals.

Metals That Contract When Cooled.   To begin the discussion of these stresses, we'll follow an aluminum rod, 2 in. in diameter and 10 ft long, through a cycle of heating and cooling. As this rod (a uniform cross section) is heated, all the dimensions expand. But consider just the length; the higher the temperature, the longer the piece becomes. At the specified heat-treating temperature, a pause ensures that the piece has a uniform temperature throughout, and then it is quenched. If the quenchant extracts heat uniformly, the rod should arrive at room temperature straight and at the same length it started.

Now consider a tapered cross section, like that shown in Figure 7-15(a). When heated, this piece lengthens like the previous rod. But when quenched, the piece loses heat more rapidly from the narrow point of the cross section than from the heavier portion. Since it is at a lower temperature, the narrow edge tends to shrink; however, the heavier (and stronger) portion, still at a higher temperature, is of a longer length. In the process of cooling and shrinking, the narrow portion actually has to stretch to stay the same length as the heavier portion.[#]

As the heavier portion approaches room temperature, it shrinks to its original length, approximately. However, the narrow portion has reached room temperature sometime previously, already longer than its original length. *Thus the piece has two extremes of length* (and all gradations in between), and it adopts the "banana" shape shown in Figure 7-15(b).

The same analysis can be done for a sphere or ball of aluminum. As the outer surface of a quenched aluminum ball cools, it has to stretch to accept the still-hot, expanded inner core. The outer surface or shell of the ball is thus bigger than it should be. When the core finally cools and contracts, it pulls inward on the outer

---

[||] For a comparison of commercial cooling rates and the Jominy test, see Roy F. Kern, Intense quenching, *Heat treating,* Sept. 1986, 19–23.

[#] It must do this in order to remain fastened to the heavier portion. In T-sections of greatly different cross-sectional thicknesses, this difference in the cooled length is sometimes resolved when the smaller section shears off completely; buckling or waving are other results of differences in thickness and temperature.

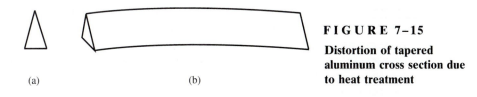

FIGURE 7–15

Distortion of tapered aluminum cross section due to heat treatment

(a)   (b)

shell, putting it in *compression*. Rather than separate the metal at its surface, these stresses compress it. Since many failures, especially fatigue failures, start with surface cracks, *compressive surface stresses are normally desirable*.

Metals That Expand When Cooled.   So far only changes due to the coefficient of thermal expansion have been considered. The prior analysis would have brought us the same results if the metal was aluminum, brass, or pearlitic steel. However, if the ball is a steel with enough hardenability that it can quench to martensite throughout (i.e., can be "through-hardened"), the situation is the exact reverse. Figure 7-16 shows the TTT diagram for such a steel, and the cooling curves for the outer shell and inner core of a ball. Note that the γ-austenite for both the outer shell and the inner core quench rapidly enough that they both reach M_s.

When austenite crosses the M_s temperature, it expands as it transforms to martensite in the BCT lattice. When the outer-shell austenite cools to M_s, its contraction due to the reduction in temperature has been accomplished, and the

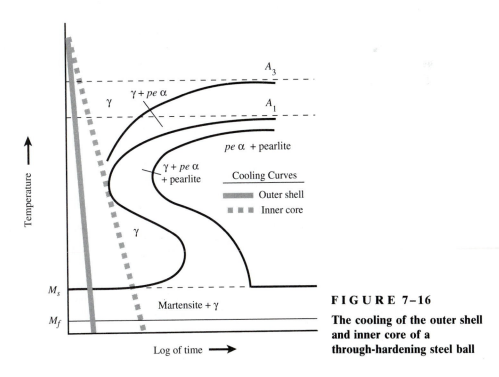

FIGURE 7–16

The cooling of the outer shell and inner core of a through-hardening steel ball

shell then tries to expand as the austenite transforms to martensite. This puts the outer shell in compression, and the inner core, still *austenitic*, adds to the compression as it continues to cool and shrink, still above M_s. But, when the core crosses M_s and converts to martensite, it expands. This puts the *outer shell in tension,* a situation not favorable to resisting cracking or fatigue failure.

Many of the highly alloyed steels capable of through-hardening to martensite readily crack when quenched severely; a 1-in.-diameter rod of 4140 usually cracks when quenched in water from an austenitizing temperature.

If the steel has lower hardenability, however, and is not capable of through-hardening, the center of the ball does not convert to martensite, it does not go through the expansion process, and thus it does not create the surface tensile or separating stresses. Instead, the surface is in compression, a desirable condition.

If the tapered section of Figure 7-15 were made from a through-hardening steel, the distortion would be reversed. The thicker portion of the cross section would convert to martensite last, then expand; the piece would be too long on the bottom, too short on the tip. As positioned in Figure 7-15 the ends would bow upward.**

7.12   TEMPERING

Martensite in the quenched condition (i.e., "as-quenched") is very hard and brittle, usually too brittle for most uses. Its toughness, and its machinability, can be improved by **tempering** or drawing, which involves holding the steel at a temperature below the A_1 critical line. This relieves some of the residual stresses of the type discussed previously. During the tempering process some of the supersaturated interstitial carbon leaves the BCT lattice, allowing it to convert to a BCC lattice; the liberated carbon precipitates as complex carbides in the α-ferrite solid solution.

Tempering is done in the range of 300 to 1200°F (150 to 650°C). In general, the higher the temperature, the lower the tensile strength and hardness and the higher the ductility become. Toughness is not so straightforward; if the steel is tempered at the wrong temperature, the toughness will not be optimized. (See Figure 7-17.) The temperature at which the low toughness occurs varies with composition but is usually in the 500-to-700°F (260-to-370°C) range.

Figure 7-17 illustrates the relationship of properties versus tempering temperature for typical carbon and alloy steels. Notice that the carbon steel has a single peak of toughness, whereas the alloy steel has two peaks. For either type steel the decision has to be made to temper for maximum hardness (low tempering temper-

**It is believed that this expansion property of martensite was used to put the curve in samurai swords. A steel, capable of quenching to martensite, was quenched with the sharp, thin edge exposed to the quenchant. The thicker back part of the blade was protected from the quenchant, so it quenched to a softer product. The cutting edge (martensite) got longer, the back edge shrank, and the blade took on the characteristic curve of the samurai blade. See Cutting Curves in Samurai Swords, *Science news, 137* (Apr. 28, 1990), 270.

FIGURE 7-17 Representative tempering curves for carbon and alloy steels

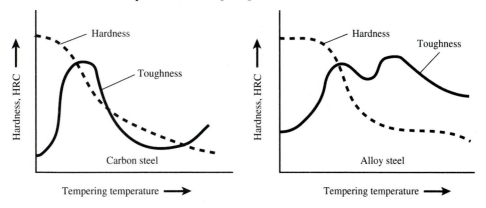

ature) or for optimum toughness (higher tempering temperature). One of the common reasons for using an alloy steel is to get the higher level of toughness, so such steels are often tempered for maximum toughness rather than maximum hardness. Under most circumstances *the low-toughness, or brittle, temperature range should be avoided.*

The tempering temperature limits the heat range in which the steel should be used. That's because use at a higher temperature essentially retempers the steel at the higher value. For example, steel tools used in an environment hotter than their tempering temperature are essentially retempered and will soften. Also, this excess temperature may be in the low-toughness range which means the tool not only softens but also becomes more brittle. Generally, alloy steels can be tempered at temperatures higher than those used for comparable carbon steels, an advantage of alloy steels.

Since the tempering process involves the internal energy of the metal, both time and temperature are important. That is, the degree of softening accomplished at one temperature may be accomplished at a lower temperature by using a longer cycle, and vice versa.

7.13   AUSTEMPERING

Figure 7-18 illustrates the **austempering** process whereby steels are austenitized, cooled fast enough to miss the pearlite nose of the TTT diagram, and then held above the M_s temperature to transform to bainite. The resulting steels have a good combination of strength, hardness, and toughness; rotary lawn-mower blades and hand shovels are examples of products made from austempered steel.

Austempered steels do not have the high hardness of quenched and tempered martensite, but with typical hardnesses of 45 to 55 HRC, they have very good properties. In addition, austempered steels have less distortion, since the austenite-to-martensite volume change does not occur and the quench is not as drastic.

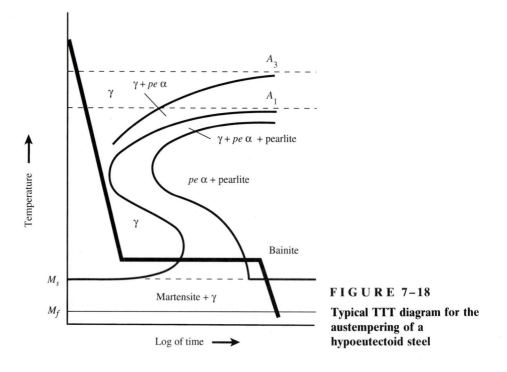

F I G U R E 7–18

Typical TTT diagram for the austempering of a hypoeutectoid steel

There is a limitation to the process, however. The interior portions of large cross sections have difficulty in following the required quenching path and may not transform to all bainite. Therefore, *section thickness is limited to about ½ in.* This limit can be raised by using alloy steels, but while this moves the TTT diagram to the right to prevent the austenite-to-ferrite transformation, it also lengthens the time the steel must be held above M_s, all of which increase the cost.

7.14   MARTEMPERING

Martempering is a process designed to minimize the effects of the severe quench often required to obtain a fully martensitic part. Figure 7-19 shows cooling curves representing the surface and center of a quenched part. You can see from the figure that if the cooling were not interrupted above M_s, the surface would completely transform to martensite before the center even got to the M_s, as shown in Figure 7-16. This would generate the residual stresses discussed previously.

The objective of martempering is to have both the surface and center portions miss the nose of the curve, but equalize in temperature before going through the martensite-to-austenite transformation. By holding these portions above the M_s temperature before the quenching is complete, they would go through the M_s–M_f region at almost the same time, thus generating much lower residual stresses.

FIGURE 7–19 **Typical TTT diagram for the martempering of a hypoeutectoid steel**

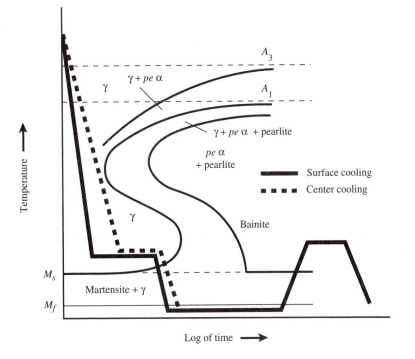

As shown in Figure 7-19, after quenching, the part would be tempered to the desired hardness. As in austempering, the part must be of a size that enables it to follow the proper cooling path. In this case it must be quenched from the holding bath (usually salt or lead) before it starts to transform to bainite.

Martempering is also known as **marquenching.**

7.15   SURFACE HARDENING

There are times when it is desirable to develop a high degree of hardness on only the surface of a part. A gear, for example, is essentially a series of cantilever beams that exert a force but must also resist wear and absorb shock loads. A martensitic surface will combat the surface wear, but the internal portion of the shock-loaded tooth root, as well as the hub, will perform better if it is of a tough material such as pearlite.

Such a combination of properties can be achieved by two different approaches:

1. Heat-treat a steel that has a reasonably high carbon content, say 0.40% C, so it is pearlitic throughout, and then austenitize and quench just the surface to martensite.

2. Use a low-carbon steel that is capable of achieving the internal toughness desired, but change the chemical composition of its surface so it has the ability to harden to the desired level when the whole part is heat-treated later.

Since these methods produce a hard exterior layer, they are commonly called **case hardening.**

Flame hardening and **induction hardening** are two methods that use the first approach. The part is first converted to all pearlite, probably by normalizing, and then the surface is reheated and quenched. Flame hardening may be done by simply heating a part with a torch and quenching it in a bucket of water, or a sophisticated device that automatically heats and quenches a small area of a part and then indexes to the next area may be used. Induction hardening is accomplished through use of a high-frequency magnetic field to induce eddy and hysteresis electric currents in the surface of the steel. The part to be case-hardened must somehow come under the influence of a magnetic field. Often this is accomplished by putting the part into, or on top of, a magnetic coil. The heating cycle is usually controlled by time, after which the part is quenched. For both methods the quench must be rapid and done immediately after heating.

Carburizing, cyaniding, carbonitriding, and **nitriding** are methods that use the second approach to surface hardening—i.e., by changing the chemical composition of the surface. These methods take advantage of the ability of small-diameter atoms such as carbon and nitrogen to form interstitial solid solutions with γ-austenite.

In carburizing, the low-carbon steel is subjected to an atmosphere containing carbon monoxide. At an elevated temperature, for example, 1700°F (925°C), carbon interstitially penetrates the surface of the steel and raises the carbon content in the outer layer, or case. Later heat-treating generates a hard martensitic case with a tougher, probably pearlitic, interior. Figure A-17 shows a section through the surface of a carburized specimen and illustrates the difference in structure.

Figure A-18 is illustrative of the reverse process, **decarburizing.** That is, carbon is removed at the surface through lengthy heating in the presence of oxygen. The layer of reduced carbon content cannot be hardened; it has lost too much carbon.

Cyaniding is accomplished using liquid salt baths; carbonitriding is done with a gas atmosphere. Both treatments have the same purpose—to increase the amount of carbon and nitrogen in the surface layer of the treated steel. Later heat-treating hardens the surface. The carbon enrichment obviously helps the hardness; the nitrogen stabilizes the austenite, slows the transformation, and thus increases hardenability.

Nitriding should not be confused with the treatment just described. Its objective is to form hard nitrides at the surface. The part is held in an atmosphere of ammonia at about 1000°F (540°C). The lower temperature means nitriding can be performed on finished tools, for example, and no subsequent heat-treating is necessary.

Tempering of surface-hardened parts is recommended.

SUMMARY

1. The slow-cooled treatments of steel are used to accomplish specific micro-structures, but all produce the equilibrium products of α-ferrite and γ-cementite.

2. Austenite transforms to:
   - pearlite at high temperatures (from $A_1/A_{3,1}$ down to about 1000°F or 540°C)
   - bainite at lower temperatures (from about 1000°F, or 540°C, down to the M_s)
   - martensite, beginning at the M_s temperature

3. The austenite-to-martensite transformation is summarized at the end of Section 7.3.

4. Hardenability is the *ability to harden,* not the *hard*ness itself. It is indicated when TTT diagrams are situated to the right, with lower critical cooling rates. Jominy curves with very low slopes also indicate hardenability.

5. The Jominy test is a relatively inexpensive way to quantitatively measure the hardenability of steels and to get data that can be used for the specification and selection of steels to meet property requirements.

6. Liquid quenchants cause cooling rates to vary through three stages. Agitation, quenchant temperature, and type of quenchant are all variables in the quenching process. The surface-to-volume ratio is a variable related to the product.

7. Rapid quenching causes distortion and residual stresses. Some of this is due to the variation in cooling rate caused by variations in the thickness of the part, and some of it is due to the fact that steels expand when they transform to martensite.

8. Various methods are used to relieve or reduce these stresses. Martensite is tempered. Austempering avoids the martensite-expansion problem by transforming to bainite instead. Controlled cooling using the martempering process attempts to reduce the effects of the martensite expansion by causing the whole piece to transform, as nearly as possible, at the same time.

9. The surface of steels may be selectively hardened by two general approaches—heat-treating only the very surface of a steel that has enough carbon content to achieve the hardness desired, or heat-treating a whole part that has a high carbon (and/or nitrogen) content only at the surface.

REVIEW QUESTIONS

1. Describe the differences in the microstructures of fully annealed and normalized hypoeutectoid and hypereutectoid steels; explain why they occur.

2. Define critical cooling rate; explain how it relates to both hardness and hardenability.

3. Of the two TTT diagrams shown in the figure, which has the faster or greater critical cooling rate, a or b? Why?

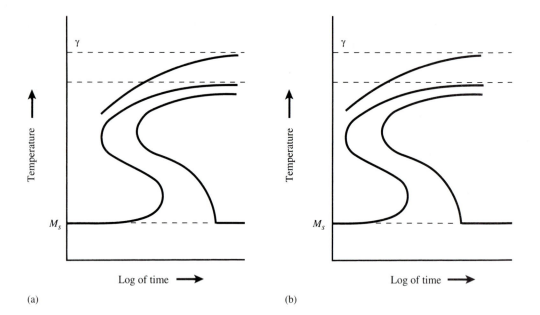

(a)   (b)

4. Assuming that the phases and microstructures were shown on the TTT diagrams of Question 3, how would you know if they were for hypoeutectoid or hypereutectoid steels?

5. Sketch a TTT diagram for a hypoeutectoid steel and on it show the cooling paths needed to get a room-temperature microstructure of all martensite, all bainite, ferrite and pearlite, and the unlikely combination of all four.

6. What are the advantages of intercritical heat-treating of low-carbon steels?

7. Use a TTT diagram to describe the tempering, austempering, and martempering processes. Why is each process used? Compare the results obtained from each.

8. Compare the surface/volume ratio of a 1-in.-diameter sphere, a 1-in. cube, and a 1-in.-diameter cylinder, 1 in. long.

9. Given the 1-in. pieces in Question 8, are they likely to crack if quenched in water from an austenitizing temperature if they are 1040 steel? 4340 steel? Explain.

10. What are the three stages of quenching? How would you increase and decrease the cooling rate in each of the stages?

11. Describe the two basic approaches to surface-hardening steels.

FOR FURTHER STUDY

Alexander, W. O., G. J. Davies, K. A. Reynolds, and E. J. Bradbury. *Essential metallurgy for engineers.* Wokingham, U.K.: Van Nostrand Reinhold, 1985, 61–71.

Avner, Sidney H. *Introduction to physical metallurgy* (2d ed.). New York: McGraw-Hill, 1974, chap. 8.

DeGarmo, E. Paul, J. Temple Black, and Ronald A. Kohser. *Materials and processes in manufacturing* (7th ed.). New York: Macmillan, 1988, 107–121.

Flinn, Richard A., and Paul K. Trojan. *Engineering materials and their applications* (3d ed.). Boston: Houghton Mifflin, 1986, secs. 6.6–6.20.

Kern, Roy F. Intense quenching. *Heat treating,* September 1986, 19–23.

Mason, Clyde W. *Introductory physical metallurgy.* Materials Park, OH: ASM International, 1947, chap. VII.

Metals handbook (9th ed.). Materials Park, OH: ASM International, 1981, vol. 4.

Reed-Hill, Robert E., and Reza Abbaschian. *Physical metallurgy principles* (3d ed.). Boston: PWS-Kent, 1992, chaps. 18 and 19.

UCON quenchants for ferrous and non-ferrous metals. Danbury, CT: Union Carbide Corporation, 1988.

C H A P T E R   8

Alloyed Steels

OBJECTIVES

After studying this chapter, you should be able to

- explain why alloying elements are used in steels
- identify the alloying elements that form carbides and those that dissolve in ferrite, i.e., describe where alloying elements will reside in steels
- state the specific function of the alloying elements commonly used in steels
- describe the processes by which alloying elements harden annealed steel
- describe the hardenability problems created by the formation of carbides
- identify the alloying elements that greatly improve hardenability
- describe the classification systems commonly used with alloy steels, tool steels, HSLA (high-strength, low-alloy) steels, and highly alloyed steels and give the general characteristics of the steels in each system
- describe the general composition of AISI/SAE* alloy steels
- identify the alloying elements commonly used in tool steels and explain why they are added
- describe the HSLA steels and explain how their properties are achieved
- describe the dual-phase steels, explain how their properties are achieved, and list the advantages they have over other HSLA steels
- describe the three types of stainless steels and their numbering system and explain the role of chromium and nickel in stainless steels
- describe Hadfield's steel and the role manganese plays in its formation
- describe the processing of maraging steels and explain why they are used
- describe the cost factors important in the selection of tool steels and explain why many accounting systems have difficulty in identifying these costs

*American Iron and Steel Institute/Society of Automotive Engineers.

178

DID YOU KNOW?

It's probable that the knife used to carve your Thanksgiving turkey is stainless steel; most likely, it's magnetic. The kitchen sink may also be stainless steel, but it will probably be nonmagnetic. That is, stainless steel is not just stainless steel; there are various types and they all have different properties and characteristics. For instance the knife would be a martensitic stainless steel, the sink an austenitic stainless steel.

Alloying elements can have small, subtle effects on steels, or they can have big effects that drastically change properties and microstructures. Without these varying capabilities of alloys the cutting tools, ball bearings, furnace parts, kitchen sinks, and contoured auto bodies so commonplace would not be possible.

Industry is still learning to use its ability to monitor and control steel-making processes through the application of programmable controllers and computers. As these capabilities mature, the use of more sophisticated steels will increase. More and more steel products will be processed through continuous operations where subtle differences in composition can be used advantageously to improve properties.

A PREVIEW

Alloying changes the characteristics of steels according to the amount of the alloying element(s), the presence of other elements, and the heat-treating process used. For example, relatively small amounts of some alloying elements will affect hardenability; very large amounts of some others may make the material nonheat-treatable—i.e., it won't even behave like a steel. Combinations of elements can give steels similar properties, and systems of alloy steels have been developed for applications that take advantage of these similar properties, e.g., tool steels and stainless steels.

This chapter first looks at why alloying elements are used and then at what each of the elements does when added to steels. Four systems of alloy steels are studied—the AISI/SAE alloy steels, the AISI/SAE tool steels, HSLA steels, and lastly the highly alloyed steels. These systems are not explored in detail here. Our purpose is simply to present the steels within each system as examples of the use of alloying elements. At the end of the chapter, we'll discuss the costs associated with the use of tool steels, which are singled out for a separate discussion because they are so difficult to determine.

8.1   PURPOSES OF ALLOY ELEMENTS IN STEELS

Hardenability. Carbon steels can be made very hard—martensite is an extremely hard material—but they lack hardenability. For a part made of carbon steel to achieve martensite throughout requires such a rapid quenching rate that the part will usually crack unless it is very small (i.e., thin).

Improving the hardenability of steels is probably the major, but not the only, reason for the use of alloying elements in steels. The very hard martensite structure is not necessarily the appropriate one for all applications of steel, and alloying elements achieve other effects when added to steels.

Formation of Carbides. Another use of alloying elements is for the formation of carbides. Although martensite is hard, it is not as wear-resistant as some of the carbides created by alloying elements such as tungsten and vanadium; these hard particles are particularly valuable in providing the wear resistance required of many parts and tools. The microstructure of these steels is martensite plus the carbide particle, so some hardenability is still required.

Corrosion Resistance. The corrosion resistance of steels is greatly improved by the addition of chromium, the alloying element common to all stainless steels.

Grain Refinement. Some elements, such as vanadium, molybdenum, and niobium form carbides and/or nitrides, which retard the rate of growth of austenite grains. The final product thus has a smaller or finer grain size, contributing to the strength and hardness of the final product.

Other Purposes. Nickel is advantageously used in many steels to improve their toughness, but it is also a key contributor to improving both the high- and low-temperature strength of steels. Some alloy additions, e.g., silicon, improve the electrical and/or magnetic properties of steels. Others, like silicon and manganese, improve the un-heat-treated strength of steels. Other characteristics that can be controlled or improved by alloying include machinability and resistance to decarburization.

8.2   DISTRIBUTION OF ALLOYING AND OTHER ELEMENTS

Metal alloys take three basic forms—as mixtures, as compounds, and as solid solutions. The mixtures themselves may be combinations of pure metals, compounds, and/or solid solutions. When elements other than iron and carbon are present in steel, they will necessarily freeze into the austenite phase, where they have the potential to form in these same three basic ways—

- in a substitutional solid solution with FCC austenite
- as compounds with iron, carbon, or another alloying element, or as a complex

compound with cementite, which would then exist as particles mixed with the austenite
- as a mixture with austenite

As the solid iron cools, it changes from austenite to ferrite, which may change these solubility relationships, because of both the change in temperature and the change in lattice.

Solid Solutions. If an alloying element is to form a substitutional solid solution with iron, the conditions for solubility described in Chapter 4 must be satisfied. In many applications the amount of the alloying element used is relatively small, so partial solubility requiring atom diameters that differ less than 15% is of interest. Thus the elements that have the potential to form solid solutions with austenite or ferrite are those that are nearby on the periodic table or that have a reasonably sized atom diameter—e.g., manganese, cobalt, chromium, nickel, vanadium, molybdenum, tungsten, titanium, silicon, and aluminum.

Compounds. Many of the same elements that can go into solution with iron are transition elements capable of forming compounds with the carbon, nitrogen, or other elements that may be present in the austenite.[†] Whether an element goes into solution or forms a compound often depends on the absence or presence of other elements. For example, aluminum has such a strong affinity for oxygen that if it is added to a steel containing oxygen it forms aluminum oxide; however, if no oxygen is present, the aluminum will dissolve in the ferrite.

Another element, say chromium, might go into solution with a low-carbon austenite, or if the carbon content is high, might instead form a carbide. However, that same element, chromium, might go into solution with a high-carbon austenite if a stronger carbide-forming element, say titanium, is also present.

Other elements, e.g., niobium, may be added to a steel, not to dissolve in the austenite but to form carbides or nitrides.

Opposing tendencies. Thus *there are these two opposing tendencies of the alloying elements—to dissolve in austenite and/or ferrite or to form other compounds, especially carbides.* Although it is not a certainty that an element will be found in a given form, there is an approximate hierarchy in which the elements tend to be found as carbides, or in the reverse order, in ferrite:

⇐ Tendency to form carbides increases as you go from right to left
Titanium   Vanadium   Molybdenum   Tungsten   Chromium   Manganese
        Tendency to dissolve in ferrite increases as you go from left to right ⇒

It should not be assumed that, just because an element to the left is present, one to the right cannot form a carbide, but the general principle is that some elements

[†]Some alloying elements are added because of their influence during the solidification process in forming slags and removing unwanted elements; these are the elements discussed in Section 6.7. They involve the formation of a compound between the element and something other than iron and carbon; the elements of interest here are those that influence the properties of heat-treated and/or non-heat-treated steels.

tend to form carbides more readily than others and some dissolve in ferrite more readily than others.

All the major alloying elements have some solubility in both austenite and ferrite, although that solubility varies widely. However, the *stronger carbide formers have the least solubility in austenite,* so that elements to the left end in the table above have the lowest solubility in austenite; solubility increases going to the right.

Silicon, nickel, and aluminum are absent from the above list because they don't form carbides but tend to convert the carbon to the graphite form.

Carbides and Hardenability. The typical purpose for using alloying elements to form solid solutions with austenite is to improve the hardenability of the steel.[‡] We have already discussed the effect of solute atoms on recrystallization. We saw (Table III-1) that the addition of solute atoms to pure metals raises the recrystallization temperature—or increases the time it takes the metals to recrystallize at the same temperature. Solute atoms do the same thing here; they *slow the ability of austenite to transform.* All alloying additions, except cobalt, slow the austenite-to-pearlite transformation; they decrease both the rate of nucleation of new grains and the growth of the individual grains. Alloy additions move the upper portion as well as the nose of the TTT diagram to the right.

The factor that stands in the way of these elements becoming solute atoms for the austenite is their potential for forming interstitial compounds. So elements such as chromium, vanadium, and titanium (1) may go into substitutional solution with austenite, (2) may form carbides or nitrides, or (3) may do both. This variability can have confusing effects. In order to improve hardenability, alloying elements must be in solution in the austenite; on the other hand, an element that forms a carbide is not in the austenite lattice for hardenability purposes, plus the austenite is robbed of the carbon needed to form the martensite. Therefore the selection of alloying elements is a matter of balancing their effects to get the properties desired.

8.3   SPECIFIC FUNCTIONS OF COMMON ALLOYING ELEMENTS

Nickel. Nickel has some fundamental advantages as an alloying element in steels. Its atomic diameter is very close to that of iron, with the same lattice (FCC) as austenite. Thus nickel has 100% solubility with austenite, improves hardenability, and doesn't form carbides.

[‡]The hardenability of a steel is improved by slowing down or retarding the ability of its austenite to transform until it gets to the M_s temperature. Although the austenite grain size is a variable in this process, the usual method of improving hardenability is the use of alloying elements that go into substitutional solid solution with the austenite. See Section 7.6 for a more extensive discussion on hardenability.

Nickel doesn't have 100% solubility with ferrite, since the two metals have different lattices, but it does have a large partial solubility, which improves strength and toughness at room temperature and below.

When nickel is added to steel it lowers the A_1 critical transformation temperature and reduces the carbon content of the eutectoid point. When used in larger amounts and combined with chromium, nickel can depress the A_1 temperature sufficiently that steels are austenitic at room temperature, as in some stainless steels.

Chromium. Chromium has enough solubility in austenite to be a strong contributor to hardenability, but it is also a relatively strong former of chromium carbides and complex carbides with iron. Chromium has a BCC lattice and an atom diameter (2.498 Å) almost the same as iron's (2.482 Å) so it is 100% soluble in ferrite and provides solid-solution strengthening to that phase.

Chromium is a popular alloying addition in regular alloy steels, is used in a majority of tool steels, and is used in all corrosion- and heat-resistant (stainless) steels.

Tungsten. Because of its diameter and/or lattice type (BCC), tungsten has only a partial solubility with austenite, but even in small quantities, it can influence hardenability. Because of its expense, however, it is not generally used for this purpose. Tungsten is primarily used because of its ability to form carbides, but it does have a large partial solubility with ferrite.

Molybdenum. Molybdenum's BCC lattice and atom size give it limited solubility in austenite, but even when used in amounts as little as 0.20 to 0.35%, it is a strong contributor to hardenability. Molybdenum has a large partial solubility in ferrite; however, if carbon is available, it is a strong carbide former. These carbides can help prevent the growth of the austenite grain size and promote fine-grained steels.

Molybdenum also raises the tempering temperature required to achieve the same amount of softening. When tempered at temperatures above 1000°F (540°C) steel containing molybdenum forms complex carbides, which further raise the steel's hardness. This process is termed **secondary hardening;** it is also effected with vanadium.

Vanadium. With a BCC lattice vanadium has a low solubility in austenite, but in small quantities, e.g., 0.10%, it has a marked effect on hardenability. Although its diameter is close to that of iron, the resulting 100% solubility with ferrite is usually of no significance because vanadium forms carbides so readily. It typically forms carbides in the austenite, which serves to restrict grain growth and promote fine-grained steels. Like molybdenum, vanadium retards tempering and contributes to secondary hardening.

Vanadium is a frequently used alloying element in tool steels and HSLA steels but is used infrequently in regular alloy steels.

Niobium.   Niobium (columbium) aggressively forms compounds with carbon and nitrogen, i.e., carbides, nitrides, and carbonitrides. As with vanadium, these compounds can reduce the growth of austenite grains and contribute to finer-grained steels. Niobium is often used in the HSLA steels.

Titanium.   Titanium is such a strong carbide former that the presence of carbon makes its limited solubility in austenite and ferrite of little importance. It will also form nitrides as well as carbides and so is used to refine grain size in the same manner as are niobium and vanadium. Titanium is also used in some HSLA steels.

Nitrogen.   Nitrogen is found in the liquid of most steels or can be purposely added to form nitrides, which act in the same manner as carbides to inhibit the growth of austenite grains. Aluminum, vanadium, niobium, and titanium are among the elements that form nitrides.

Manganese.   As a general alloying element, manganese is used in lesser amounts as a deoxidizer and desulfurizer in many steels, but it is also an inexpensive and effective hardenability agent.

Manganese dissolves readily in austenite, and when used in amounts over 1.0% it contributes to hardenability—and thus to hardness and strength. It has some solubility in ferrite[§] but is a weak carbide former. Manganese lowers the critical transformation temperatures and when used in larger amounts, e.g., over 10%, can make the steel completely austenitic at room temperature. This explains its role in Hadfield's steel, discussed in Section 8.8.

Silicon.   Silicon tends to graphitize carbon so it doesn't form carbides,[||] but it does dissolve in ferrite and, to a lesser extent, in austenite. For most alloy steels, at low levels of composition its prime role is that of deoxidizer. However, it contributes solid-solution strengthening when dissolved in ferrite and improves strength and toughness. Silicon also reduces the distortion caused by the austenite-to-martensite volume change. At higher levels, about 3.00%, silicon is used in steels intended for magnetic and electrical applications.

Aluminum.   In relatively small amounts aluminum is used to **"kill,"** or degas, many steels. However, aluminum is also capable of forming nitrides, which can serve to help control austenite grain size. If not tied up as an oxide or nitride, it can dissolve in ferrite and contribute to solid-solution strengthening.

[§]The elevated-temperature closest-approach distances of γ-iron and β-manganese are 2.579 and 2.37 Å, respectively; manganese is thus within 8% of the atom diameter of austenite. At room temperature the closest-approach distances of α-iron and α-manganese are 2.482 and 2.24 Å, respectively; manganese thus differs 10% from the atom diameter of ferrite.

[||]Silicon is important in the formation of the graphite that gives the gray, ductile, and malleable cast irons their characteristics, as discussed in Chapter 9.

Other Elements.   An alloying element that doesn't dissolve in iron or form a carbide (i.e., if it only forms a mixture with the steel) will do little to alter the basic properties or characteristics of the steel. For example, sulfur is commonly used to improve the machinability of steels. As discussed in Chapter 6, sulfur tends to embrittle steels by forming FeS, which has a melting temperature within the austenite region and makes the steel appear "hot-short." Manganese is added to the steel to encourage the formation of MnS, which forms a slag. However, if the MnS has a small particle size and is well dispersed throughout the steel, it improves machinability by acting as a chip breaker. The MnS is a compound, but not a compound with either the iron or the carbon forming the steel, since it is insoluble to them.

Lead is another insoluble that, when dispersed as small particles, can encourage chip formation during machining.

Other Effects.   *All alloying elements that dissolve in austenite tend to lower the carbon content at which the eutectoid occurs;* i.e., the eutectoid point moves to the left on the phase diagram. Most alloying elements *raise* the eutectoid, or A_1, temperature. The significant exceptions are nickel and manganese which can *lower* the A_1 enough that austenite can exist at room temperature. This is important in stainless steels and is discussed later in this chapter.

8.4   THE ALLOY STEEL SYSTEMS

In order to achieve the diverse properties required by the broad uses made of steels, a number of systems of alloy steels have been devised. These are covered by standards that have been established by organizations such as ASTM, SAE, and AISI.[#] We'll discuss these systems and their steels briefly here by way of illustrating the use of specific alloying elements and providing examples of the different characteristics and properties that can be achieved in steels. Complete details about these systems are available in the annual publications of the organizations cited and in many other handbooks.[**]

*These systems fall into two general categories according to how properties are achieved—through heat treatment or through chemical composition—*but there are overlaps that limit precise classification. Within each of the systems there are also major divisions and subdivisions.

In general, the alloy steels and the tool steels of the AISI/SAE systems are intended to be heat-treated. The high-strength, low-alloy (HSLA) steels normally

---

[#]The American Society for Testing Materials, The Society of Automotive Engineers, and the American Iron and Steel Institute.

[**]For example, the 1990 *SAE Handbook,* vol. I, secs. 1 and 2, defines the types and compositions of alloy steels, gives Jominy curves and property data, cites the general purposes of the alloying elements, recommends heat treatments, and gives much useful information.

improve properties through chemical composition and process control; however some do develop properties by heat treatment or through special processing that is, in effect, heat-treating.

The stainless and other highly alloyed steels combine heat treatment and composition, and perhaps processing, to achieve mechanical properties; some of these steels can be heat-treated, while others are used in the cold-worked or annealed condition.

8.5   AISI/SAE ALLOY STEELS

Designation System. Perhaps the most familiar designation system is that used for AISI/SAE **alloy steels.** This system uses a four-digit identification number to *classify the steels by composition;* the first two digits identify the major alloying element or elements (e.g., 1 = carbon; 2 = nickel; 3 = nickel, chromium; 4 = molybdenum; 43 = nickel, chromium, molybdenum), and the last two give the carbon content in points (e.g., 10 points of carbon = 0.10% C). The system is illustrated in Table VIII-1.

Variations in Chemical Composition. A quick inspection of Table VIII-1 shows that the carbon content of most of these steels is relatively low (less than 0.60% C);[††] that only nickel, chromium, and molybdenum are frequent additions;[‡‡] and that silicon, manganese, tungsten, and vanadium are added to only a few of the series.

The previous discussion on alloying elements has provided some generalities on the properties and characteristics of these steels. Those high in nickel will be tough at normal and low temperatures; the presence of chromium will mean good hardenability, with some possibility for formation of carbides; molybdenum is a reasonably strong carbide former that can also help control austenite grain size and contribute to secondary hardening during tempering. Obviously, these characteristics will vary greatly with the alloying composition and the amount of carbon, but the AISI/SAE alloy steels are used primarily where strength, hardness, and toughness are important.

For purposes of illustration, compositions for individual alloys and alloys series are shown in Table VIII-1. A specific alloy is identified when all four digits of its number are given, e.g., 4340. When the carbon digits are shown by "xx," e.g., 86xx, the range of nominal carbon contents used with that whole series is shown in parentheses.

Variations in Properties. With the many combinations of carbon and alloy content available in this system, these steels can be heat-treated to tensile and yield strengths as high as 280 and 230 ksi, respectively, with lower values possible by

---

[††]See Section 7.3; maximum martensite hardness is achieved with about 0.6 to 0.8% carbon.
[‡‡]Manganese and silicon are also present, but for the reasons described in Section 6.7 rather than as alloying elements.

T A B L E VIII–1 AISI/SAE steel numbering system, with major alloying elements and carbon ranges for some series, and data for individual steels

AISI/SAE Number	Carbon (nominal)	Nickel	Chromium	Molybdenum	Other	UNS Designation
Carbon Steels						
10xx	(0.05–0.95)	1.00% Mn max.				G10xx0
11xx	(0.10–0.52)	Resulfurized				G11xx0
12xx	(0.15 max.)	Rephosphorized & resulfurized				G12xx0
15xx	(0.13–0.90)	High-Mn carbon steel				G15xx0
Alloy Steels						
13xx	(0.30–0.40)				1.60–1.90 Mn	G13xx0
23xx	(0.17–0.45)	3.25–3.75				G23xx0
25xx	(0.12–0.17)	4.75–5.25				G25xx0
31xx	(0.15–0.45)	1.10–1.40	0.55–0.75			G31xx0
32xx	(0.32–0.50)	1.50–2.00	0.90–1.25			G32xx0
33xx	(0.10–0.16)	3.25–3.75	1.40–1.75			G33xx0
34xx	(0.15–0.50)	2.75–3.25	0.60–0.95			G34xx0
40xx	(0.23–0.47)			0.20–0.30		G40xx0
41xx	(0.30–0.50)		0.80–1.10	0.15–0.25		G41xx0
4340	0.38–0.43	1.65–2.00	0.70–0.90	0.20–0.30		G43400
4422	0.20–0.25			0.35–0.45		G44220
4620	0.17–0.22	1.65–2.00		0.20–0.30		G46200
4720	0.17–0.22	0.90–1.20	0.35–0.55	0.15–0.25		G47200
4820	0.18–0.23	3.25–3.75		0.20–0.30		G48200
51xx	(0.20–0.60)		0.70–0.90			G51xx0
E52100	0.98–1.10		1.30–1.60			G52986
6150	0.48–0.53		0.80–1.10		0.15 V min.	G61500
7260	0.50–0.70		0.50–1.00		1.50–2.00 W	G72600
86xx	(0.15–0.45)	0.40–0.70	0.40–0.60	0.15–0.25		G86xx0
8720	0.18–0.23	0.40–0.70	0.40–0.60	0.20–0.30		G87200
8822	0.20–0.25	0.40–0.70	0.40–0.60	0.30–0.40		G88220
9259	0.56–0.64				1.80–2.20 Si	G92560
9840	0.38–0.43	0.85–1.15	0.70–0.90	0.20–0.30		G98400

Source: *SAE Handbook*. Warrendale, PA: Society of Automotive Engineers, 1990, vol. 1.

slower quenching. The actual steel chosen also depends on other factors, such as type of fabrication, severity of quench, and need for weldability.

In any series of these steels the alloying elements are constant or very close to constant. *Since alloying elements, as opposed to carbon, are the major contributors to hardenability, any one series of alloy steels has Jominy curves with about the same general shape and slope; they are displaced up and to the right as the carbon content increases.* For example, in the 86xx series, the nominal carbon varies from 0.15 to 0.60%, but the alloying elements (0.40–0.70% Ni, 0.40–0.60% Cr, 0.15–0.25% Mo) are specified to be the same for all carbon contents. The 43xx, 47xx, 87xx, and 88xx series also contain nickel-chromium-molybdenum alloying elements, but in differing amounts, so that families of steels with a variety of Jominy curves, or hardenabilities, result.

Particular Steels. Steel 4340 is one of the most popular steels for general applications requiring a balance of hardenability, strength, hardness, and toughness. The hardenability of 4340 is so good that it achieves hardnesses of over 50 HRC in the center of a 3-in.-diameter rod quenched in water; this steel can often be air-cooled to usable hardness in smaller parts. The various 4xxx steel families are used for a wide variety of automotive and aircraft parts, including shafts, gears, pins, and landing gears.

The 5xxx series of chromium steels is used at two general levels of carbon. The lower level, less than 0.60% C, is usually carburized, which gives a product with a tough low-carbon interior and a very hard exterior containing chromium carbides. The higher-carbon varieties, E51100 and E52100 steels, with 1.00% C, are often encountered because of their popularity for use in ball and roller bearings; they produce a very hard, wear-resistant steel. The high-quality steel required is produced in the electric furnace, indicated by the prefix "E."

The 71xxx and 72xx series use tungsten, with chromium; some steels use levels of 12 to 18%, but 7260, shown in Table VIII-1, has only 1.50 to 2.00% W. The main contribution of tungsten is to form carbides that give strength and hardness at high temperatures. In tool steels, tungsten is used at much higher levels in the grades referred to as "high-speed steels."

The 11xx and 12xx carbon steels contain sulfur and sulfur plus phosphorus to aid in machining and improve surface finish. The sulfur forms manganese-sulfide inclusions that interrupt the homogeneous nature of the metal and act as chip breakers. The inclusions may also serve to lubricate the surface, also improving the finish. Phosphorus, being partially soluble in ferrite, strengthens the ferrite, which helps to break the chip into smaller pieces and avoid stringiness.

H Steels. These AISI/SAE alloy steels are also produced as H steels, e.g., 4340H, made to slightly broader chemical-composition specifications but guaranteed to generate a Jominy curve that will lie between standard limits. The use of Jominy curves to select H steels to meet specifications is discussed in Appendix E.

Deleted Steels. The societies that maintain these systems and standards often "delete" a steel or series of steels from their official listing. This merely means that the production volume of that steel was below a certain minimum; it does not necessarily mean that the steel cannot be purchased. However, the prudent designer would check the availability of the steel at a reasonable price before committing to its use.

8.6   AISI/SAE TOOL STEELS

Designation System. Table VIII-2 illustrates the system used to categorize tool steels; a brief inspection of the table shows that tool steels are classified by more than just their chemical composition, the system used to classify the alloy steels.

Type Tool Steel	AISI/SAE Designation	C	Mn	Si	Ni	Cr	V	W	Mo	Co	UNS Designation
Water hardening	W2	0.60–1.40	0.25	0.25		0.50	0.25				T72302
	W5	0.60–1.40	0.25	0.25		0.25					T72305
	W6	0.60–1.40	0.25	0.25			0.25				T72306
Shock resisting	S1	0.50	0.25	0.70		1.50		2.50	0.40		T41901
	S5	0.50	0.80	2.00							T41905
Cold work, oil hardening	O6	1.45	0.70	1.00					0.25		T31506
Cold work, air hardening	A5	1.00	3.00	0.30		1.00			1.00		T30105
	A7	2.25	0.50	0.30		5.25	4.50		1.00		T30107
Cold work, high C & Cr	D2	1.50	0.40	0.40		12.00			1.00		T30402
	D7	2.35	0.40	0.40		12.00	4.00		1.00		T30407
Hot work, chromium	H12	0.35	0.30	1.00		5.00	0.40	1.50	1.50		T20812
	H14	0.40	0.35	1.00		5.00		4.50			T20814
Hot work, tungsten	H21	0.30	0.28	0.30		3.25	0.45	9.25			T20821
	H23	0.30	0.28	0.37		12.00	1.00	12.00			T20823
	H26	0.50	0.28	0.28		4.00	1.00	18.00			T20826
Hot work, molybdenum	H42	0.60	0.28	0.32		4.00	2.00	6.00	5.00		T20842
High-speed, tungsten	T1	0.70	0.30	0.30		4.00	1.00	18.00			T12001
	T2	0.85	0.30	0.30		4.00	2.00	18.00	0.85		T12002
	T8	0.80	0.30	0.30		4.00	2.00	14.00		5.00	T12008
	T15	1.50	0.30	0.30		4.00	5.00	12.00		5.00	T12015
High-speed, molybdenum	M1	0.80	0.30	0.30		4.00	1.10	1.50	8.50		T11301
	M6	0.80	0.30	0.30		4.00	1.50	4.00	5.00	12.00	T11306
	M15	1.50	0.30	0.30		4.00	5.00	6.50	3.50	5.00	T11315
Special-purpose, low alloy	L2	0.80	0.50	0.30		0.95	0.20				T61202
Carbon-tungsten	F2	1.25	0.30	0.30		0.30		3.50			T60602
Mold steel, low-carbon	P3	0.10 max.	0.40		1.25	0.60					T51603

Source: *SAE Handbook*. Warrendale, PA: Society of Automotive Engineers, 1990, vol. 1.

The identification designations—sequential numbers with a letter prefix—are simpler than the alloy steel system, but they separate the steels into overlapping categories that are more complicated. Instead of being related simply to composition, the designations are dependent on

- type of quench, e.g., water, oil, or air
- major alloying element, e.g., tungsten, molybdenum, or chromium
- use or application, e.g., cold work, hot work, or molds
- properties or characteristics, e.g., shock resistance or high speed

Variations in Compositions. A look at the compositions shown in Table VIII-2 gives some idea of the major ways in which tool steels differ from alloy steels:

- Their carbon contents are generally higher.
- Chromium is obviously the major alloying element.
- In general, more alloying elements are used and in broader ranges of content.
- Elements like vanadium and tungsten are used more often.

Variations in Microstructures and Properties. In broad terms, the *alloying of tool steels is done to provide hard, wear-resistant carbides.* Tools need to be strong— some have tensile strengths approaching 300 ksi—but the yield strength is the real measure of strength; having tools that will deform plastically is of little value. (Who would want a bent drill or a twisted thread tap?) That is, *resilience* is of more importance in tool steels; *toughness* is of more importance in the alloy steels.[§§]

As a general comparison, the microstructures of AISI/SAE alloy steels have a hard, tough martensite, while the tool steels generate a hard martensite containing wear-resistant carbides. Generating martensite is usually no problem for the tool steels because they have hardenability from all the alloying elements used, and these same elements easily provide the carbides if the carbon content is high enough.

The following examples illustrate these general principles.

Chromium hot-work steels. In tool steels chromium generally contributes hardenability, although it is also a carbide former. Chromium, in combination with other hardenability contributors, makes it possible to heat-treat tools with a large cross section to HRC 50 or higher when air-cooled. For example, H13 steel, with 5% chromium, can be air-cooled in thicknesses of 10 in. or more and achieve optimum properties.

Tool steels of this type have high room-temperature tensile and yield strengths; when tempered to 52 HRC, these steels have strength values of approximately 260 and 220 ksi. They also retain their strength at high temperatures; at a test temperature of 800°F the tensile strength can be as high as 220 ksi. These steels are often used in tooling for operations requiring high temperatures, e.g., forging, hot extrusion, and die casting.

[§§]See Section B.6 for a discussion of toughness and resilience.

High-speed steels. The tungsten (T) and molybdenum (M) **high-speed steels** are of particular interest because of their exceptional properties and widespread use. Some are capable of achieving a *compressive* yield strength of 500 ksi and a tempered hardness of 64 HRC. Tools made from these steels have many uses—as punches, dies, etc.—but their name and primary use come from their ability to operate at speeds where they are heated until red; that is, they achieve "red-hardness." Drills, taps, and milling cutters are frequently made from these steels.

The high carbon content in the presence of the carbide-forming elements chromium, vanadium, tungsten, and molybdenum creates many hard carbides in the microstructure. In some versions cobalt is added to harden ferrite by solid-solution strengthening; it is probable that this provides hardness at higher temperatures where carbides may tend to redissolve in austenite.

The best known of the high-speed steels are T1 and T2, commonly known as 18-4-1 and 18-4-2 because of their contents of tungsten, chromium, and vanadium. The tungsten and molybdenum high-speed steels have slightly different properties and characteristics. The tungsten variety is more wear-resistant and is therefore favored for abrasive applications, but it is more brittle. The molybdenum steels are somewhat tougher and are favored for interrupted cuts.

8.7 HIGH-STRENGTH, LOW-ALLOY (HSLA) STEELS

A group of low-carbon steels has been developed with strengths higher than plain carbon steels yet containing low levels of alloying elements. The term *HSLA* is used to identify these steels, but they are not a homogeneous, well-defined system of steels. Some of these HSLA steels are hot-rolled, some are later cold-rolled, some are quenched and tempered, and some are processed and become dual-phase (ferrite and martensite) steels.

Microalloying. The most common denominator among the HSLA steels is that they usually contain small, or **microalloying,** amounts of elements intended to help control grain size during hot-processing of the steel. Rather than being subjected to solid-solution strengthening or heat-treating, these steels gain their strength through their fine grain. So, where a plain carbon steel, heated to a rolling temperature in the austenite region, might undergo grain growth, the carefully added elements of a HSLA steel prevent grain growth and thus bring greater strength even though the carbon content of the steel is lower.

The alloying elements used to accomplish this are niobium, vanadium, aluminum, zirconium, titanium, and nitrogen—elements that can form carbides or nitrides in the austenite and prevent the growth of the austenite grains. In some alloys the solubility for these particles decreases as the temperature falls, and a *strengthening also occurs due to precipitation.*

Thus there are two major results of this alloying: the yield strength is raised, often to within 10 ksi of the tensile strength, and the improved properties are accomplished at a carbon content less than that for plain carbon steels with comparable properties.

Designation System.   The designation systems for these steels commonly use the minimum yield strength in ksi (thousands of psi) in the identifying number, with additional numerals and/or letters tagged on.[‖‖] Table VIII-3 illustrates four SAE grades and a dual-phase steel typical of the HSLA category. Specified or typical compositions and properties are shown. In examples 1, 2, and 3 properties are determined by alloying elements and process control. The grades of examples 1 and 2 are specified to have a 10-ksi spread between yield and tensile strengths, but other relationships are available, as shown in example 3. Stress-strain diagrams typical of the steels of Table VIII-3 are shown in Figure 8-1.

SAE Grades.   Example 1, grade 050X (Table VIII-3), is a specification for hot- and cold-rolled HSLA steel sheet with a minimum yield strength of 50 ksi; the "X" means it contains small amounts of elements such as niobium, vanadium, and/or nitrogen. Example 2, grade 50F, is specified for hot-rolled steel plates, bars, and shapes with a minimum yield strength of 50 ksi; it also contains small amounts of elements designed to control grain size. The "F" designates it as a formable grade.

Example 3, grade 950X, is a general specification for a microalloyed HSLA steel with a yield strength of 50 ksi, minimum. The specification for this group of steels is being replaced by those for the steels of examples 1 and 2, but it is still in use. Although the properties of examples 1 through 3 are not identical, they are generalized by one curve in Figure 8-1.

In examples 1, 2, and 3 note the presence of a yield point with discontinuous yielding. Note also that the yield point is close to the value of the tensile strength and that the elongation value is relatively high.

[‖‖]For example, K = killed steel, A = carbon and manganese only, C = carbon, manganese, and nitrogen, etc. But the meaning of the letters is not consistent among specifications. A, B, C, etc. are also used to indicate weldability, formability, etc. X can mean microalloying, but X may also indicate that there is a 10-ksi spread between YS and TS (Y indicates a 15-ksi spread, Z a 20-ksi spread).

T A B L E VIII–3 Compositions and properties for some HSLA and plain carbon steels

Example Number	SAE Grade or Name	Carbon, %	Manganese, %	Control Grain Size	Yield Strength or Point, ksi	Tensile Strength, ksi	Elongation, %
1	050X	0.13 max.	0.90 max.	√	50	60	22
2	50F	0.18 max.	1.65 max.	√	50	60	20
3	950X	0.23 max.	1.35 max.	√	50	65	22
4	Q980	0.20 max.	1.35 max.	√	80	95–115	18
5	Dual-phase	0.10	0.40 Si	√	64	101	17
6	1023	0.20–0.25	0.30–0.60		31	56	25
7	1047	0.43–0.50	0.70–1.00		47	85	15

Source: *SAE Handbook.* Warrendale, PA: Society of Automotive Engineers, 1990, vol. 1.

FIGURE 8-1 Typical stress-strain curves for the steels of Table VIII-3

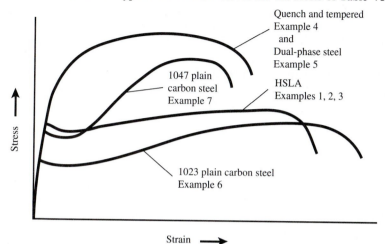

Example 4, grade Q980, is a low-carbon, low-alloy steel that achieves properties by being quenched and tempered. Although no microalloying elements are specified, the steel is produced to a fine grain and other elements are added to achieve certain strength levels. Table VIII-3 indicates that this steel achieves comparatively high yield and tensile strengths with a reduction in elongation; it would have a stress-strain diagram similar to that for example 5.

Dual-Phase Steels. Example 5 is typical of a dual-phase steel. **Dual-phase steels** are low-carbon, low-alloy steels that achieve their properties by being cooled from above the A_1 critical line, but not above the A_3; that is, they are cooled from the region of α-ferrite and γ-austenite, or intercritically.[##] The result of this treatment is a microstructure similar to that seen in Figure A-23, a matrix of soft ductile α-ferrite with hard martensite laced through it.

The intercritical treatment results in the dual-phase steels exhibiting **continuous yielding behavior;** i.e., there is no yield point followed by a period of discontinuous yielding, as is often exhibited by hot-worked, low-carbon steels. Instead, the 0.2% offset yield strength is relatively low, but the metal work-hardens rapidly (see Figure 8-1) so that a small amount of later cold work will raise the yield strength appreciably. The steels are also characterized by having high tensile strengths and elongations.

Dual-phase steels are relatively new but are being used in many products that require forming, e.g., automobile wheels.

Plain Carbon Steels. For comparison purposes, the chemical compositions and typical mechanical properties for two plain carbon steels are also shown in Table

[##]See Section 7.8 for a discussion of intercritical heat-treating. Heating a low-carbon steel to just above A_1 transforms the pearlite to austenite of about 0.80% C; the ferrite does not transform to austenite at that temperature.

VIII-3 and Figure 8-1. Note that examples 1, 2, 3, and 7 all have yield points of about 50 ksi, but that the plain carbon steel, example 7, requires twice the carbon to achieve that yield point. The plain carbon steel of example 6 has as much or more carbon than examples 1 through 3, but has a much lower yield point and tensile strength.

The point is that the careful alloying and processing of the HSLA steels leads to a great improvement in properties over plain carbon steels.

8.8   HIGHLY ALLOYED STEELS

So far in this chapter we have discussed the use of relatively small amounts of alloying elements. There are some tool steels that contain significant amounts of alloying elements, e.g., D7, T1, M6, but generally the alloy steels we've talked of to this point have about 5% or less alloying elements. In this section we'll discuss steels where something other than iron accounts for 25% or more of their composition. In some alloys the amount of carbon is also very low, which almost makes the use of the name "steel" inappropriate.

Effect of Nickel and Chromium on Iron. Nickel and chromium both form alpha (α) and gamma (γ) phases with iron, but each has opposite effects on the location and shape of the phase regions. The phase diagrams for these alloy systems are seen in Figures 8-2 and 8-3.

At high temperatures, e.g., 2200°F (1200°C), nickel and iron have 100% solubility in the gamma (γ) phase. At lower temperatures, and a nickel content of less

F I G U R E 8–2 Iron-nickel phase diagram

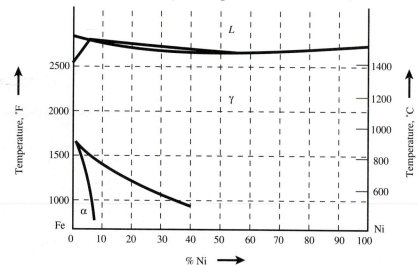

Source: *Metals handbook* (8th ed.). Materials Park, OH: ASM International, 1973, vol. 8, 304.

FIGURE 8-3   Iron-chromium phase diagram

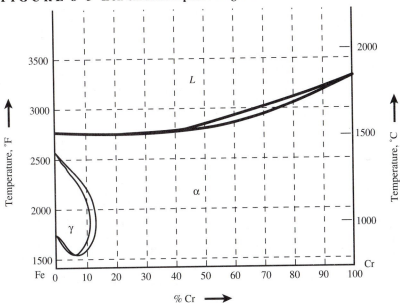

Source: *Metals handbook* (8th ed.). Materials Park, OH: ASM International, 1973, vol. 8, 291.

than 10%, an alpha (α) phase forms, but otherwise the *gamma* (γ) *phase exists down to room temperature* for all other compositions.

The iron-manganese phase diagram is not shown here, but iron and manganese react much as do iron and nickel, with large amounts of the manganese gamma (γ) phase extending down to room temperature. Since most steels contain some manganese, it is available to assist nickel in this role of depressing the gamma (γ) phase to room temperature.

Of opposite influence is *chromium, which stabilizes the alpha* (α) *phase;* although it forms a gamma (γ) phase, it exists only up to about 13% chromium and above 1562°F (850°C). The gamma (γ)-phase loop can be seen in Figure 8-3.

By varying the kind and amount of these major alloying element(s)—nickel and chromium—stainless, austenitic-manganese, precipitation-hardened, and maraging steels are developed. In the remainder of this section we'll discuss each of these types in turn; their compositions and properties are shown in Table VIII-4.

Stainless Steels.   One of the important groups of highly alloyed steels are the heat- and corrosion-resistant steels, more commonly referred to as *stainless steels. Chromium is the common denominator of all stainless steels.* When 12 to 13% chromium is in solution in the iron—i.e., not as a compound with carbon or other elements—a protective oxide coating is formed that is highly resistant to chemical attack. If some of the chromium is used to form carbides or other compounds, then additional chromium must be added to protect the corrosion resistance of the steel.

TABLE VIII–4 Comparison of compositions and properties for highly alloyed steels*

Type	AISI No.	Carbon, %	Manganese, %	Silicon, %	Chromium, %	Nickel, %	Other, %	Condition	Tensile Strength, ksi	Yield Strength, ksi	Elongation, %	Brinell Hardness	Source
Ferritic stainless	430	0.12 max.	1.00 max.	1.00 max.	16.00–18.00			Annealed	75	45	30	155	1,2
Martensitic stainless	440C	0.95–1.20	1.00 max.	1.00 max.	16.00–18.00		Mo. 0.75 max.	Oil Q/T @ 600°F†	285	275	2	580	1
	410	0.15 max.	1.00 max.	1.00 max.	11.50–13.50			Oil Q/T @ 1000°F†	157	145	13	300	2
	501	0.10 min.	1.00 max.	1.00 max.	4.00–6.00		Mo. 0.40–0.65.	Annealed / Q/T†	70 / 145	30 / 110	28 / 18	160 / 300	3 / 3
Austenitic stainless	302	0.15 max.	2.00 max.	1.00 max.	17.00–19.00	8.00–10.00		Annealed / Approx. 15% c.w.‡	92 / 125	38 / 75	68 / 12	162 / 253	2 / 2
	304	0.08 max.	2.00 max.	1.00 max.	18.00–20.00	8.00–10.50		Annealed	85	30	60	148	2
	310	0.25 max.	2.00 max.	1.50 max.	24.00–26.00	19.00–22.00		Annealed	92	40	47	180	2
	201	0.15 max.	5.50–7.50	1.00 max.	16.00–18.00	3.50–5.50	N. 0.25 max.	Annealed	115	55	55	183	1
Hadfield's steel		1.20 typ.	12.00 typ.	0.60 typ.				Cast. h.t.‡	120	55	50	200	1
Precipitation-hardened steel	17-7PH	0.09 max.	1.00 max.	1.00 max.	16.00–18.00	6.50–7.75	Al. 0.75–1.50	h.t.‡ / precip.-hard.	230	217	6	452	2
Maraging steel		0.03 max.	0.10 max.	0.10 max.		18.00 typ.	‡	h.t.‡ / precip.-hard.	310	300	5		2

*Compositions shown are specification limits unless marked "typ" (typical). Mechanical properties shown are typical values.

†Q/T = quenched and tempered; c.w. = cold-worked; h.t. = heat-treated.

‡Approx. Co = 8.0, Mo = 4.0, Ti = 0.5, Al = 0.10.

Sources: 1. *Metals handbook* (8th ed.). Materials Park, OH: ASM International, 1961, vol. 1.
2. *SAE Handbook*. Warrendale, PA: Society of Automotive Engineers, 1990, vol. 1.
3. Sidney H. Avner. *Introduction to physical metallurgy* (2d ed.). New York: McGraw-Hill, 1974.

Ferritic stainless steels. The ferritic stainless steels take advantage of the limited size of the gamma (γ) region in the chromium-iron system (Figure 8-3). If the amounts of chromium in solution are *greater* than the limit of the gamma (γ) region, the steel will be ferritic; if carbon is present, the chromium content has to be increased to offset the amount tied up in carbides. In Table VIII-4, steel 430 is given as a ferritic type; note that its chromium content is greater than 13% (16 to 18%) to offset the maximum carbon content of 0.12%. The ferritic stainless steels are characterized by their high chromium content; their magnetic nature; and, in the annealed condition, their low strength and hardness and their high ductility.

Since these steels are ferritic, they cannot be heat-treated, and their properties can be altered only by cold-working and annealing. Steels of this type are used where the degree of forming is not too severe and high-temperature corrosion resistance is important; automotive exhaust systems and oil-burner nozzles are examples. These steels are also used where appearance is important in a corrosive environment, e.g., in auto trim and windshield wipers.

The AISI designation for ferritic stainless steels is a three-digit number beginning with a 4; the second two digits are merely assigned identifiers and contain no information.

Martensitic stainless steels. If the chromium in solution in the iron is reduced *below* the 13% limit of the gamma (γ)-phase loop of Figure 8-3, then the gamma (γ) phase can exist at a temperature of about 1900°F (1040°C). If the steel also contains carbon, the gamma (γ) can be quenched to martensite, just as if it were a carbon or alloy steel, and a hard, corrosion-resistant steel is formed. Being martensitic, the steels are magnetic.

Steel 440C (Table VIII-4) is a popular version of this type steel. Because the carbon content is high, chromium carbides are formed, lowering the chromium in solution so that the steel is in the gamma (γ) region at high temperatures. The chromium and molybdenum help the hardenability of the steel so that it can be quenched to martensite; martensite plus chromium carbides gives 440C the highest hardness and wear resistance of any of the corrosion- or heat-resistant steels.

The 440 steel is also produced in two versions with lower carbon content (440A, B) and two with variations of other elements (440F, F Se). Steel 410, with less carbon and chromium, produces a steel with lower properties in the quenched-and-tempered condition. Steel 501 is an example of the 5xx stainless steels that use a lower level of chromium and add other elements to aid hardenability.

In general, the hardenability of the martensitic group of stainless steels is so high that oil quenching, and often only air cooling, is sufficient to achieve maximum properties.

The AISI designations for martensitic stainless steels use 4xx for the higher-chromium versions and 5xx for the lower.

The martensitic stainless steels are used where high hardness and strength are required. Typical applications are bearing races and balls, cutlery, spray nozzles, and surgical instruments.

Austenitic stainless steels. When nickel is added to iron, the gamma (γ) phase is expanded. Although Figure 8-2 indicates that a large amount of nickel is required to maintain gamma (γ) down to room temperature, other elements, e.g., manganese, also help to lower the gamma (γ)-to alpha (α) transformation temperature. Thus nickel in the ranges shown in Table VIII-4 is sufficient to make these steels austenitic. Being austenitic, they have the more ductile FCC, rather than the BCC, lattice; they are nonmagnetic in the annealed condition, but some may be slightly magnetic when cold-worked.

These steels are used in the cold-worked or annealed condition, and as shown in Table VIII-4, have good ductility and can achieve high mechanical properties with cold-working.

The corrosion- and heat-resistance properties of these alloys are dependent on keeping chromium in solution with the gamma (γ)-iron. If carbon is too readily available, chromium carbides may form, robbing the grains of the protective chromium. Thus although not hardenable, the austenitic steels are usually cooled rapidly from the annealing temperature to prevent chromium from coming out of solution in the gamma (γ) and forming carbides.

Another obvious way to prevent the formation of carbides is to reduce the carbon content. Note that the carbon content of grade 302 is 0.15% C, compared with only 0.08% C for type 304; grade 304L has only 0.03% C maximum as a further attempt to reduce this problem. The lower-carbon variants are preferred for welding to reduce the tendency for these alloys to precipitate carbides when cooling from the welding temperature.

Another approach to avoiding the formation of chromium carbides is to add another alloying element that is a stronger carbide former than chromium. Thus "stabilized" grades with chemistries similar to those of 304 add titanium (321) and niobium and tantalum (347) to form carbides, leaving the chromium free in the gamma (γ).

Grade 302 is the most commonly used of the corrosion- or heat-resistant steels; the stabilized variations of 304 are popular for welded uses. Grade 310 and others like it are noteworthy for their use at higher temperatures; note that with higher carbon content, additional chromium is specified to maintain the corrosion resistance.

The AISI system uses 2xx and 3xx to designate the austenitic stainless steels. As can be seen in Table VIII-4, these steels contain large amounts of nickel, whereas the others do not. In the 2xx type some of the expensive nickel is replaced with manganese.

This family of stainless steels is used in a wide range of applications such as food and chemical processing equipment, cookware, hubcaps, hydraulic tubing, jet engine parts—anywhere that appearance, plasticity, corrosion resistance, and/ or heat resistance are required.

Hadfield's (Austenitic-Manganese) Steel. Manganese in the quantities used in these steels, about 12%, greatly delays the austenite-to martensite transformation; it is not eliminated, but at room temperature it is slowed appreciably. As the FCC austenitic steel is deformed in use, it slowly transforms to martensite and its properties go up. The data of Table VIII-4 show that its yield strength is not very high, but with deformation (in this case the tensile test itself) the properties increase. Steels of this type are used in construction equipment and railroad rail joints, where the hardness after use can be twice as high as that shown in Table VIII-4.

Precipitation-Hardened Stainless Steels. The steels of this group are the 18-8 (#302) type of stainless steel modified by the addition of precipitate forming elements such as aluminum, niobium, molybdenum, and titanium. They are usually identified by their chromium and nickel content, followed by "PH." As can be seen from Table VIII-4, they are capable of quite high strengths; they develop their maximum properties by being solution-heat-treated at about 1900°F, cooled to room temperature, and then given a low-temperature (900°F) precipitation treatment. These steels are used for jet engine parts, wire, springs, and whip antennas.

Maraging Steel. **Maraging steels** use large amounts of nickel, 18 to 25%, and iron to produce a soft martensite, which is then further hardened by the precipitation of intermetallic compounds created by the addition of alloying elements such as molybdenum, titanium, aluminum, and cobalt. There are three significant advantages to these steels:

- They are easily formed and fabricated; the solution heat treatment is essentially a 1500°F anneal, after which cold work or machining can be done. The strength is then gained by heating to only 900°F for the precipitation-hardening or maraging treatment. Because of the low temperature, distortion and scaling are at a minimum.
- They have extremely high tensile and yield strengths.
- Nevertheless they have good toughness.

These steels are usually identified by composition and strength level, e.g., 18Ni(300). The combination of high strength and toughness makes maraging steels useful for press tools, where a typical tool steel might shatter if a misalignment occurs. Because they can be heat-treated with little distortion, they are useful for intricate machine parts requiring high strength and close tolerances.

8.9 SELECTION OF TOOL STEELS

Product and Tooling Costs. It is important to properly account for *all* material costs, steel or otherwise; however, tool steels are singled out in our discussion here because of the difficulty in allocating the costs of tools to the product. Often, if not usually, carbon and alloy steels, stainless steels, HSLA steels, and other materials are used as part of the product. That is, their cost as a material is directly included in the product cost. The total product cost is then the accumulation of labor, machine, and material costs, with overhead expenses being allocated on some basis, e.g., labor hours, machine hours, or sales dollars.

Tools are somewhat different. Any particular tool may be used to produce a number of products, so the cost of the tool related to a particular product is not usually known; the normal accounting system is not set up to determine such costs. Instead, these are usually buried in the overhead charge, and the data normally available to engineers in evaluating the cost-effectiveness of tools are not available.

Costs Associated with the Use of Tools.   There are a number of costs to consider when selecting a tool steel, but the most important is the *overall* cost to produce the product. Some of the costs that should be included in this analysis are given below.

- *The cost of production:* These are the identifiable costs that show in the normal accounting reports—labor and machine costs. A properly selected tool produces more output per time with fewer "problems"; an improper tool reduces output per time. Since a tool is often a very small part of the total production process, it is sometimes difficult to identify its impact on the production costs.
- *The cost of quality:* The attributes that determine the product quality can be controlled better with certain types of tools than with others. This often shows up in the next factor.
- *The cost of maintaining, repairing, or altering the tool:* If, for example, it is expected that the part design will be changed, steels can be chosen that will facilitate the changes; a steel that produces a large number of pieces per hour but has to be resharpened more often may not be the best choice.
- *The cost of safety to workers and the workplace:* Hard, brittle tools may wear better and give better quality, but if they can shatter and endanger the lives of workers and/or their workplace the cost of preventing that hazard must be considered by prudent engineers.
- *The cost to protect the environment:* If a steel is chosen that requires a lubricant, or quench media, that is hazardous to employees or involves special disposal provisions, then those costs must be considered.
- *The cost of the tool itself:* Sometimes this cost is significant, but in most cases it is not the most important factor to be considered.***

One of the major problems in dealing with these costs is that accounting systems rarely accumulate data related to tools at all, let alone by these categories, and few tool suppliers are capable of citing actual data on the quality, maintenance, environmental, or safety costs of tools made from their steels. Even the cost of production will be highly dependent on differences in machinery, lubricants, and tool design, so comparisons are often very difficult. Usually the engineer must resort to trial runs and accumulate data outside the normal accounting system.

***In producing the tool itself, the most significant cost factors are usually distortion and cracking due to heat-treating.

Comparisons of Tool Steels. The amount of steel in a tool is usually rather small, so tool steels are not produced in as large a volume as are many of the other alloy steels. Instead, there are a few manufacturers who produce relatively small volumes *of their own versions* of the AISI/SAE steels. Care must be taken in selecting tool steels, especially when attempting to replace a satisfactorily performing steel with one that has a lower selling price; steels may be identified as the same AISI/SAE type, but subtle differences can have a significant influence on their performance in a particular operation.

The cost of tools is often the concern and responsibility of a buyer or purchasing agent. The desire to save money on the tool itself must be weighed against the overall cost to produce the part; this can usually be determined only by objective comparisons made in the workplace. The cost of the *tool* is not as important as the *total* cost of the *use of the tool.*

SUMMARY

1. One of the major reasons alloying elements are added to steels is to improve hardenability, but alloying elements will improve a number of other properties of steels as well, including

Toughness	Hardness
Wear resistance	Corrosion resistance
Machinability	Grain refinement
Strength at ordinary temperatures	Magnetic and electrical properties
Strength at both high and low temperatures	

2. Elements near iron on the periodic table will most easily go into solution with austenite and/or ferrite. Those somewhat farther away, e.g., titanium and niobium, have less solubility and are more apt to generate interstitial compounds with small-diameter elements like carbon and nitrogen.

3. Going from *right to left,* the following elements increase in their tendency to form carbides:

   Titanium Vanadium Molybdenum Tungsten Chromium Manganese

   On the other hand, reading from *left to right,* these same elements have an increasing tendency to dissolve in ferrite.

4. Nickel, silicon, and aluminum don't form carbides but tend to graphitize the carbon.

5. The alloying elements must be dissolved in the austenite to affect hardenability. If the element forms a carbide or nitride, it is not available for hardenability purposes, and if a carbide is formed, the carbon is not available to harden the martensite.

6. There are four major groups or types of alloy steels:
  a. The AISI/SAE alloy steels (see Table VIII-1)
   - These steels are classified by their composition.
   - Their classification is indicated by a four-digit identification number, e.g., 4340.
   - The first two digits of that number indicate major alloying element(s), the last two the carbon content in points.
   - The characteristic alloying elements of these steels are chromium, nickel and molybdenum.
   - These steels are generally intended to be hard and tough.
  b. The AISI/SAE tool steels (see Table VIII-2)
   - These steels are classified in overlapping categories based on the type of quench, the alloying element, the application, and specific properties or characteristics.
   - They are identified by letters and numbers, e.g., H13.
   - Their characteristic alloying elements include chromium plus other carbide formers like vanadium, tungsten, and molybdenum.
   - These steels are generally intended to be hard and wear-resistant and to have resilience rather than toughness.
  c. The HSLA steels (see Table VIII-3)
   - These steels are loosely classified, usually by application.
   - They are identified by their yield strength in ksi, e.g., 050.
   - Some types are given their properties by controlling grain size during hot processing; others are heat-treated. Dual-phase steels are heat-treated intercritically and have continuous, rather than discontinuous, yielding.
   - Characteristically these steels have low carbon and alloy content and are microalloyed with carbide or nitride formers (e.g., vanadium, niobium, aluminum, or nitrogen) to control grain size.
  d. Highly alloyed steels (see Table VIII-4) which are generally classified by composition but are not a well-defined grouping
   (1) The AISI stainless steels
    - Their identification system uses a three-digit number; the first digit is meaningful, but the last two are merely identifiers.
    - The three types of AISI stainless steels are classified by both microstructure and composition:
      * austenitic-FCC, which is nonmagnetic when annealed, is non-heat-treatable, is the most corrosion-resistant of the three types, and is designated as either 2xx (chromium + nickel + manganese) or 3xx (chromium + nickel)
      * ferritic-magnetic, which is non-heat-treatable and is designated as 4xx (high chromium content)
      * martensitic-magnetic, which is heat-treatable, is the hardest and the least corrosion-resistant of the three types, and is designated as either 4xx (high chromium + carbon) or 5xx (low chromium + carbon)
    - The characteristic alloying element of these steels is chromium.

    (2) Hadfield's (austenitic-manganese) steel
- The various versions of this steel are identified by composition.
- This steel is noted for its austenitic structure, which becomes martensitic with work-hardening.
- The characteristic alloying elements are large amounts, e.g., > 10%, manganese with high carbon, e.g., > 1.00%.

    (3) Precipitation-hardened stainless steels
- These steels are identified by their chromium-nickel composition, e.g., 17–7PH.
- They harden by precipitation.
- Their characteristic alloying elements are chromium and nickel, with other elements added to create the precipitate.

    (4) Maraging steels
- The various versions of this steel are identified by composition.
- The characteristic alloying element is large amounts (18 to 25%) of nickel.
- This steel heat-treats to a soft, tough martensite, which can be formed or machined.
- The steel's strength is obtained by a low-temperature aging treatment.

7. The costs of all materials should be known and monitored, but the real costs associated with the use of a tool made with any particular steel is often difficult to determine since normal accounting systems do not collect data that permit analysis. The emphasis of cost control should be on the overall cost to manufacture the product.

REVIEW QUESTIONS

1. Where on the periodic table (Table I-1) are the elements that are most likely to alloy with iron? What is the percentage difference between the diameters of their atoms and the atomic diameter of iron? Which elements are most likely to form carbides? Where are they in relation to iron on the periodic table?

2. Where can alloying elements be found in steels?

3. Describe the common designation systems used with steels, especially those used for carbon steels, alloy steels, tool steels, and "stainless" steels.

4. What are the major reasons the following alloying elements are added to steels? What characteristic(s) do they impart to steels?

    a. Manganese Silicon   d. Tungsten Molybdenum Vanadium
    b. Nickel   e. Titanium Niobium
    c. Chromium   f. Aluminum
                           g. Nitrogen

5. Use the iron-nickel and iron-chromium phase diagrams to explain how nickel and chromium influence the properties of highly alloyed steels.

6. Describe the mechanisms by which alloying elements contribute to the strength or hardness of steels.

7. Discuss the four major groups of alloy steels described in this chapter. How are they similar in terms of alloying elements used? How do they differ? How are their properties similar? How do they differ? How do these differences relate to their intended uses?

8. List as many reasons as you can for adding alloying elements to steels.

FOR FURTHER STUDY

Avner, Sidney H. *Introduction to physical metallurgy* (2d ed.). New York: McGraw-Hill, 1974, chaps. 9 and 10.

Bain, Edgar C. *Functions of the alloying elements in steel*. Cleveland: American Society for Metals, 1939.

Carpenter matched tool and die steels. Reading, PA: Carpenter Technology Corp., 1985.

DeGarmo, E. Paul, J. Temple Black, and Ronald A. Kohser. *Materials and processes in manufacturing* (7th ed.). New York: Macmillan, 1988, 133–49.

HSLA steels technology and applications. Metals Park, OH: American Society for Metals, 1984.

Kot, R. A., and B. L. Bramfitt (eds.). *Fundamentals of dual-phase steels*. New York: Metallurgical Society of AIME, 1981.

Payson, Peter. *The metallurgy of tool steels*. New York: Wiley, 1962.

SAE Handbook. Warrendale, PA: Society of Automotive Engineers, 1990, vol. 1, secs. 1, 2, 7, and 10.

Smallman, R. E. *Modern physical metallurgy* (4th ed.). London: Butterworth, 1985, secs. 12.7.1–12.7.5.

C H A P T E R   9

Cast Irons

OBJECTIVES

After studying this chapter, you should be able to

- define and describe the types of cast irons
- describe the processes and/or treatments used to obtain each type
- describe the microstructures (carbon form and matrices) of each type and how the matrices can be heat-treated
- describe the properties, general characteristics, and major uses of each type

DID YOU KNOW?

The metal-casting process has been known for thousands of years, and it is still one of the most important industrial processes used today. In the mid-1980s the United States had over 3000 foundries producing approximately 14 million tons of castings per year in all metals. In the 1970s the volume of cast iron was as high as 17 million tons, though environmental and market problems have reduced that to the current level of about 10 to 12 million tons per year.* Although the volume is down, this is still a large output, and most ground vehicles and industrial machines still make extensive use of iron castings.

A PREVIEW

Cast iron is introduced as a steel **matrix** containing carbon in different amounts and forms than contained in steel itself. We find that the various types of cast iron are dependent on the form the carbon takes in the steel matrix. We discuss the processes used to obtain each of the forms and subsequent treatments given them, and we relate the properties and uses of cast irons to their microstructures.

Metals handbook (9th ed.). Metals Park, OH: ASM International, 1988, vol. 15, 37.

9.1   PHASE DIAGRAM FOR CAST IRONS

The iron–iron carbide phase diagram was discussed in Chapter 6; a review of Section 6.2 will probably be helpful before you proceed with this chapter.

Although the diagram shown in Figure 9-1 (Figure 6-2 modified for cast irons) is commonly considered to indicate phases stable in its various fields under equilibrium conditions, cementite is actually unstable over long periods at elevated temperatures and tends slowly to decompose, giving carbon as graphite and iron. The temperatures and compositions of the true iron-graphitic carbon diagram are very close to those shown in Figure 9-1, so that figure will be used.

The decomposition of cementite is negligible in relatively pure iron-carbon alloys and steels, but it is greatly encouraged by the presence of silicon—a common ingredient of pig iron that is largely removed in steel-making furnaces but remains or is added to cast irons. *Cast irons should be considered ternary alloys of iron, carbon, and silicon* rather than just binary alloys of iron and carbon.

F I G U R E 9–1   Iron–iron carbide phase diagram labeled for white cast iron

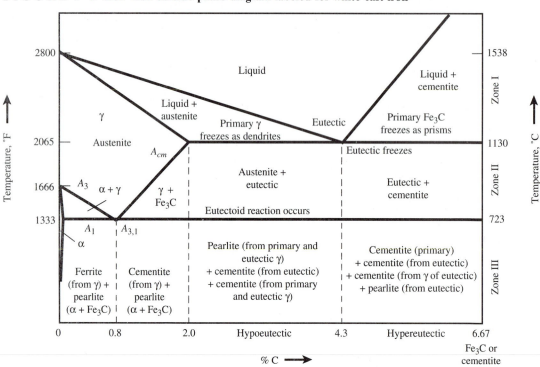

Source: Clyde W. Mason. *Introductory physical metallurgy.* Materials Park, OH: ASM International, 1947, fig. 50, 86. Used by permission.

9.2 DIFFERENCES AND SIMILARITIES BETWEEN CAST IRON AND STEEL

Most cast irons have about 2.5 to 3.5% carbon; it can be seen from the phase diagram that their freezing range is greater than that of steels, and since they contain the eutectic, their minimum freezing temperature is lower than that for steels. Cast irons are thus very fluid in the molten state; they are cast directly into molds and can form complicated shapes. As the name suggests, they form cast, not wrought, products.[†]

Although carbon is soluble in liquid iron, solid solubility occurs only in the α-ferrite and γ-austenite regions. This means that all the carbon will *not* be in solution for any iron that freezes with a carbon content over the 2.0% maximum solubility of γ-austenite. This is the characteristic that gives cast irons the properties that differentiate them from steels. In steels carbon can only be in the form of Fe_3C-cementite or "combined carbon." *In cast irons carbon can be Fe_3C-cementite, or graphite, or both.*

Figure 9-1 shows that, upon freezing, the *hypoeutectic* cast irons will contain the eutectic mixture and dendrites of the γ-austenite solid solution (see Figure A-25). So while there are differences between steel and cast iron, there are also similarities: Both contain γ-austenite upon solidification.[‡] If the γ-austenite of steel can be alloyed and heat-treated to different structures, so can that of cast irons. Cast irons should be thought of starting as an austenite (or steel) matrix *mixed* with carbon in various forms or shapes.

9.3 TYPES OF CAST IRONS

There are four general types of cast irons—white, gray, nodular (or ductile), and malleable. Each of the types relates to the form of the carbon in the steel matrix. **White cast iron** gets its name from the presence of large quantities of Fe_3C-cementite, which is white in appearance. In **gray iron** the carbon takes the form of flakes, which give the metal an overall gray color. **Nodular iron** is **inoculated** immediately before casting so that the carbon forms into spheroids, or nodules. **Malleable iron** is produced by heat-treating white cast iron; holding white iron at temperatures above $A_{3,1}$ for long periods of time causes the Fe_3C-cementite to form into jagged clumps, or **temper carbon.**

Figure 9-2 is an overview of how the four types are formed. Notice that whether an iron is white, gray, or nodular is determined during freezing; only malleable iron is formed by later treatment. Also notice the interrelationship of carbon content, silicon content, and cooling rate.

[†]Wrought products are rolled, forged, extruded, and drawn.
[‡]Commercial cast irons are hypoeutectic, but the general category includes hypereutectic irons as well.

FIGURE 9–2 The cast-iron portion of the phase diagram, showing the phases and structures that exist in various zones, along with the room-temperature microstructures

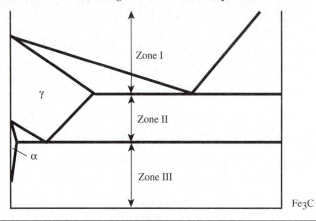

Given	Low carbon Low silicon Fast cool	Medium carbon Medium silicon Moderate cool		Higher carbon Higher silicon Slower cool	
Zones	White CI	Pearlitic		Ferritic	
		Gray	Nodular	Gray	Nodular
I	$\gamma + L$	$\gamma + L$	$\gamma + L + \text{Mag}$	$\gamma + L$	$\gamma + L + \text{Mag}$
II	$\gamma + \text{Cem}$	$\gamma + G_f$	$\gamma + G_n$	$\gamma + G_f$	$\gamma + G_n$
III	$P + \text{Cem}$	$P + G_f$	$P + G_n$	$\alpha + G_f$	$\alpha + G_n$

Malleable		Given	Fast cool	Slow cool
Reheat: Hold for long period in Zone II		Zones	Pearlitic	Ferritic
		II	$\gamma + G_t$	$\gamma + G_t$
		III	$P + G_t$	$\alpha + G_t$

P   = Pearlite
Cem = Cementite
G_f = Graphite flakes
G_n = Graphite nodules
G_t = Graphite temper carbon

Source: William G. Moffatt, George W. Pearsall, and John Wulff. *The structure and properties of materials*, vol. 1: Structure. New York: Wiley, 1964. Reprinted by permission.

9.4 WHITE CAST IRON

Formation. If silicon is low (1% or less), and cooling fairly fast, cast irons show the phases indicated by the iron–iron carbide diagram (Figure 9-1). A melt of, for example, 2.5% C begins to freeze as dendrites of the solid-solution γ-austenite (zone I of Figure 9-2). As freezing continues, the melt is enriched in carbon and its composition varies toward the eutectic, **ledeburite** (see Figure A-27). The rapid diffusion rate of carbon in γ-austenite at such a high temperature prevents coring to any great extent.

At the eutectic temperature, the remaining melt solidifies between the dendrites of γ-austenite as a fine-grained mixture of eutectic and γ-austenite; the zone II phases are γ-austenite and Fe_3C-cementite. As cooling continues, the γ-austenite undergoes the changes already described for high-carbon steels; that is, as the end of the tie line follows A_{cm} toward the eutectoid point, carbon is rejected from the γ-austenite as Fe_3C-cementite, adding to the Fe_3C-cementite that originally formed part of the eutectic. When the temperature reaches $A_{3,1}$ (zone III) the remaining γ-austenite transforms to pearlite, and the structure is made up of Fe_3C-cementite and pearlite. Figure 9-1 identifies the sources of the structures.

Cementite. The photomicrographs in Figures A-26, A-27, and A-28 show the presence of the Fe_3C-cementite, which has such an influence on the properties of the white cast irons. Because of the large amounts of Fe_3C-cementite, these irons are very hard (500 HB), brittle (0% elongation in tension), and not machinable. They are weak in tension (30,000 psi TS), but strong in compression (225,000 psi). They break with a bright fracture and are thus termed "white irons."

The principal use of white cast iron is as a starting material for malleable iron, which will be discussed later.

9.5 GRAY CAST IRON

Formation. If enough silicon (1.5% or more) is added to the 2.5% C iron considered previously, *the carbon precipitated during the freezing process forms as graphite flakes rather than as Fe_3C-cementite.* The freezing process (zone I) starts as in the white cast iron, with γ-austenite dendrites being formed. The FCC γ-austenite takes carbon into interstitial solid solution just as it did in the white cast iron, but now at the eutectic temperature graphite flakes are formed instead of Fe_3C-cementite. If any Fe_3C-cementite is formed, it is quickly graphitized under the influence of the silicon.

At the eutectic temperature the only phases present are γ-austenite and graphite. Since the solubility of carbon in γ-austenite decreases with temperature, as cooling proceeds (zone II) the γ-austenite must give up its carbon. Because of the silicon, the liberated carbon either attaches to the already formed flakes or forms new flakes. Thus the metal arrives at the eutectoid temperature, $A_{3,1}$, as γ-austenite and graphite flakes.

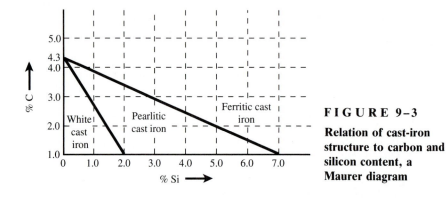

F I G U R E 9–3

Relation of cast-iron structure to carbon and silicon content, a Maurer diagram

Pearlitic Gray Cast Iron. At $A_{3,1}$, γ-austenite must transform to α-ferrite, which has almost zero solubility for carbon. Since the carbon cannot go into the α-ferrite, there are only two other phases available—*graphite* and *Fe_3C-cementite*. If cooling proceeds at a relatively rapid rate, pearlite (α-ferrite and Fe_3C-cementite) is formed (zone III). The gray cast iron thus has a microstructure of pearlite surrounding graphite flakes and is called *pearlitic gray cast iron* (see Figure A-29).

Ferritic Gray Cast Iron. However, if cooling through the $A_{3,1}$ region is slow and the silicon content is high, formation of the Fe_3C-cementite in the pearlite is arrested and additional graphite flakes are formed. The microstructure will consist of nothing but graphite flakes surrounded by α-ferrite and is termed *ferritic gray cast iron* (see Figures A-30 and A-31).

Microstructure and Properties. From the above discussion it is evident that the final microstructure of gray cast iron is determined by alloy content *and* cooling rate. Figure 9-3 shows the relationship between structure and carbon and silicon content for relatively thin sections (i.e., those having fast cooling rates) for gray irons. The strongest gray irons are those that lie in the middle of the diagram; for example, 2.75% carbon plus 1.5% silicon produces a pearlitic gray iron with a tensile strength of over 40,000 psi. If carbon and silicon content are reduced, say to 2.5 and 0.5%, respectively, a brittle white cast iron is produced with a tensile strength under 30,000 psi. If the carbon and silicon contents are increased, say to 3.0 and 4.5%, respectively, a softer ferritic gray iron with a tensile strength of about 25,000 psi is produced. Gray irons have elongations of 1% or less; they are not very ductile.

9.6   NODULAR CAST IRON

Commercially nodular cast iron is usually called **ductile iron;** the name *nodular* is used here because it better describes the microstructure and the process by which the iron is made. The term *ductile* is used commercially because it identifies the iron by its major characteristic, ductility. Although nodular cast irons are not as ductile as low-carbon steels, elongations of 18% at tensile strengths of 60,000 psi are possible, which is much more ductility than possible with white or gray irons.

Formation. A study of Figure 9-2 shows that the results of freezing and cooling a nodular iron are the same as achieved for a gray iron, except that the form of the graphite is nodular, or spheroidal or "roundish," instead of in flakes. As in gray iron, silicon is used to graphatize the carbon; the nodular shape of the carbon is accomplished by inoculating the melt with a controlled amount of magnesium[§] immediately before casting.

Other than this difference in the form of carbon, the nodular and gray irons go through the same processes in freezing and cooling. High silicon content and slow cooling tend to produce a ferritic matrix; lower silicon and higher cooling rates tend to produce a pearlitic matrix.

Because the elements used to form the nodules have a strong affinity for sulfur the melt must be "desulfurized" before inoculation.

9.7   MALLEABLE CAST IRON

In this discussion we've considered the Fe_3C-cementite of white cast iron to be a stable phase, and under normal conditions it stays in this form. However, at elevated temperatures it is less stable; this instability is the basis of malleable iron.

Pearlitic Malleable Cast Iron. When a white cast iron is heated to $A_{3,1}$, the pearlite transforms to γ-austenite; initially the phases present above $A_{3,1}$ are γ-austenite and Fe_3C-cementite. If the cast iron is held for a long period of time, e.g., 30 h, at about 1700°F (925°C), i.e., zone II of Figure 9-2, the Fe_3C-cementite breaks down into its iron and carbon elements. As a result of this first stage of the annealing process, the carbon of the Fe_3C-cementite forms irregularly shaped "clump graphite" or temper carbon.[‖] If the iron is cooled rapidly to room temperature, the γ-austenite will change back into pearlite. The resulting structure, which consists of graphite temper carbon in a pearlite matrix, is called *pearlitic malleable iron.*

Ferritic Malleable Cast Iron. It is more typical that, instead of the rapid cool, a second step is used in the anneal. From 1700°F (925°C) the iron is cooled at a very slow rate through the $A_{3,1}$ temperature; this causes the carbon that is rejected from the cooling γ-austenite to join the previously formed temper carbon. That is, because of the slow cooling rate, instead of forming pearlite when $A_{3,1}$ is crossed,

---

[§]Magnesium is generally used, though cerium, calcium, lithium, and sodium, among others, will also spheroidize carbon. See *Metals handbook* (8th ed.). Metals Park, OH: American Society for Metals, 1961, vol. 1, 379.

[‖]It is called "clump" because of its irregular shape, "temper" because it forms during the tempering or annealing process.

this carbon becomes graphite and joins the previously formed temper carbon. Such a two-stage annealing treatment results in a completely ferritic matrix, and the product is called *ferritic malleable iron* (see Figure A-32).

Microstructure. To control the properties of malleable irons, it is important that the original casting be *all white iron*. If during casting heavy sections cool too slowly, or if the carbon and/or silicon content are too high, some parts of the casting may be gray iron and be weaker after malleablizing. The requirement that the *whole* casting be initially white iron places limitations on the thickness of malleable castings.

9.8   ANNEALING, HEAT-TREATING, AND ALLOYING CAST IRONS

Since cast irons have a matrix of steel, the thermal processes used with steels also apply to cast irons. Castings are annealed to reduce residual stresses and/or to soften the casting for machining, for example. Castings are heated into the austenitic range and cooled at various rates to achieve annealed, normalized, or quenched structures and properties.

Proper alloying and quenching can form martensitic structures, which are tempered to reduce brittleness. Alloying elements that contribute to the hardenability of steels do the same thing for cast irons, and Jominy curves for a cast iron can be used to evaluate hardenability just as in steels.

Stress-Relief Annealing. Stress-relief annealing is perhaps the most common heat treatment of cast irons. The casting process and casting design cause cooling rates that vary throughout the casting, prompting residual stresses similar to those discussed in the section on quenching in Chapter 7. Heating to 1000 to 1200°F (540 to 650°C), holding, and slow-cooling relieves these stresses. This can be important if the casting undergoes later machining or is used in a corrosive environment.

Annealing and Normalizing. Annealing and normalizing require the casting be heated above $A_{3,1}$, to about 1600 to 1700°F (870 to 925°C). Whether annealing or normalizing occurs depends on the cooling rate from that temperature, annealing being a slower (furnace) cool and normalizing being an air cool. Since the production of malleable cast irons involves a reheat and slow cool of the white iron casting, this material should not need annealing.

Hardening. With suitable hardenability all the cast irons can be quenched to achieve harder, stronger structures of pearlite, bainite, and martensite. It must be recognized that an unalloyed cast iron has the hardenability of a plain carbon steel, but with alloying elements the matrix can be transformed to all martensite if desired. This process involves heating the casting to above $A_{3,1}$, e.g., 1600 to 1700°F (870 to 925°C), and cooling in a medium that achieves the desired structure.

The same concepts and principles that were of use in understanding the heat-treating of steel can be applied to cast irons. TTT diagrams and the Jominy test, for example, pertain to cast irons as well as steels.

An austempered nodular iron is a recent product of this technology. Given the name **austempered ductile iron,** or **ADI,** it was introduced in the 1950s and has become a very popular form of cast iron. The structure is essentially that of a bainitic steel matrix with nodular carbon. The austempered product can have a tensile strength twice as high as regular ductile iron with the same toughness. Those irons that go through the austenite-to-bainite transformation at higher temperatures have higher ductility, toughness, and fatigue strength and are useful in structural applications. Those that transform at lower temperatures, closer to M_s, are harder and stronger and are useful for such things as gears.

9.9 PROPERTIES AND USES OF CAST IRONS

As the various types of cast irons were discussed, some mechanical property data was presented for comparison purposes. Now we'll summarize that information, present some additional data, and identify some of the uses of cast irons. The information is summarized in Table IX-1.

In discussing mechanical properties it is helpful to remember that the graphitic cast irons are essentially steels with variously shaped holes filled with soft graphite. The graphite contributes nothing to strength or hardness; those properties are dependent on the steel matrix. On the other hand, the shape of the graphite, e.g., flake versus sphere, can create stress concentrations that affect the fatigue properties of the irons.

Modulus of Elasticity. The modulus of elasticity of gray iron is not linear; it varies with the tensile strength. By convention it is measured as the slope of the line that intersects the stress-strain curve at 25% of the tensile strength. For the weaker ferritic-type irons, it can be about 10×10^6 psi; for the stronger alloyed gray irons it can be over 20×10^6 psi.

The elastic region of the stress-strain diagram for white, nodular, and malleable irons is linear; all three have a modulus of elasticity of about 22 to 28×10^6 psi.

Resilience. Since the modulus of resilience is inversely related to the modulus of elasticity (see Section B.6), cast irons will be better *elastic* energy absorbers than steels. One of the major uses of gray iron is as a machine base because it is a good vibration damper. Nodular and malleable iron are not as good as gray iron in this regard, but they are still better than steel. The graphite-filled openings in cast irons interrupt the transmission of the vibrations.

Tensile Properties. The tensile strengths of white and gray irons are not very high, being in the range of 40 to 60 ksi. Cast irons exhibit no yield point, as do some low-carbon steels, so the yield strength must be found using the 0.002 in./in.

TABLE IX–1 Typical properties of white, gray, nodular (ductile), malleable, and special cast irons

Type & Designation	Composition, %	$E \times 10^6$ psi	Compressive Strength, ksi	Tensile Strength, ksi	Yield Strength, ksi	Elongation, %	Hardness, HB	Typical Uses	Source
White Unalloyed	3.5 C, 0.5 Si	24–28	225	40	40	0	500	Wear-resistant parts; as starting stock for malleable cast iron	2, 3
Gray Class 20 Ferritic, a.c.*	3.5 C, 2.4 Si	12	83	22	18	0.4	160	Pipe; sanitary ware; small, thin castings	2, 3
Class 40 Pearlitic, a.c.*	3.2 C, 2.0 Si	18	140	40	30	0.4	220	Machine tools, motor blocks, gear blanks	2, 3
Quenched Bainitic	3.2 C, 2.0 Si 1.0 Ni, 1.0 Mo			70	70	0	300	Camshafts	2
Quenched Martensitic	3.2 C, 2.0 Si 1.0 Ni, 1.0 Cr, 0.4 Mo			80	80	0	500	Wearing surfaces	2
Nodular 60-40-18 Ferritic, ann.*	3.0 C, 2.5 Si	24.5	50†	66.9	47.7	15	160	Spun heavy-duty pipe	1, 2
80-55-06 Pearlitic, a.c.*	3.5 C, 2.2 Si	24.4	56†	81.1	52.5	11.2	190	Crankshafts	1, 2

Type / Grade	Composition							Application	Sources
120-90-02 Mart. Q/T*	3.50 C, 2.2 Si	23.8		141	125.3	1.5	331	High-strength machine parts	1, 2
Malleable 35018 Ferritic	2.4 C, 1.25 Si	25		53	38	18	130	General heat resistance, machinability	1, 3
45010 Pearlitic	2.3 C, 1.25 Si	25.5–28	220	65	45	10	180	Couplings	1, 3
80002 Mart. Q/T*	2.2 C, 1.0 Si	~25		100	80	2	250	Connecting rods, U-joint yokes	2, 3
Special Austenitic Gray, a.c.*	3.0 C, 2.0 Si, 20 Ni, 2.0 Cr			30	30	2	150	Exhaust manifolds	2
Austenitic Ductile, a.c.*	3.0 C, 2.0 Si, 20 Ni			60	30	20	160	Pump casings	2
Martensitic White, h.t.*	2.7 C, 0.8 Si, 20 Cr, 2 Mo, 1 Ni			80	80	0	600	Wear-resistant parts, liners	2

*a.c. = air-cooled; ann. = annealed; Q/T = quenched and tempered; h.t. = heat-treated.

†Yield strength.

Sources: 1. *ASM metals reference book* (2d ed.). Materials Park, OH: ASM International, 1983, 166–76.

2. Richard Flinn and Paul Trojan. *Engineering materials and their applications* (3d ed.). Boston: Houghton Mifflin, 1986, table 6.6, 270.

3. *Metals handbook* (8th ed.). Materials Park, OH: ASM International, 1961, vol. 1, 349–406.

or 0.2% offset. For most white and gray cast irons there is very little deformation before fracture, so the yield strength and tensile strength are often shown as the same, with zero elongation. A rule of thumb for as-cast gray iron is that the yield strength will be 85% of the tensile.

The tensile strength of as-cast nodular and annealed malleable irons is much higher than that of white or gray irons; the former irons are as much as twice as strong, i.e., 50 to 80 ksi. Since their elongations are higher (10 to 20% compared with zero), they show deformation before fracture and have yield strengths that are 60 to 80% of their tensile strengths.

Hardness.   The hardness of cast irons is directly related to the tensile strength, although the relationship is not as linear as it is for the low-carbon steels. Hardness readings can vary from 150 to 600$^+$ HB. Because so much of the microstructure consists of a soft material, graphite, an averaging-type hardness reading like the Brinell may not give a good indication of the "effective" hardness. For example, a 400-HB reading for a graphitic cast iron may hide the fact that the hard matrix is actually 600 HB. Thus many cast iron parts wear very well. The imbedded soft graphite can also act as a lubricant for the hard matrix, which is a contributing reason for the frequent use of gray cast iron in engine piston rings.

Classification.   The graphitic cast irons are generally classified by mechanical properties. Gray irons are divided into classes by the ASTM system; e.g., class 25 has a minimum tensile strength of 25 ksi. The SAE system classifies this iron as G2500.

The ASTM classification systems for ductile irons uses the minimum tensile and yield strengths (ksi) and elongation, respectively, e.g., 60-40-18. In the SAE system the same properties are designated by "D" for ductile; this system uses only the yield strength and elongation: D4018.

The systems for malleable irons use only the minimum yield strength and elongation but without the hyphens; a malleable iron with a yield strength of 32 ksi and 10% elongation would be referred to as 32510 by ASTM and M3210 by SAE.

Applications.   As mentioned previously, the principal use of white cast iron is as a starting material for malleable iron. As white iron it is used for any wear application, or as a chilled area of another type of iron where wear is to occur.

Gray iron is best known for its use in engine blocks; its vibration damping and self-lubricating properties, plus its low cost, make it well suited to this use. Gray iron is also in common use in a wide variety of applications—machine tools, sanitary pipe, fire hydrants, firebox and stoker parts, brake drums, clutch plates, pistons, and piston rings.

Ductile or nodular iron is used for spun-cast pipe and engine parts, e.g., crankshafts, gears, and cams. It is replacing wrought or cast steels and other metals in many applications because of its good strength-to-weight ratio, damping capacity, machinability, and castability. Because ductile iron can be made in thicker sections, it is being applied where malleable iron cannot be used. ADI, or austempered ductile iron, is increasingly used for gears and other critical parts where steels were used previously.

Although ductile iron is advantageous for thick sections, malleable iron is good for thin sections and where machinability is critical; the maximum thickness produced is dependent on the foundry but is usually from 1-½ to 4 in. Malleable iron is often used for connecting rods, valves, pipe, and general hardware; it has many of the same uses as ductile iron.

Many uses are found for alloyed versions of all the types of cast iron where heat and corrosion must be resisted.

SUMMARY

1. Cast irons are tertiary alloys of iron, carbon, and silicon. Most cast irons contain 2.5 to 3.5% carbon.

2. Cast irons are essentially steels containing more carbon that γ-austenite can dissolve. The excess carbon precipitates within the γ-austenite as Fe_3C-cementite or graphite; the form of the precipitated carbon determines the type of cast iron.

3. In white cast iron the excess carbon is in the form of Fe_3C-cementite.

4. In gray and nodular (ductile) cast irons the carbon is in the form of graphite flakes and nodules, respectively.

5. Which of these three forms the cast iron takes is determined when the cast iron freezes. (See Figure 9-3.)
   a. If carbon and silicon contents are low and the cooling rate is fairly fast, the iron will be white.
   b. If carbon and silicon contents are higher, the iron will be gray.
   c. If magnesium, for example, is added to an iron of higher carbon and silicon content, the iron will be nodular (ductile).

6. The cooling rate and alloy content determine the matrix, or background structure, in which the excess carbon exists. Above $A_{3,1}$ this matrix is γ-austenite.
   a. If carbon and silicon contents are moderate and the cooling rate moderately fast, the γ-austenite will transform into pearlite at $A_{3,1}$ and the matrix will be pearlitic.
   b. If carbon and silicon contents are higher and the cooling rate slow, the γ-austenite will precipitate its carbon on the flakes or nodules already formed, and the matrix will be α-ferrite.
   c. If alloying elements are added to increase the hardenability of the γ-austenite and it is then properly quenched, the matrix can become martensitic or bainitic.

7. Malleable cast iron is produced by heat-treating white cast iron.
   a. Heating above $A_{3,1}$ transforms the white iron into γ-austenite and Fe_3C-cementite; holding above $A_{3,1}$ allows the Fe_3C-cementite to break down

into iron and carbon, with the carbon taking the form of irregularly shaped clumps called *temper carbon.*

   b. If the iron is cooled rapidly to room temperature, the γ-austenite will transform to pearlite at $A_{3,1}$ and the matrix will be pearlitic.

   c. If the iron is cooled slowly through the $A_{3,1}$ temperature, the γ-austenite will also break down into iron and temper carbon, and the matrix will be α-ferrite.

   d. One limitation of the malleable iron process is that the starting material must be white iron; any carbon not initially produced as Fe_3C-cementite will not be converted to temper carbon.

8. Cast irons serve a number of useful purposes; some of these are outlined in Table IX-1 and are discussed in Section 9.9.

9. Austempered ductile iron, or ADI, is a relatively new treatment of ductile iron that shows much promise. It yields products with very high strength, hardness, and ductility.

REVIEW QUESTIONS

1. Graphitic cast irons are ternary alloys of iron, carbon and what other element? What is the purpose of adding this other element?

2. How are cast irons classified?

3. Describe the processes or treatments used to produce each of these classifications.

4. Explain, in terms of processes and microstructures, how gray cast iron and nodular or ductile cast iron are similar and how they are different.

5. What are the similarities and differences between malleable and nodular cast iron in terms of processes and microstructures?

6. Which portion of the microstructure of the graphitic irons determines their *mechanical* properties? Explain.

7. What are the advantages of ADI, austempered ductile iron, over other cast irons?

8. Which should be capable of producing the greater section thickness, malleable cast iron or nodular cast iron? Why?

9. What are the characteristics of gray cast iron that make it so popular for automotive engine blocks?

10. Which of the types of cast iron would you choose in producing a manhole cover, an automobile crankshaft, a base for a machine tool, a woodworking plane, a wear surface on a machine tool? Explain your choices.

FOR FURTHER STUDY

Alexander, W. O., et al. *Essential metallurgy for engineers.* Wokingham, U.K.: Van Nostrand Reinhold, 1985, secs. 2.16 and 2.16.1.

ASM metals reference book (2d ed.). Metals Park, OH: American Society for Metals, 1983, 166–76.

Avner, Sidney H. *Introduction to physical metallurgy* (2d ed.). New York: McGraw-Hill, 1974, chap. 11.

Flinn, Richard A., and Paul K. Trojan. *Engineering materials and their applications* (3d ed.). Boston: Houghton Mifflin, 1986, secs. 6.29–6.34.

Iron by Brillion, Bulletin No. 486. Brillion Iron Works, 200 Park Avenue, Brillion, WI, 54110, 1987.

Metals handbook (8th ed.). Metals Park, OH: American Society for Metals, 1961, vol. 1, 349–406.

Metals handbook (9th ed.). Metals Park, OH: ASM International, 1988, vol. 15, 627–97.

Nonferrous Metals and Alloys

OBJECTIVES

After studying this chapter, you should be able to

- list applications where nonferrous, rather than ferrous, metals would be used and give reasons for their use
- describe the strengthening methods possible with nonferrous alloys
- describe the phase-diagram configuration that enables the solution heat-treating of heat-treatable nonferrous alloys and give the steps used in the treatment
- describe the principal engineering properties of copper and copper alloys, how levels of these properties are designated, and how the various alloys and tempers are identified
- describe the strengthening processes used for α and α-β brasses and bronzes
- describe the principal engineering properties of aluminum and aluminum alloys and the systems for identifying the various alloys and tempers
- describe how aluminum alloys are strengthened
- explain what occurs during precipitation-hardening, or aging, and the difference between natural and artificial aging
- explain what special controls are used to prevent incipient eutectic melting when heat-treating aluminum alloys
- describe the differences between the aluminum and magnesium alloy designation systems
- contrast the advantages and disadvantages of aluminum and magnesium alloys
- describe the characteristics of nickel and nickel alloys that make their use worth the extra expense

There are many possible nonferrous alloys; this text is able to treat relatively few of them. Many products, especially in the electronics field, are made from unusual nonferrous metals. In the manufacturing, production, and transportation world of engineering, unusual nonferrous metals may be encountered in many products, but most of the equipment, structures, and tooling used will be alloys of steel, copper, aluminum, nickel, and/or magnesium. This chapter concentrates on copper and aluminum, while giving some information on magnesium and nickel. Nickel is discussed in more detail in Chapters 12 and 13, which deal with failure at low and high temperatures, respectively.

A PREVIEW

We'll first cover, in a general way, the applications of nonferrous metals and the methods used to achieve their properties. Then we'll go on to discuss copper and the types of copper alloys and to describe how the alloys and tempers are identified. We'll then use composition and property data from commercial alloys to investigate the metallurgy behind the alloys and processes.

We'll follow approximately the same pattern for aluminum and its alloys— how the alloys and tempers are identified, how the alloys are worked and heat-treated, and how composition and property data can illustrate the principles involved. We'll discuss applications of copper and aluminum alloys where it seems appropriate rather than devote a separate section to a description of their uses.

Aluminum is a good introduction to the section on magnesium; the same format is used in this later section, but we'll go into less detail with magnesium. The discussion on nickel and nickel alloys places emphasis on applications because the metallurgical principles involved are similar to those already discussed.

10.1   USES AND PROPERTIES OF NONFERROUS METALS

Ferrous metals are those based on iron. Thus nonferrous metals are all the metal elements in the periodic table other than iron. Often ferrous metals contain large percentages of "nonferrous" metals, but as alloying additions. In similar fashion, some nonferrous metals may contain iron, but as an alloying element.

Given the large number of metals considered nonferrous, it should be expected that they would find a broad range of applications. Observations made in a home or hardware store, at an airport, or on the highway immediately identify some of these applications.

Below you'll find two lists of properties that suggest reasons for the use of the very adaptable nonferrous metals. The list on the left contains properties where the application may require either a high or low value. For example, nonferrous metals can be made with either very high or very low electrical conductivity; i.e., they can be made very conductive or very resistive. The list on the right contains properties that are desirable as stated. These lists are not complete but they are indicative of the many reasons nonferrous metals are used.

Property, desirable at varying levels	Property, desirable as stated
Specific gravity (density)	High strength-to-weight ratio
Magnetic capability	High strength at low temperatures
Coefficient of friction	High strength at high temperatures
Electrical conductivity	Ease of alloying
Thermal conductivity	Nonsparking
Chemical activity	Machinable
Thermal expansion	Easily formed, joined, fabricated
Modulus of elasticity	Pleasing appearance
	Low melting point
	Liquid at room temperature

Since ferrous metals are generally less expensive than nonferrous, nonferrous metals are usually only considered if a ferrous metal won't do the job. For many applications low-carbon steels are very hard to beat; however, when special properties are required, nonferrous metals are often the cost-effective approach.

10.2   CONTROLLING PROPERTIES IN NONFERROUS METALS

Lack of Allotropy. In Chapter 6 we found that ferrous metals, e.g., steel, are unusual in that they undergo an allotropic change that is central to their ability to be heat-treated for a broad range of uses. Some nonferrous metals exhibit allotropic change, but this capability does not play an important part in their use. Thus *the mechanical properties of nonferrous metals are controlled primarily through cold-working, annealing, alloying, and/or solution heat-treating and precipitation-hardening.*

Cold-Working, Annealing, and Alloying. A number of wrought nonferrous alloy systems are hardened through a combination of cold-working and alloying. Because of their commercial importance the copper and aluminum systems are discussed in detail later in this chapter. The general metallurgical principles behind these treatments were covered previously—strengthening by cold-working in Chapter 2, annealing in Chapter 3, and solid-solution strengthening in Section 4.8.

Solution Heat-Treating and Precipitation-Hardening. To be heat-treatable, the nonferrous alloy systems must be of the *partially soluble–insoluble* type; that is, the alloys must have the potential to *form supersaturated solid solutions* by rapid quenching from a one-phase region into a mixture region. This means that the solvus line must slope so that the solubility of the precipitating phase decreases with temperature and that the precipitating phase must be such that it hardens the metal. This subject was dealt with in Sections 5.4 and 5.10; see especially Figure 5-23.

Copper, aluminum, nickel, and magnesium all form alloy systems with other metals that are partially soluble–insoluble; some of these systems are commercially important and their heat treatment and properties are discussed later in the chapter.

10.3 COPPER ALLOYS

General Characteristics. Copper is a reddish-color metal with approximately the same density as iron; it is ductile and easily formed (FCC atom lattice) and joined. It has very good electrical and thermal conductivities, is nonmagnetic, alloys easily with many other metals, is corrosion-resistant in normal atmospheric and marine environments, and can be used where appearance is important.

Ability to Alloy. Copper has complete solubility (see Figures 5-1 and 5-2) with those FCC metals that have nearly the same atom diameters, e.g., nickel, rhodium, platinum, palladium, and gold. The properties of these binary alloys can only be altered by cold-working, annealing, or changing the ratio of the two metals.

Copper has significant partial solubility (see Figures 5-6 and 5-7) with a number of other elements; some of these are listed in Table X-1. Because of their partial solubility, these alloys have the potential to create supersaturated solid solutions by solution heat-treating.

Brasses. The brass family, both wrought and cast, is composed of alloys made of copper and zinc, with other elements designated if used, e.g., tin brass, leaded brass, and silicon brass. Some cast yellow brasses are known as manganese bronze or leaded-manganese bronze, although the major alloying element is still zinc.

T A B L E X–1 Maximum solubilities of elements in copper

Element	Maximum Solubility, %	Element	Maximum Solubility, %
Beryllium	2.7	Silver	7.9
Cadmium	3.7	Aluminum	9.4
Titanium	4.7	Tin	15.8
Silicon	5.3	Zinc	39.0
Antimony	6.7		

Source: *Metals handbook* (8th ed.). Materials Park, OH: ASM International, 1973, vol. 8.

Bronzes.   At one time the term *bronze* was used to designate alloys of copper and tin. Today the term identifies, with the exception of the brasses just mentioned, an alloy of copper and a metal other than zinc, in both wrought and cast products. Because of the variety of metals used with copper, the other metal is used as a prefix to bronze, e.g., tin bronze, aluminum bronze, silicon bronze, etc. Some bronzes involve copper plus two or more alloying elements, e.g., tin and phosphorus (phosphor bronze); tin, lead, and phosphorus (leaded-phosphor bronze); and tin and nickel (nickel-tin bronze).

Others.   Other alloy systems are formed by copper and other major alloying elements, e.g., nickel (copper nickels), nickel-zinc (nickel silvers), and lead (leaded coppers).

Manufacturing Processes.   Sand, centrifugal, and continuous casting methods are commonly used with copper and its alloys; wrought processes used include rolling, drawing, stamping, forging, and extruding.

10.4   COPPER ALLOYS: ALLOY-TEMPER DESIGNATION SYSTEMS

Standard Designation System.   There are a number of systems used to identify the compositions of copper and copper alloys. Table X-2 outlines the **UNS,** or unified numbering system, the standard used in North America. This system, with a C prefix, expands a previously existing three-digit system to five digits. The table shows the range of numbers that define a category of alloys, the major alloying elements in those alloys, and the common name used to describe that category. The commonly used names for specific alloys often relate to their composition (70-30 brass, tin bronze), color (yellow brass), or application (naval brass, cartridge brass). Wrought alloys have the lower range of number; cast alloys start at C80xxx.

T A B L E X–2 **Unified Numbering System (UNS) for wrought and cast copper and copper alloys**

Copper Alloy Number	Alloying Metals	Common Name
Wrought Alloys		
C10xxx to C19xxx	Cu	Coppers & high-copper alloys
C20xxx to C28xxx	Cu-Zn	Brasses
C31xxx to C38xxx	Cu-Zn-Pb	Leaded brasses
C40xxx to C49xxx	Cu-Zn-Sn	Tin brasses
C50xxx to C52xxx	Cu-Sn	Phosphor bronzes
C53xxx to C54xxx	Cu-Sn-Pb	Leaded-phosphor bronzes
C55xxx	Cu-P & Cu-Ag-P	Copper phosphorus alloys
C60xxx to C644xx	Cu-Al	Aluminum bronzes
C647xx to C661xx	Cu-Si	Silicon bronzes
C664xx to C69xxx	Cu-Zn	Misc. copper-zinc alloys
C70xxx to C72xxx	Cu-Ni	Copper-nickel alloys
C73xxx to C79xxx	Cu-Ni-Zn	Nickel silvers
Casting Alloys		
C80xxx to C82xxx	Cu	Coppers & high-copper alloys
C83xxx to C85xxx	Cu-Sn-Zn-Pb	Brasses & leaded brasses
C86xxx	Cu-Zn-Mn-Al-Fe-Pb	Manganese bronzes
C87xxx	Cu-Si	Silicon brasses & bronzes
C90xxx to C91xxx	Cu-Sn	Tin bronzes
C92xxx to C945xx	Cu-Sn-Pb	Leaded-tin bronzes
C947xx to C949xx	Cu-Sn-Ni	Nickel-tin bronzes
C95xxx	Cu-Al-Fe/Al-Fe-Ni	Aluminum bronzes
C96xxx	Cu-Ni-Fe	Copper nickels
C97xxx	Cu-Ni-Zn	Nickel silvers
C98xxx	Cu-Pb	Leaded coppers
C99xxx	Cu + ?	Special alloys

Source: *Standards handbook, wrought and cast products.* Greenwich, CT: Copper Development Assn., 1988.

Strain-Hardening Temper Designation System. Many of the copper alloys, like the **alpha (α) brasses,** and many of the bronzes, use cold-working to achieve their hardness or strength. These properties are stated as a fraction, in units of a quarter, of a "full-hard" condition—quarter-hard, half-hard, three-quarters hard—and are referred to as *tempers*. Beyond the full-hard and extra-hard temper are five levels of *spring* tempers. The fractional tempers are written ¼ hard, ½ hard, etc. Table X-3 lists the percent cold work required to achieve the tempers shown.

Heat-Treating Temper Designation System. Some of the wrought and cast copper alloys are capable of being strengthened by solution heat-treating and precipitation-hardening. To indicate that a copper alloy has been solution heat-treated

T A B L E X–3 Cold-working designations used for copper and copper alloys

Condition	% Cold Work
Annealed*	0
Quarter hard	10.9
Half hard	20.7
Three-quarters hard	29.4
Hard	37.1
Extra hard	50.0
Spring	60.5
Extra spring	68.7

*Grain size often shown in mm.

Source: *Metals handbook* (8th ed.). Materials Park, OH: ASM International, 1961, vol. 1, 1006.

only the letter "A" is used. If the heat-treating is followed by a precipitation-hardening treatment the letter "T" is added, i.e., AT.

Some copper alloys are further hardened by cold-working after the solution heat treatment and before the precipitation-hardening. In those cases, the cold-working designator is added. A copper alloy that was solution heat-treated, cold-worked to the ½ hard temper, and then precipitation-hardened would be designated ½HT.

Some alloys are quenched and then tempered or stress-relieved; these are referred to as *quenched and tempered.*

10.5   COPPER ALLOYS: HEAT TREATMENTS, PROPERTIES, AND USES

Heat-Treatable Cast and Wrought Copper Alloys. Table X-4 gives the mechanical properties of four cast copper alloys. Although the first one, high-strength yellow brass, is not heat-treatable, it still has the highest properties. The as-cast and heat-treat data for the nickel-tin and aluminum bronzes show the improvement in strength possible by heat-treating.

Figure 10-1 shows the copper-rich end of the copper-aluminum phase diagram. Aluminum bronzes of the type shown in Table X-4 are dependent on this diagram. With compositions of about 10% aluminum, the alloys convert to the beta (β) phase when heated. Because of the sloping solvus line, quenching creates a supersaturated solid solution, but the metal also goes through the eutectoid reaction. (Note the "vee" at the bottom of the beta (β) region.) Thus this nonferrous metal goes through a *martensitic-type reaction,* and in the resulting structure of the alpha (α) and beta (β) phases becomes the needlelike structure seen in Figure A-33, not unlike that of Figure A-21, which is for steel. These bronzes are tempered after quenching.

T A B L E X–4 **Approximate compositions and minimum mechanical properties for selected cast-copper alloys**

UNS No.	Alloy Name	Approximate Composition + Cu	Condition	Tensile Strength, ksi	Yield Strength, ksi	Elongation, %
		Nonheat-Treatable Alloys				
C86200	High-strength yellow brass	25% Zn, 3% Fe, 3.75% Mn, 4% Al	As-cast	90	45*	18
C92500	Phosphor bronze	11% Sn, 0.3% P	As-cast	35	18[†]	10
		Alloys Capable of Heat Treatment				
C94700	Nickel-tin bronze	5.75% Sn, 5.75% Ni, 1.25% Zn	As-cast	45	20[†]	25
			Q/T[‡]	75	50[†]	5
C95300	Aluminum bronze	10% Al, 1.15% Fe	As-cast	65	25[†]	20
			Q/T[‡]	80	40[†]	12

*At 0.2% offset.

[†]0.5% extension under load.

[‡]Quenched and tempered.

Source: *SAE Handbook*. Warrendale, PA: Society of Automotive Engineers, 1990, vol. 1, sec. 10.

The brass alloy system also has a solvus line whose solubility varies with temperature that separates the alpha (α) and beta (β) regions. (See the copper-zinc phase diagram, Figure 5-10.) If the alloy composition is about 40% zinc, the alloy can be heated from the alpha (α)-beta (β) mixture region into an all-beta (β) region; if quenched from this single phase, a supersaturated solid solution of alpha (α) in beta (β) can be created.

Table X-5 gives the compositions and mechanical properties of typical wrought-copper alloys. Those at the top of the table are heat-treatable alloys; those at the bottom are the cold-worked type. Note that the aluminum bronze alloy has a composition similar to the cast alloy listed in Table X-4. (Their properties are not directly comparable, however; Table X-4 data are minimum values, while Table X-5 data are typical values.)

The beryllium-copper alloys* are used in both cold-worked and heat-treated tempers; their properties are compared in Table X-5. When solution-heat-treated and precipitation-hardened (the AT condition), they can achieve quite high strengths; adding cold-working between the two thermal treatments can raise the strength even more (½HT indicates solution heat treatment, ½ hard temper cold

---

*The toxicity of beryllium requires that it be used with care and that it not be used where it may be ingested, e.g., in cooking utensils.

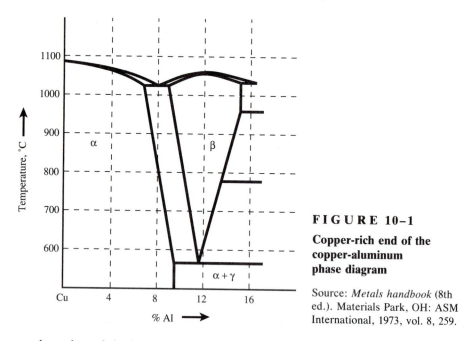

FIGURE 10–1

Copper-rich end of the copper-aluminum phase diagram

Source: *Metals handbook* (8th ed.). Materials Park, OH: ASM International, 1973, vol. 8, 259.

work, and precipitation-hardening.) Note that when heat-treated only they have a higher elongation; i.e., they are a little less brittle.

Nonheat-treated cast bronzes are generally used in applications where their appearance and corrosion resistance are an advantage, e.g., plumbing fixtures, furniture hardware, and radiator fittings. The heat-treated cast and wrought alloys find applications that take advantage of their hardness, higher strength, and wear resistance, e.g., gears, worm wheels, screw conveyors, roller-bearing cages, electric contacts, valve seats, coil and leaf springs, diaphragms and bellows, and spark-resistant tools.

Cold-Worked Brasses.   Perhaps the most familiar use of copper is in the cold-worked brasses. As seen in Figure 5-10, and discussed in Section 5.5, there is a large alpha (α) solid solution area that extends up to approximately 33% zinc—to the left of the heat-treatable two-phase region just discussed. Within this range of compositions, *properties of these brass alloys can be changed only by mechanical working and annealing.*

As would be expected from the discussion in Section 4.8, as zinc solute atoms are added to copper its lattice is strained and the tensile strength increases. Figure 4-4 of that section of the text shows the strengthening effect of combining copper and nickel atoms, which are completely soluble in each other. The effect of adding zinc to copper is similar, but the zinc lattice is CPH, not FCC, so although the diameter of the zinc atom is very close to that of copper there cannot be complete solubility.

Figure 10-2 shows how increasing the number of solute atoms in the alpha (α) solid solution increases both the annealed tensile strength *and* elongation. The

T A B L E   X–5 **Approximate compositions and typical mechanical properties for selected wrought-copper alloys**

UNS No.	Alloy Name	Approximate Comp. + Cu	Condition	Tensile Strength, ksi	Yield Strength, ksi	Elongation, %
Alloys Capable of Heat Treatment						
C62400	Aluminum bronze	11% Al, 3% Fe	Q/T	100		12
C17200	Beryllium copper	1.9% Be	Hard*	110	104	5
			AT[†]	178	155	6
			½HT[‡]	200	182	4
C17500	Beryllium copper	0.5% Be, 2.0% Co	Hard	78	70	5
			AT	110	90	12
			½HT	120	108	8
Cold-Working Alloys						
C21000	Gilding	5% Zn	¼ hard*	42	32	25
			Spring*	64	58	4
C23000	Red brass	15% Zn	¼ hard	50	39	25
			Spring	84	63	3
C26000	Cartridge brass	30% Zn	¼ hard	54	40	43
			Spring	94	65	3
C26800	Yellow brass	34% Zn	¼ hard	54	40	43
			Spring	91	62	3

*Full hard, spring hard, etc., by cold work; see Table X-3.

[†]AT = solution-heat-treated, precipitation-hardened.

[‡]½HT = solution-heat-treated, ½ hard cold worked, precipitation-hardened.

Source: *SAE Handbook*. Warrendale, PA: Society of Automotive Engineers, 1990, vol. 1, sec. 10, tables 1 and 10B.

tensile strength curve increases rather smoothly to its maximum value at about 30% zinc, dips slightly, and then goes up abruptly when the two-phase alpha (α)-beta (β) region is reached.

Although the data of Figure 10-2 are for annealed brasses, data for cold-worked brasses are similar and approximately parallel. For example, the lower

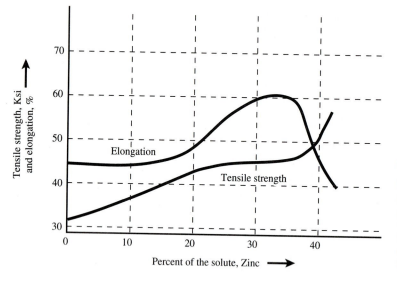

F I G U R E   10–2

Influence of solute atoms on the mechanical properties of annealed brass

Source: *Metals handbook* (8th ed.). Materials Park, OH: ASM International, 1961, vol. 1.

portion of Table X-5 shows the properties of four compositions of brass, each in the quarter-hard and spring tempers. Note that for either temper, as percent zinc goes up the properties increase; *it is especially unusual that the elongation increases along with the tensile strength.*

Applications of the brasses typically take advantage of their ductility, or formability, corrosion resistance, and pleasing appearance. Gilding brass has many ornamental and trim applications; it is used as a base for vitreous enamel jewelry. Red brass is used for weather stripping, fire extinguishers, electric sockets, and radiator and plumbing hardware. Cartridge brass is used in radiator tanks and tubing, light reflectors, ammunition cases, and lamp bases.

Modulus of Elasticity.   The modulus of elasticity for copper and brasses up to 15% zinc (red brass) is 17×10^6 psi. As percent zinc increases, the modulus goes down, e.g., with 20% (low brass) and 30% zinc (cartridge brass), it reduces to 16×10^6 psi; at 35% (high brass) and 40% (Muntz metal) the modulus decreases to 15×10^6 psi. The modulus of elasticity of the cast aluminum bronze of Table X-4 is 16.5 to 17.0×10^6 psi, depending on the solution heat treatment and temper used.

10.6   **ALUMINUM ALLOYS**

General Characteristics.   Aluminum is a silvery colored metal about one-third the density of copper and iron. It is easily formed (FCC atom lattice), nonmagnetic, corrosion-resistant in normal atmospheric environments if used properly, and alloys with a number of other metals. When properly alloyed and heat-treated, it can have a high strength-to-weight ratio. Aluminum has good thermal conductivity, and its electrical conductivity per weight is better than that of copper. Aluminum and its alloys can be used where appearance is important, and its appearance and corrosion resistance can be enhanced by a process called *anodizing.*

Ability to Alloy. Aluminum forms alloys with a number of metals, but commercially the most important of these are magnesium, copper, zinc, manganese, and silicon. These elements are often used in combination. There has been strong interest in the aluminum-copper-lithium system in recent years, but the alloys are still not widely used. Other elements—bismuth, lead, chromium, zirconium, titanium, and boron—are used in lesser amounts for special purposes.

Cast and Wrought Products. Aluminum and its alloys are used in both cast and wrought products. Cast products are produced by sand, permanent mold, and die-casting methods; they are used in the as-cast and heat-treated conditions.

Wrought products are used in both heat-treated and nonheat-treated (or cold-worked) conditions. These products include sheet, plate, foil, extrusions, wire, rod, bar, and tubing; the processes used include hot- and cold-rolling, extrusion, impact extrusion, drawing, and forging.

Applications. The nonheat-treated wrought alloys are used in many applications very familiar to the general public: Kitchen foil is made from essentially unalloyed aluminum; beverage containers, pots and pans, small boats and canoes, roofing and siding are made from alloys containing magnesium or manganese. Alloys high in magnesium are very weldable and are often used in storage tanks and ships.

Cast and wrought alloys capable of heat treatment are usually required for structural and machine applications—engine blocks; pistons; cylinder heads, aircraft, truck, and trailer frames and parts; military hardware; lawn furniture; ladders. The principal alloying elements of these alloys are copper, magnesium and silicon, and zinc. All of these systems contain regions of partial solubility, with a solvus line that decreases in solubility with temperature; thus the alloys are capable of generating supersaturated solid solutions when cooled rapidly.

10.7 ALUMINUM ALLOYS: ALLOY-TEMPER DESIGNATION SYSTEMS

Alloy Designation Systems. The alloys of aluminum used for wrought products are designated by a four-digit number; the first digit indicates the principal alloying element (see Table X-6), the second digit indicates whether the alloy is a modification of another alloy, and the last two digits are merely identifiers. Thus alloy 2017 is a high-copper alloy; 2024 is another high-copper alloy, and 2224 is a modification of 2024. In other examples, 6061 is a magnesium-silicon alloy, 6063 is another magnesium-silicon alloy, and 6463 is one of the modifications of that alloy.

T A B L E X – 6   Aluminum alloy designation systems

Wrought Alloy Composition	Alloy Designation		Cast Alloy Composition	Alloy Designation	
	AA/ANSI*	UNS		AA/ANSI*	UNS
Aluminum ≥ 99%	1xxx	A91xxx	Aluminum ≥ 99%	1xx.x	
Alloys, major element			Alloys, major element		
Copper	2xxx	A92xxx	Copper	2xx.x	A02xxx
Manganese	3xxx	A93xxx	Silicon, w/Cu &/or Mg	3xx.x	Ax3xxx
Silicon	4xxx	A94xxx	Silicon	4xx.x	Ax4xxx
Magnesium	5xxx	A95xxx	Magnesium	5xx.x	A05xxx
Magnesium & silicon	6xxx	A96xxx	Zinc	7xx.x	A07xxx
Zinc	7xxx	A97xxx	Tin	8xx.x	
Other	8xxx		Other	9xx.x	
Unused series	9xxx		Unused Series	6xx.x	

*AA = Aluminum Association, ANSI = American National Standards Institute.
Source: *Aluminum standards and data.* Washington, D.C.: Aluminum Association, 1984, 7 and 8.

A four-digit numbering system is also used for casting alloys, except here one of the digits is a decimal. As with the wrought system (Table X-6), the first digit indicates the major alloying element(s); the second two digits are numerical identifiers of the alloy, and the decimal indicates the form of the cast product (xxx.0 = casting, xxx.1 = ingot); modifications of alloys are indicated by alphabetical prefixes beginning with A. For example, 319.0 indicates a casting made from a high-silicon alloy with copper, and B319.0 is that same alloy modified by the addition of magnesium. A table of chemical compositions must be consulted to find out what the modification is.

Heat-Treated Temper Designation System. Both wrought and cast heat-treatable alloys use the T system shown in Table X-7. The "T" and appropriate numbers follow the alloy number, separated by a hyphen.

Since wrought products are by definition "worked," their annealed designator is part of the cold-working designation system discussed next and shown in Table X-8. Since castings are not worked, they have a special annealing designator, T2. The T3 and T4 tempers both involve solution heat-treating and natural aging, but T3 is used only with wrought alloys that have cold work after heat-treating. T4 is used with both cast and wrought products.

The T5 temper is used with cast and wrought alloys that are capable of achieving a specified level of properties when cooled from an elevated temperature during processing. For example, most extrusions in alloy 6063, exiting the die in the alpha (α) region, can achieve supersaturation by cooling in forced air, and if thin enough, in still air. They are then artificially aged.

T6 is a commonly used temper where the metal is formed by casting or wrought processes, solution heat-treated, and then artificially aged. T8 is similar except it is used only with wrought products since it involves cold-working after heat-treating and before aging.

T A B L E X – 7 Commonly used thermally treated tempers for aluminum alloys

Symbol	Condition
T2	Annealed-cast products only
T3	Solution-heat-treated, cold worked, natural aged
T4	Solution-heat-treated, natural aged
T5	Cooled from an elevated temperature shaping process, artificial aged
T6	Solution-heat-treated, artificial aged
T7	Solution-heat-treated, stabilized
T8	Solution-heat-treated, cold worked, artificial ageed

Source: *Aluminum standards and data.* Washington, D.C.: Aluminum Association, 1984, 11.

T7 is also similar to T6 except that the artificial aging is usually an "overaging" process to stabilize some characteristic of the alloy. (Overaging is discussed in the next section.)

Strain-Hardening Temper Designation System. There are two temper designations commonly applied to both heat-treated and strain-hardened alloys: "O" for annealed and "F" for as-cast or as-fabricated products.

Alloys that develop their properties by strain-hardening (cold-working) are designated by a system using an "H" followed by two or three digits that describe the level of properties and/or the fabricating process used; a hyphen separates the alloy and temper designations. The first digit following the H identifies the basic

T A B L E X – 8 Cold-work designations used for aluminum and aluminum alloys

Condition	Symbol	% Cold Work
As-cast or as-fabricated	F	Unspecified
Annealed	O	0
Quarter hard	HX2	15*
Half hard	HX4	35
Three-quarters hard	HX6	55
Full hard	HX8	75

X = 1, 2, or 3 depending on thermal treatment. See text.

*Extrapolated.

Source: *Aluminum standards and data.* Washington, D.C.: Aluminum Association, 1984, 10.

fabricating process: H1 = strain-hardened only, H2 = strain-hardened and partially annealed, H3 = strain-hardened and stabilized. Table X-8 outlines the system and gives the approximate percent cold work required to achieve the tempers.

The second digit indicates the quartile into which the properties fall. For
example, H12 means the metal is *cold-worked only,* and at a level that is approximately 25% of the strength possible for the alloy; H24 achieves about 50% of the
strength level possible by a **partial anneal** of cold-worked metal; H38 is in the full-
hard condition achieved by cold-working followed by a *stabilizing thermal treatment,* i.e., a recovery-type anneal.

When the third digit is used, it indicates some variation from the two-digit
temper specified, e.g., closer property limits than standard.

10.8   ALUMINUM ALLOYS: HEAT TREATMENT AND PROPERTIES

Solution Heat-Treating and Precipitation-Hardening. Figure 10-3 is the aluminum-rich half of the aluminum-copper phase diagram. To the left is the kappa (κ)
partial–solid solution region with a sloping solvus line that indicates that solubility
decreases as temperature decreases; to the right is the theta (θ) intermediate
phase.

A number of high-copper aluminum alloys, both wrought (2xxx) and cast
(2xx.x and some 3xx.x), are based on this diagram. These alloys often contain

F I G U R E 10–3 Aluminum-copper phase diagram at the aluminum-rich end

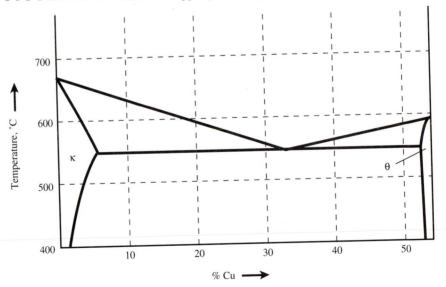

Source: *Metals handbook* (8th ed.). Materials Park, OH: ASM International, 1973, vol. 8, 259.

about 4% copper, varying amounts of magnesium, and various other elements. The heat-treating process begins by heating the part into the kappa (κ) region, holding long enough at the elevated temperature that that kappa (κ)-theta (θ) mixture is completely converted to kappa (κ), and then cooling rapidly, usually in water, so that the kappa (κ) retains the theta (θ) in supersaturated solid solution at room temperature.

Figure 10-4 is an enlarged view of the kappa (κ) region with the above process shown, including schematic micrographs. Micrograph 1 assumes a deformed microstructure of a cold-worked or fabricated part. The part is heated (solid arrow) across the solvus line into the kappa (κ) region, so that at point 2 it has recrystallized and become a *single phase*. It is then quenched (dotted arrow) to room temperature; at point 3 the microstructure is still a single-phase solid solution, but it is *supersaturated;* with the passage of time at room temperature, a submicroscopic precipitate begins to come out of solution, strengthening and hardening the alloy. Depending on the alloy, this **natural aging** may or may not achieve optimum properties, in which case the alloy is **artificially aged;** i.e., it is held for a comparatively long period at a relatively low temperature (10 to 20 h at 250 to 350°F, or 120 to 175°C).

Figure 10-5 shows the aluminum-rich end of the magnesium-silicon wrought alloy system (6xxx). In this case, the precipitating phase is a compound of the two

F I G U R E 10–4 Kappa (κ) region of the aluminum-copper phase diagram

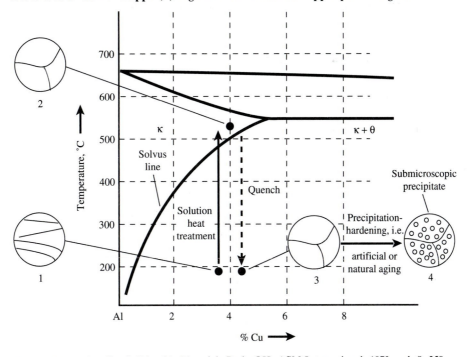

Source: *Metals handbook* (8th ed.). Materials Park, OH: ASM International, 1973, vol. 8, 259.

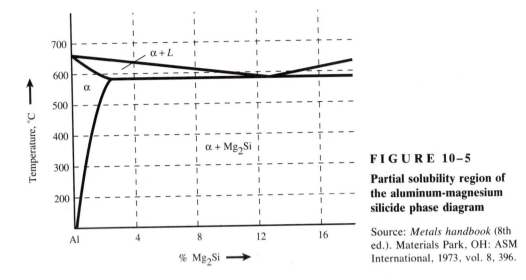

FIGURE 10–5

Partial solubility region of the aluminum-magnesium silicide phase diagram

Source: *Metals handbook* (8th ed.). Materials Park, OH: ASM International, 1973, vol. 8, 396.

chief alloying elements, magnesium and silicon—magnesium silicide, Mg_2Si. When alloyed so that the compound amounts to just under 2%, the metal can be quenched from the alpha (α) region, creating the supersaturated situation described previously.

Precipitation-Hardening, or Aging.   Figure 10-6 contains the "aging curves" for changing alloy 2014 from the T4 to the T6 condition; that is, previously solution-heat-treated metal is aged for various temperatures and times. The yield strength that results is plotted against time for the various test temperatures; the curves for tensile strength and elongation would have similar shapes. The curves lead to a number of conclusions about aging that can be applied to all precipitation-hardening alloys:

- Aging temperatures that are very high or very low, relatively speaking, do not contribute to an increase in strength; low temperatures (e.g., 0°F or −18°C) retard the hardening process; high temperatures (e.g., 650°F or 340°C) cause "overaging," which becomes an annealing treatment if carried on long enough.
- For a range of temperatures between these extremes, the level of the strength varies with the length of the aging treatment and reaches optimum levels; i.e., the curves have "humps."
- These optimum levels of strength occur at times that vary inversely with the temperatures; i.e., higher temperatures cause the optimums to occur in less time, and lower temperatures require a longer treatment to achieve the optimum strength.
- Aging at lower temperatures for longer periods of time achieves a higher strength level, and aging at higher temperatures for shorter periods of time achieves a lower strength level.

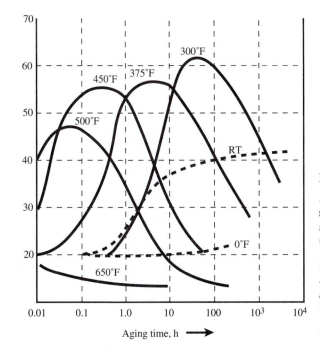

FIGURE 10–6

Aging curves for 2014 alloy, showing yield strength vs. aging time for various temperatures

Source: John E. Hatch (ed.). *Aluminum: Properties and physical metallurgy.* Materials Park, OH: ASM International, 1984, fig. 2, 138. Used by permission.

- Note that room-temperature aging (or natural aging along the room-temperature curve in Figure 10-6) does improve strength with the passage of time, but this alloy requires artificial aging to achieve its maximum properties. Those alloys that are used in the naturally aged condition, such as 2024, come very close to their maximum properties after a period of a week or so. Other alloys, such as the zinc-magnesium and zinc-magnesium-copper (7xxx) alloys, do not reach a stable heat-treated condition and require artificial aging.
- Also note that a high yield strength can be achieved by heating to only 300°F (149°C), but this requires approximately 40 h in the furnace. For most products this is not an economical use of an expensive furnace; a more likely commercial treatment for 2014-T6 is 315 to 325°F (157 to 163°C) for 16 to 20 h.

Overaged Tempers. Figure 10-7 gives an interesting view of what happens during aging. The plot is of aging temperature versus aging time for alloy 7075; the curves are constant values of yield strength, with English units at the bottom left and international standard units (ISU) at the upper right. Note that the curves are hairpin-shaped; thus at a given temperature there are two times that give the same yield strength, and vice versa.

   If solution-heat-treated 7075 is aged at 250°F (121°C) for 20 h (the shaded box labeled "T6 aging" in Figure 10-7), the yield strength reaches a maximum of about 80 ksi. However, if the temperature is raised to 325°F (163°C), a somewhat lower

FIGURE 10-7   Aging temperature vs. time for the yield strength of solution-heat-treated 7075 alloy

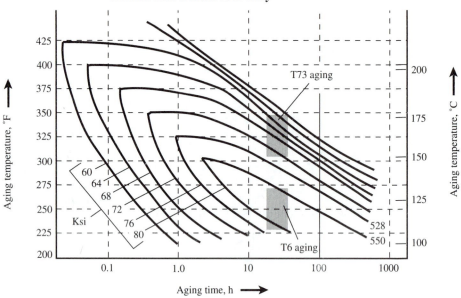

Source: John E. Hatch (ed.). *Aluminum: Properties and physical metallurgy*. Materials Park, OH: ASM International, 1984, fig. 48, 186. Used by permission.

level of strength is achieved in the same length of time (the shaded box labeled "T73 aging"). In effect, the metal has been "**overaged,**" but the result is that the distribution of the precipitate changes so that resistance to corrosion is improved. For many critical applications, e.g., aircraft structures, the reduced strength level is offset by the advantage of increased corrosion resistance.

Eutectic Melting.   When alloys are frozen at cooling rates common in commercial processes, equilibrium conditions do not exist. As discussed in Section 5.8, this lowers the solidus lines on phase diagrams. Figure 10-8 depicts a solid-solution region under the influence of rapid freezing and cooling. The original phase diagram is in heavy lines, the displaced diagram is in light lines, and the cooling of the alloy is represented by the vertical dotted line.

Under slow (approximately equilibrium) cooling the alloy freezes as alpha (α) solid solution at point *x* (Figure 10-8). As slow-cooling continues across the solvus line, the alpha (α) begins to precipitate the beta (β) phase, and the metal becomes a mixture of the two phases. This is what happens to alloy number 1 in Figure 5-23. If such a microstructure were reheated into the alpha (α) region for the purposes of solution heat-treating, as long as it was heated below point *x*, it would not melt and a product of acceptable quality would result.

However, if the alloy originally froze by a rapid-cooling method, the phase

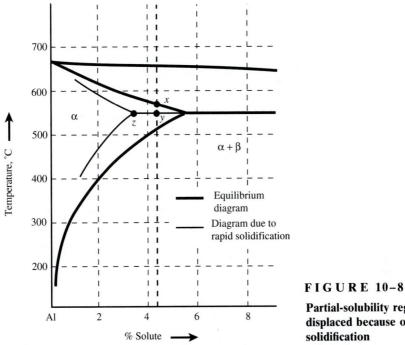

FIGURE 10–8

Partial-solubility region displaced because of rapid solidification

diagram is shifted to the left; the metal will not be all frozen alpha (α) at point x, but will remain liquid down to point y, the eutectic temperature. (There is a eutectic in this alloy system out of view to the right.) The short tie line y-z represents the amount of the liquid that froze as eutectic. Now, if this alloy is reheated for the purposes of solution heat treatment, and if it is heated above point y, that portion that froze as eutectic will melt. This is termed *incipient eutectic melting*.[†]

Although the y-z tie line is probably very short, and thus the amount of the incipient melting very small, the result is that the final heat-treated product has small areas within it that are essentially of an annealed structure. When these are scattered throughout a structural member in a critical application, like an aircraft wing spar, they are cause for great concern. The photomicrograph in Figure A-34 shows incipient eutectic melting in overheated 2014 alloy.

Producers can take two steps to minimize this problem: homogenize the original cast structure to reduce the segregation caused by the rapid solidification (see Section 5.8) and reheat the product to below the eutectic temperature when solution heat-treating. Unfortunately, at the compositions used in many alloys the temperature range (vertical dimension) of the partial-solubility region is very

---

[†]"Incipient" in that eutectic melting is just beginning.

small. This requires that the temperature be controlled within a few degrees, which can be difficult when heat-treating large products in large furnaces.[‡]

Typical Properties of Representative Alloys. Table X-9 gives typical chemical compositions and mechanical properties of some frequently used wrought alloys. A comparison of the top and bottom sections of the table shows that heat-treated alloys are capable of higher strength levels at high levels of elongation. The table shows the change in properties among heat-treated alloys as treatments progress from annealing, to solution heat-treating, to precipitation-hardening. It is clear that aging has a larger influence on the yield strength than it does on the tensile strength and that aging reduces the elongation.

The low yield strength and high elongation of the T4 condition are sometimes used to advantage by fabricators; the heat-treated temper is purchased, the forming is done, and then the metal is aged to the T6 strength. The rivets of a naturally aging alloy are an extreme example of this process: They are heat-treated, refrigerated until used, and naturally aged to the higher-strength condition after being upset.

As can be seen in Table X-9 the wrought alloys discussed above are high in copper, magnesium-silicon, or zinc. In contrast, the cold-working alloys are unalloyed or are alloys high in manganese or magnesium.

Among the cold-worked alloys it can be seen that:

• Increasing the cold work for any alloy increases the strength values and lowers the elongation, as should be expected.
• For comparable amounts of strain-hardening, higher alloying content produces higher strength properties; alloy 1100 is almost pure aluminum, alloy 3003 adds 1.2% manganese, and alloy 5154 has 3.5% magnesium, and their strengths are in that order.

A comparison of the composition and property data of Table X-10 shows the influence of heat-treating and alloying in casting alloys. The high-copper alloy 201.0 in the heat-treated T7 condition has high properties. The difference between the as-cast and heat-treated conditions in other alloys is also evident. A comparison of alloys in the as-cast F condition shows the influence of alloying elements.

There is one obvious difference between the wrought and cast alloys—the large amounts of silicon used in many casting alloys. Silicon, with a lower specific gravity than aluminum, can result in lighter-weight castings. In binary alloys, in hypoeutectic and hypereutectic amounts (see the Al-Si phase diagram in Figure 5-5), it provides a large freezing range to improve castability.

Modulus of Elasticity. The room temperature modulus of elasticity of aluminum is nominally 10×10^6 psi, but it does vary with alloy content. One reason for the recent interest in the aluminum-copper-lithium alloys is that a relatively small amount of lithium raises the modulus to 11×10^6 psi. This could mean that

---

[‡]A wing spar for a Boeing 747 is just over 100 ft long when heat-treated.

T A B L E X–9 Typical chemical compositions and mechanical properties for selected wrought-aluminum alloys in the heat-treated and cold-worked conditions

Alloy	Alloying Elements, %						Temper	Tensile Strength, ksi	Yield Strength, ksi	Elongation, %	HB, 500 kg
	Si	Cu	Mn	Mg	Cr	Zn					
Heat-Treated Alloys											
2014	0.8	4.4	0.8	0.5			O	25	10	21	45
							T4	61	37	22	105
							T6	63	60	10	135
2024		4.4	0.6	1.5			O	27	11	20	47
							T3	70	50	18	120
6061	0.6	0.28		1.0	0.2		O	18	8	25	30
							T4	35	21	22	65
							T6	45	40	12	95
6063	0.4			0.7			O	13	7		25
							T4	25	13	22	
							T5	27	21	12	60
							T6	35	31	12	73
7075		1.6		2.5	0.23	5.6	O	33	15	17	60
							T6	83	73	11	150
Cold-Work Alloys											
1100		0.12					O	13	5	35	23
							H12	16	15	12	28
							H14	18	17	9	32
							H16	21	20	6	38
							H18	24	22	5	44
3003		0.12	1.2				O	16	6	30	28
							H12	19	18	10	35
							H14	22	21	8	40
							H16	26	25	5	47
							H18	29	27	4	55
5154				3.5	0.25		O	35	17	27	58
							H32	39	30	15	67
							H34	42	33	13	73
							H36	45	36	12	78
							H38	48	39	10	80

Source: *Aluminum standards and data.* Washington, D.C.: Aluminum Association, 1984, tables 1.1 and 2.1.

T A B L E   X – 1 0   Typical chemical compositions and mechanical properties in heat-treated and as-cast conditions for selected aluminum casting alloys

Alloy	Alloying Elements, %					Temper	Tensile Strength, ksi	Yield Strength, ksi	Elongation, %
	Si	Fe	Cu	Mg	Zn				
201.0*			4.6	0.35		T7	68	60	5.5
357.0	7.0			0.55		F	25	13	5.0
						T6	50	43	2.0
390.0	7.0	1.3 max.	4.5	0.55		F	40.5	35	1.0
						T5	43	38.5	1.0
A444.0	7.0					F	21	9	9.0
						T4	23	9	12.0
413.0	12.0	2.0 max.	1.0 max.			F	43	21	2.5
520.0				10.0		T4	48	26	16.0
711.0		1.0	0.5	0.35	6.5	F†	35	18	8.0

*Plus 0.7 Ag, 0.35 Mn.

†Tested 30 days after casting.

Source: *Aluminum standards and data*. Washington, D.C.: Aluminum Association, 1984, 322–33, tables 2, 4, 5, and 6.

structural members that are designed against buckling (dependent on section modulus and modulus of elasticity rather than just strength) can be made lighter; this could be important in reducing the weight of aircraft.

In the casting alloys the amounts of alloying elements used are often much larger than in the wrought alloys, so the variation in modulus of elasticity is rather large. The modulus for alloy 390.0 shown in Table X-10 is 11.9×10^6 psi; on the other hand, alloy 520.0 is only 9.5×10^6 psi.

Specific Gravity. Since aluminum alloys are often used where the strength-to-weight ratio is important, specific gravity can be an important factor in alloy selection. As might be expected, alloys with "heavier" alloying elements, e.g., zinc, have higher specific gravities and those with "lighter" alloying elements, e.g., lithium, magnesium, and silicon, have lower specific gravities.

The specific gravity for high-zinc alloys 7075 and 711.0 is about 2.8; it is 2.77 and 2.8 for high-copper alloys 2024 and 201.0, respectively. The other wrought alloys of Tables X-9 and X-10 have specific gravities very close to 2.7. The casting alloys, often with larger amounts of alloying elements, have a wider range of values; alloy 520.0, a high-magnesium alloy, has a specific gravity of only 2.57.

The conversion of the approximate specific gravity of aluminum alloys to English units of density comes out to the useful and easily remembered value of 0.10 lb/in.^{3}

10.9 MAGNESIUM AND MAGNESIUM ALLOYS

General Characteristics. Magnesium is a silvery-colored light metal that has a specific weight of about 1.8; in English units it has a density of about 0.065 lb/in.^{3} Thus it is about two-thirds the unit weight of aluminum and approximately one-fourth that of iron and copper, which have specific gravities of about 7.9 and 9.0, respectively.

Strengthening Mechanisms. Magnesium, with a CPH atom lattice, is rather difficult to cold-work; with the exception of rolled products, it is not commonly produced in cold-worked tempers. Other products are used in the as-cast or as-fabricated (F temper)§ or solution-heat-treated and precipitation-hardened (T5 and T6) tempers. Magnesium is frequently alloyed with aluminum, zinc, and zirconium; it has maximum solubilities of 12.7, 6.2, and ≈2%, respectively. Magnesium also forms partially soluble–insoluble systems with manganese, lead, tin, and thorium.

Alloy Designation System. The alloy designation system uses two letters, two numerical digits, and a letter to indicate variations of the alloy. The first two letters identify the two principal alloying elements; the two digits are the rounded percent of those elements. For example, alloy AZ31B contains about 3% aluminum and 1% zinc and is the second variation of the alloy; alloy ZK61A contains about 6% zinc and 1% zirconium and is the first variation of the alloy. The letters used to indicate alloying elements are generally the first letter of the element, with substitutions made for duplications; e.g., K is used for zirconium.

Heat-Treating. Magnesium is an extremely reactive metal so solution heat-treating is normally done in a controlled-atmosphere furnace to prevent oxidation. The rapid cooling required to create the supersaturated solid solution is provided by still or forced air, depending on the thickness of the metal; in a very few alloys water quenching is used.

Safety Precautions. Magnesium alloys themselves can catch fire at high temperatures. Just the heat from a sawblade revolving at too high a rate has been known to cause a fire. These fires must be extinguished by smothering the flames with gases or dry chemicals. Under no circumstances should water be used; water merely aids in the generation of hydrogen and creates an even more dangerous situation.

---

§Magnesium alloys use the same designation system for cold-worked and heat-treated tempers as are used for aluminum alloys. See Tables X-7 and X-8.

Representative Compositions and Properties. Table X-11 illustrates the properties possible with cast- and wrought-magnesium alloys. The increase in strength with precipitation-hardening can be seen. Alloy AZ31B is shown in a cold-worked temper produced by rolling and as-fabricated by extrusion. Note that at one-fourth the weight of steel and copper, alloy ZK60A has a very high strength-weight ratio.

Applications. The modulus of elasticity of magnesium alloys is only 6.5×10^6 psi. This can be a disadvantage in that structures built of magnesium alloys will lack the rigidity of, say, steel. However, from the principle of the modulus of resilience (discussed in Section B.6) we know that a *low* modulus of elasticity improves the modulus of resilience as long as the yield strength is kept high:

$$\text{Modulus of resilience} \approx \frac{\sigma_y^2}{2E}$$

Thus magnesium alloys are used in applications where the product must be lightweight but must also absorb shock loading without permanent deformation, e.g., hand luggage, snowshoes, and portable equipment. Magnesium's strength-to-weight relationship also makes it ideal for any application where weight is a concern—ladders, dockboards, portable hand tools, transit and camera tripods, and automobile transmission housings.

Newer aluminum-manganese (AM) alloys with a low aluminum content, e.g., AM20, are being developed that fail in a ductile rather than brittle mode. They are

T A B L E X–11 **Typical chemical compositions and mechanical properties in various temper conditions for selected magnesium cast and wrought alloys**

Alloy	% Alloy Elements			Temper	Tensile Strength, ksi	Yield Strength, ksi	Elongation, %	UNS No.
	Al	Zn	Zr					
Cast Alloys								
AZ63A	6	3		F	29	14	6	M11630
				T5	29	15	4	
				T6	40	19	5	
ZK61A		6	0.8	T6	40	26	5	M16610
Wrought Alloys								
AZ31B	3	0.9		H24	40	29	17	M11311
				F	38	29	15	
ZK60A		5.5	0.55	F	49	37	14	M16600
				T5	52	43	12	

Source: *Magnesium mill products and alloys.* Midland, MI: Dow Chemical Co., 1982, tables 2, 3, and 5.

being considered for die castings in automobiles that will provide a weight saving but will also absorb more energy in a crash.

10.10   NICKEL AND NICKEL ALLOYS

General Characteristics.   Nickel is a heavy (sp. gr. = 8.9), ductile (FCC lattice), silver-colored metal with good resistance to corrosion and oxidation. In the periodic table it occurs just before copper, with which it has complete solid solubility, and is immediately preceded by cobalt and then iron; iron, cobalt and nickel are all ferromagnetic.

The modulus of elasticity of nickel itself is 30×10^6 psi, but since nickel is alloyed with such a wide variety of metals, in such a wide variety of compositions, the moduli of the alloys vary greatly, and may be as low as 18×10^6 psi.

Strengthening Mechanisms.   The strengthening mechanisms used with the nickel alloys are the same ones previously discussed in this chapter—solid-solution strengthening with solute atoms, cold-working, and precipitation-hardening after solution heat-treating. But nickel is an expensive metal, usually on the order of 4 times as expensive as copper, and is not used just for its hardness or strength but for one or more of its other characteristics: corrosion resistance, strength at high and low temperatures, and unusual magnetic or electrical properties.

Common Alloys.   Nickel-2% beryllium alloy is capable of being precipitation-hardened to very high strength (200 ksi) and hardness, yet it has good ductility. The alloy has good fatigue and impact strengths and can be used up to 800°F (425°C).

Monel is an alloy of 70% Ni-30% Cu that has good corrosion resistance and strength. If 2.75% aluminum is added, the alloy can be solution-heat-treated. (Aluminum forms partial-solubility regions with both copper and nickel.)

Nickel and iron in varying combinations create a sequence of alloys noted for their magnetic permeability and/or low thermal expansion. Some of these alloys are designed so that their coefficients of thermal expansion are such that they can seal with glass. Invar, 36% Ni-64% Fe, is noteworthy; it has the lowest-known coefficient of thermal expansion for a metal up to 300°F (150°C).

Nickel-chromium and nickel-chromium-iron alloys form sequences of alloys, along with other alloying elements, that are useful at high temperatures as heating elements, thermocouples, and other such hardware. Trade names like Chromel, Nichrome, and Inconel are sprinkled through this group of alloys.

Nickel-iron-cobalt alloys also have thermal expansions that are low and constant with a constant modulus of elasticity. This makes them useful in applications where temperatures vary, e.g., rocket and turbine engines.

Nickel-chromium-molybdenum, along with cobalt-tungsten-copper and other elements, forms a group of alloys that are used at temperatures up to about 1800°F (1000°C). The use of nickel in "superalloys" for extreme high-temperature applications will be discussed further in Chapter 13.

SUMMARY

1. Nonferrous metals are chosen for applications that take advantage of special characteristics as well as their mechanical properties, but they must be cost-effective in meeting special requirements.

2. Properties in nonferrous metals are achieved through cold-working, alloying, and solution heat-treating/precipitation-hardening.

3. Alloying improves strength through solid-solution strengthening.

4. Alloying between metals that are partially soluble–insoluble creates the potential for solution heat treatment. If the solubility of the solid solution varies with temperature, quenching to a supersaturated solid solution is possible.

5. Precipitation-hardening, or aging, of the supersaturated solid solution naturally or artificially usually improves strength.

6. Copper is a nonferrous metal often used because it is thermally and electrically very conductive; is corrosion-resistant; has a good appearance; and yet is easily worked, alloyed, and heat-treated to have high mechanical properties.

7. The major alloy groups are the brasses, alloys of copper and zinc, and the bronzes, alloys of copper and other metals. Some alloys of both groups can be solution-heat-treated and precipitation-hardened for use in cast or wrought products.

8. The cold-worked tempers of copper alloys are described in quarter-fractions of a "hard" condition and as being of "spring" temper or beyond.

9. A unified numbering system (UNS) is used to identify specific alloy compositions, but in general, copper alloys are known by a name related to their color, composition, or application.

10. Aluminum and its alloys have good strength-to-weight relationships and corrosion resistance. Products can be cast and wrought, cold-worked or solution-heat-treated and precipitation-hardened.

11. Wrought- and cast-aluminum alloys are designated by four-digit numbers, with the first digit identifying the major alloying element(s). (The fourth digit of the cast alloy number is a decimal.)

12. The heat-treatable wrought-aluminum alloys, the ones primarily used in structural applications, are 2xxx (copper), 6xxx (magnesium-silicon), and 7xxx (zinc).

13. For aluminum the heat-treated tempers are designated by a T and a number. Frequently seen tempers are T3 (solution-heat-treated, cold-worked, naturally aged) and T6 (solution-heat-treated and artificially aged).

14. The cold-worked aluminum tempers are designated by an H followed by a digit that indicates the treatment used, followed by another digit that indicates the quartile of cold work done.

15. Rapid freezing can cause solid-solution alloys to contain metal of eutectic composition. To avoid incipient melting of this eutectic composition,
    • solution heat-treating temperatures must be controlled below the eutectic temperature
    • the original cast metal should be homogenized to allow the composition to equalize, i.e., to reduce segregation

16. Solution-heat-treated alloys will naturally age at room temperature. For some alloys like 2024, this process will achieve optimum properties.

17. Other alloys require artificial aging, or heating at a higher temperature, e.g., 250° to 360°F (120° to 180°C). Maximum strength occurs at an optimum combination of temperature and time.

18. Very low temperatures retard the aging process; very high temperatures overage the material and are essentially annealing treatments.

19. Higher properties are achieved by aging at lower temperatures for longer times, but for obvious economic reasons the temperature is usually chosen so that the furnace is not tied up for more than a day.

20. If aluminum is carefully aged at too high a temperature, i.e., overaged, strengths are somewhat reduced, but other characteristics like corrosion resistance can be improved.

21. Magnesium alloys have very good strength-to-weight ratios and are used where light weight is required. Magnesium has a low modulus of elasticity and therefore can have a good modulus of resilience.

22. Magnesium alloys are solution-heat-treated and precipitation-hardened very similar to aluminum alloys; their heat-treating temper designation system is also the same.

23. The cold-working temper system for magnesium alloys is also similar to that of aluminum, but magnesium alloys do not cold-work well because of their CPH atom lattice.

24. The magnesium-alloy designation system uses two letters to indicate the major alloying elements (e.g., A = aluminum, Z = zinc, K = zirconium) and two digits that round off the percent composition of the two main alloying elements.

25. Magnesium is a very reactive metal and easily catches fire; at high temperatures it reacts with the water, so water cannot be used to extinguish magnesium fires.

26. Nickel is used to achieve unusual electrical, magnetic, thermal expansion, and corrosion properties and strength at high and low temperature.

27. Nickel alloys are capable of being solution-heat-treated and precipitation-hardened or cold-worked.

28. Some significant moduli of elasticity are given below:
- Steel = 30×10^6 psi.
- Copper = 16×10^6 psi; alloys vary between 15 and 17×10^6 psi.
- Aluminum and alloys of low content are approximately 10×10^6 psi; some alloys can be as high as 11 to 12×10^6 psi.
- Magnesium alloys = 6.5×10^6 psi.
- Nickel = 30×10^6 psi; alloys vary to as low as 18×10^6 psi.

29. Some significant specific gravities are given below:
- Iron = 7.87
- Copper = 8.96, with values generally decreasing as alloy content increases
- Aluminum = 2.699, with alloys up to 2.8 and down to 2.57, depending on alloy content
- Magnesium = 1.74, with alloys about 1.8
- Nickel = 8.902, with alloys at lower values

REVIEW QUESTIONS

1. Describe some circumstances where it would be desirable to have a metal that (a) was chemically reactive/chemically inactive; (b) had a high modulus of elasticity/low modulus of elasticity; (c) had high electrical conductivity/low electrical conductivity; (d) had high thermal expansion/low thermal expansion.

2. Describe the strengthening mechanisms used with nonferrous metals.

3. What are the major families or types of copper alloys? How are they different? Similar? How are they designated? How do their treatments differ and how are the treatments designated? How do cast and wrought alloys differ and how are they designated?

4. Answer Question 3 for aluminum and then for magnesium.

5. What are similarities and differences between copper, aluminum, and magnesium alloys?

6. Using appropriate phase diagrams, describe the sequence of events that could lead to having small areas of eutectic material in an otherwise heat-treated (supersaturated solid solution) piece of metal.

7. Why might baseball and softball bats be made from wood or aluminum and not from magnesium or steel?

8. Describe how a martensitic structure might be found in a nonferrous metal.

9. Assume the alloy system shown in Figure 10-3. If the alloy contains 5% copper and is quenched from just under the eutectic temperature to 300°C, approximately what percent of the microstructure could be θ?

FOR FURTHER STUDY

Aluminum standards and data. Washington, D.C.: Aluminum Association, 1984.

Avner, Sidney H. *Introduction to physical metallurgy*. New York: McGraw-Hill, 1974, chap. 12.

Hatch, John E. (ed.). *Aluminum: Properties and physical metallurgy*. Metals Park, OH: ASM International, 1984.

Metals handbook (9th ed.). Metals Park, OH: ASM International, 1979, vol. 2.

Metals handbook (9th ed.). Metals Park, OH: ASM International, 1988, vol. 15.

1990 SAE Handbook. Warrendale, PA: Society of Automotive Engineers, 1990, vol. 1, sec. 10.

HOW METALS FAIL

In part 3 we won't consider *all* modes by which metals fail, but only those most commonly experienced by engineers. Metals break (fracture), they corrode, and they "wear out." Chapter 11 addresses the general area of fracture, and Chapters 12 and 13 discuss failures that occur at low and high temperatures. Chapters 14 and 15 then cover corrosion and wear. Specifically, in Chapter 11 we'll look at ductile, brittle, and fatigue failures; in Chapter 12 we'll consider failure at low temperatures, especially in BCC metals; in Chapter 13 we'll study high-temperature failure, in Chapter 14 failure from corrosion, and in Chapter 15 failure from wear.

Although some strength of materials concepts are discussed in Chapters 11 and 12, the main focus of Part 3 is on what can be done *metallurgically* to avoid these failures. Because of the close relationship between design and metal selection, some design concepts are also discussed, but in a general way.

C H A P T E R **11**

Failure Due to Fracture

OBJECTIVES

After studying this chapter, you should be able to

- describe the stress distribution(s) that contribute to ductile failures in tension and torsion and the factors that can reduce their occurrence
- describe the stress distribution(s) that contribute to brittle-type failures in otherwise ductile metals
- explain how a notch creates a complex stress distribution
- discuss differences between ductile and brittle failures in terms of loading, fracture appearance, shape of the stress-strain diagram, and energy absorbed during failure
- sketch typical *S-N* curves for a steel and a nonferrous alloy and identify the influences of carbon content and percent martensite structure on the steel diagram
- differentiate between fatigue strength and endurance limit
- describe how a fatigue crack is initiated and how it propagates
- describe the possible alternating stress patterns and how the pattern of loading influences the *S-N* curve and fatigue strength
- identify metallurgical and design steps that can be taken to reduce occurrence of the various types of failures
- describe the major nondestructive inspection techniques and their purposes

DID YOU KNOW?

Steel specimens subjected to alternating stresses can only be expected to carry stresses up to one-half of their tensile strength for an indefinite number of stress cycles. Nonferrous metals cannot carry stresses indefinitely, but at large numbers of stress cycles their load-carrying ability is one-third to one-half their tensile strength. So, although metals can be made strong and hard and capable of carrying large loads and of resisting failure in many types of environments, they can be "brought to their knees" by such things as small cracks, scratches, or tool marks, combined with vibrating or cyclical loads; under these conditions, metals can fail suddenly and with catastrophic results.

Complex stress distributions can cause ductile metals to fail as if they were a very brittle material, and this also can happen suddenly and without warning.

A PREVIEW

How a metal is loaded has a lot to do with how it fails, so this chapter deals with types of stresses and how they are distributed in ductile and brittle failures. Our main emphases here, however, are on how ductile metals fail as if they were brittle metals and on fatigue failures. The chapter begins with a brief discussion of the general subject of quality and its relation to failure. We'll then look at the mechanisms by which stress concentrations or notches create complex stress distributions and the importance of stress distribution in brittle and fatigue failures. We'll conclude by exploring the strategies, especially metallurgical strategies, that can be used to avoid failures.

11.1   QUALITY

Because metals are used in so many structural applications, the first thing that comes to mind when the word "failure" is heard is that something "broke"—the metal wasn't strong enough or hard enough or tough enough to perform in the manner intended. This type of failure can be classed as mechanical failure, and it will be the subject of the remaining chapters of this text. You should realize, however, that in the broad context of quality, there is a failure any time a specification is not met. The failure may be that the metal's conductivity is too high or too low, its magnetic properties are not as specified, its surface is too rough or smooth, or its appearance is not acceptable. Thus, by this definition, failure has many faces and takes many forms.

In the still broader context of total quality management (TQM), a failure may occur because the specification itself is inappropriate for the part, component, or piece of equipment. In satisfying engineering needs it is important that specifications be developed that meet the stated requirements and that then those specifications be met, i.e., that quality is maintained throughout the whole manufacturing process—in marketing and sales, in design, in purchasing, in production, and in testing.

11.2   MODES OF FAILURES

Mechanical failure of a metal occurs because of a load or stress, but the failure is also strongly affected by the way the load is applied, the environment in which the metal is used, and/or the characteristics of the metal. A stress applied suddenly has different consequences from one applied slowly; results differ depending on whether the use involves a high or low temperature or the presence of a liquid that will contribute to corrosion; results also vary if the metal is aluminum or steel or brass. All the many possible outcomes may require different approaches to design, and, more importantly for this study, different considerations in the selection of metals, manufacturing processes, and/or heat treatments.

The modes of failure discussed in the next five chapters—(1) ductile, (2) brittle, (3) fatigue, (4) low temperatures, especially in BCC metals, (5) high temperatures, (6) corrosion, and (7) wear can, in a general way, be separated into broad categories. For instance, failure modes 1, 2, 3, and 7 relate primarily to the manner in which the load is applied; failure modes 4, 5, and 6 are related to the environment in which a metal is used; and failure modes 1, 2, and 6 are influenced by the general characteristics of a metal.

The categories outlined above, however, are not straightforward. For example, fatigue failures often occur in environments favorable to corrosion, and corrosion often contributes greatly to wear failures. It is not uncommon for failures to be traced to multiple causes or through a sequence of causes; e.g., an unexpected leak causes corrosion in a bearing, which causes it to wear, which causes an imbalance in a shaft, which causes an eventual fatigue failure. Incorrect manufacturing processes, e.g., faulty heat treatment or machining marks, often cause stress concentrations that combine with other factors to cause fatigue failures.

One of the difficulties in performing an analysis of failure of the type just described is that the evidence is often destroyed in the process of failure. A good example of this is seen in rotating members that break; before the power source can be stopped, the mating fractured surfaces rub against each other and wipe out information useful in determining the original cause of failure.

<h2>11.3 DUCTILE FAILURES</h2>

The point is made in Chapter 2 (and in Appendix B) that **ductile** metals are so called because they deform easily; thus a ductile failure or fracture is characterized by deformation.

Single-Crystal Slip. For single crystals deformation takes place as shown in Figures 2-3, 2-4, and 2-5, where layers of atoms slide or slip over one another in response to a stress that has a component exceeding the critical resolved shear stress of that metal.

Single crystals of CPH metals deform under relatively low stresses when the stresses are applied at 45° to the (0001) plane; at that angle the maximum shear stresses coincide with the major slip plane. However, if the (0001) plane is 90° to the applied stress, slip must occur on a secondary slip system, which would require stresses higher than the critical resolved shear stress.

A single FCC crystal, oriented with its (001) face plane perpendicular to the applied stress, has its {111} principal slip planes at 45° angles to the stress, and slip and deformation occur easily.

Polycrystalline Slip. Slip in polycrystalline metals occurs first in those grains whose major slip planes are favorably oriented to the shearing stresses. As deformation occurs, other grains rotate so that they are oriented favorably for slip, and the process proceeds throughout the microstructure.

When **simply loaded** in tension—that is, the load is applied on one axis, as seen in Figure 11-1—shearing stresses cause ductile metals to go through uniform elastic and plastic deformation, and often localized deformation or necking as well. In compression, ductile metals become barrel-shaped and some are smashed flat if loaded sufficiently. In torsion (Figure 11-2) the applied torque and resulting shearing stresses may cause the shaft to twist off at the point of failure, with surfaces perpendicular to the axis of rotation. In all of these ductile failures there is movement of metal occurring at an angle to the applied stress, i.e., movement in the direction of the shearing stresses.

Movement of Metal in Deformation. In Section 1.4 we discussed the closest-approach distance of atoms. During slip, atoms move relative to each other, but

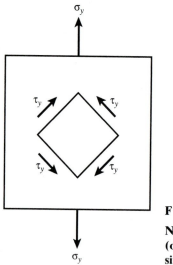

FIGURE 11-1

Normal or separating stresses (σ_y) and shear stresses (τ_y) in simple tension

the closest-approach distance still basically exists between them. Deformation by tension, compression, or torsion may cause atoms to relocate somewhat and take on slightly higher energy locations, but not in the amount necessary for the reduction in diameter or width that occurs in a tension specimen of a ductile metal, for example. In order for the specimen to deform, i.e., change shape, metal must move at an angle. In simple tensile loading, the metal generally moves at 45° to the stress, the angle of the maximum shear stresses, which are represented by τ in Figures 11-1 and 11-2.

Ductile Fracture. In tension, when a ductile polycrystalline metal can no longer move at an angle—i.e., all its slip planes are exhausted, or all the available dislocations are used (see Section 2.3)—the final failure begins. It usually starts with small cracks that originate in the center of the specimen and travel out to the periphery, where a shearing action leaves a cup and/or cone at 45°. In the final rupture the cohesive (atomic) forces that hold the metal together are overcome; the extensive deformation that occurs makes the center portion rough, giving it a dull-gray color.

Such a failure is illustrated in Figure 11-3. Both halves of the specimen are shown necked; the right portion has failed with a 45° cup, the left portion with a

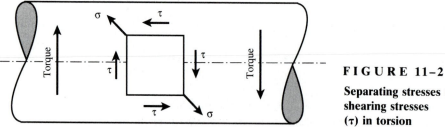

FIGURE 11-2

Separating stresses (σ) and shearing stresses (τ) in torsion

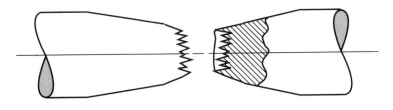

FIGURE 11-3

Schematic representation of the cup (right) and cone (left) failure of a ductile metal in tension. (Cup and cone are oriented vertically during testing.)

45° cone. Each half has a rough center portion that will have a dull-gray appearance.

Note that because the metal could no longer slip and move at an angle to the direction of the applied stress it pulled apart. This is an important concept: *When a ductile metal can no longer deform, which requires movement of the metal in a direction at an angle to the applied stress, it will fail as if it were a brittle metal,* although it is, in fact, a ductile failure.

In torsion the maximum shear stresses, τ in Figure 11-2, cause ductile metals to twist off like taffy and/or form sheared surfaces that are parallel to the axis.

Avoiding Ductile Failures. Ductile failures are usually not of major concern to designers and users of metal parts or equipment. Design is usually based on the yield stress adjusted by a suitable factor of safety. In practice, ductile failures are usually not catastrophic; deformation precedes failure and if detected it serves as a warning. A bent beam or a stretched crane cable has obviously been overloaded and, as long as the overload is below the tensile strength, corrective action can be taken.

Also, ductile failures can often be avoided by installing protective devices; for example, circuit breakers, clutches, couplings, and shear pins can be used to limit the load-carrying capacity of mechanical systems and thus reduce the opportunities for overloading.

To reduce ductile failures in mechanical systems, the normal approach is to increase the load-carrying ability. That is, the machine elements or parts are made stronger, either by using metals with higher yield strengths or by using bigger members. (Care must be taken that increasing the capacity of one part of a mechanical system doesn't just cause another portion to become the "weak link.")

Unfortunately, this solution often doesn't reduce failures, especially in machines and equipment, because loads seldom generate simple one-axis stresses. More typically, failures involve complex, or multiaxial loading, which causes the metal to fail as if it were brittle, even though it is a ductile metal.

11.4   MULTIAXIAL STRESS DISTRIBUTIONS AND BRITTLE FAILURES

Brittle Metals. **Brittle** metals fail because of normal or separating stresses, i.e., cleavage stresses that separate the metal. In tension the maximum separating stress is the applied tensile force acting on the minimum area of the specimen (see

Section 2.4), so the failure occurs at 90° to the force, i.e., the specimen breaks off square. In torsion the maximum separating stress is at 45° to the axis (Figure 11-2), so in torsion brittle metals have a twisted, screw-thread-like failure; breaking a piece of writing chalk by twisting it is a good example, as is the example seen in Figure A-35.

In completely brittle failures little energy is absorbed because the individual grains cleave along similar crystallographic planes. This is in contrast to the high amount of energy required to overcome the atomic bonds of the ductile failure. When the fracture surface of the brittle specimen is turned in the light, the cleaved crystallographic planes make it sparkle; the ductile-fracture surface appears a dull-gray color because of the extensive deformation of the final rupture.

White and gray cast irons are examples of brittle metals used for engineering purposes. They are *cast* irons because they are not capable of being formed by any process other than casting. Their stress-strain diagrams terminate essentially at their proportional limits; there is no plastic deformation region. They are brittle, as are glass, pottery, some tool steels, and carbides.

These two cast irons, white and gray, achieve their brittleness in two different ways: The white cast iron contains a large amount of the very hard and brittle compound, cementite; the gray cast iron contains many soft graphite flakes that are stress concentrations and interrupt the matrix (pearlite, ferrite, etc.) and prevent it from making its normal contribution to mechanical properties. Although the study of brittle materials can be important, the major concern of this chapter (and the next) is with normally ductile metals that behave *as if* they were brittle because of multiaxial, or complex, stress distributions.

Multiaxial Stress Distributions. **Multiaxial,** or **complex, stress distributions** are generated when metals cannot deform as they would normally. That is, when metal is pulled, it should elongate and reduce in area. When metal is pulled one way, it should be able to move in a direction at some angle to the pulling force; when it is constrained from moving sideways, it is not in a simple stress-strain relationship and a complex distribution exists.

Large tensile forces or stresses are required in rupturing or pulling metal apart in order to overcome the basic cohesive forces that bond the atoms together. Since ductile metals slip readily, it is difficult for these large stresses to build up; maximum shear stresses form easily at 45° to the tensile stress (Figures 11-1 and 11-2), and the metal slips.

However, if a specimen of the same ductile metal of Figure 11-1 is subjected to stresses that are equal and perpendicular (the biaxial stress distribution seen in Figure 11-4), then these shear stresses are canceled. There remain no shear stresses to cause slip; if the tensile stresses (σ_x and σ_y) become large enough, the metal will fail as if it were a brittle material: It will simply pull apart.*

---

*Note that if the direction of one of the tensile stresses is reversed, the shear stresses add, which increases the stresses causing slip, thus increasing the possibility for slip; this is analagous to applying tension to a strip of sheet metal going through a rolling mill.

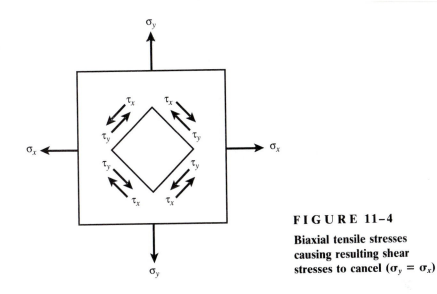

FIGURE 11–4

**Biaxial tensile stresses
causing resulting shear
stresses to cancel ($\sigma_y = \sigma_x$)**

If it is assumed that the specimen in Figure 11-4 has stresses perpendicular to the page on a z-axis, then the metal is subject to a triaxial stress distribution.

In typical machine and structural members the stresses in the various directions are not necessarily equal, but they can still generate multiaxial distributions. Brittle-type failures due to these complex stress distributions are very common in otherwise ductile metals. Often a "chevron" pattern can be seen in the fractured surface, the pattern radiating from where the fracture started.

11.5   STRESS CONCENTRATIONS

One of the more common ways that a complex or multiaxial stress is created is through a *notch,* or **stress concentration.** Figure 11-5 shows a round specimen in tension, with a notch on its circumference. To explain its effect, the notch is enlarged in the view on the right. There the notch has been drawn as a series of thin wafers of different diameters; the wafer at the base, or root, of the notch is shown shaded. For any given tensile force, the stress at the root of the notch is the highest in the specimen since its cross-sectional area is the least.

The normal consequence of a stress applied in the y-direction would be to elongate the shaded wafer in the y-direction (ε_y) and, because of the diagonal slip discussed earlier, *reduce* its diameter along the x- and z-axes, or ε_x and ε_z. However, immediately above and below the shaded wafer are wafers at lower stress levels (since they have larger areas), which do not change in diameter as much as the shaded wafer would like to; these wafers apply stresses in the x- and z-directions that oppose the movement of the shaded wafer. Therefore, because of

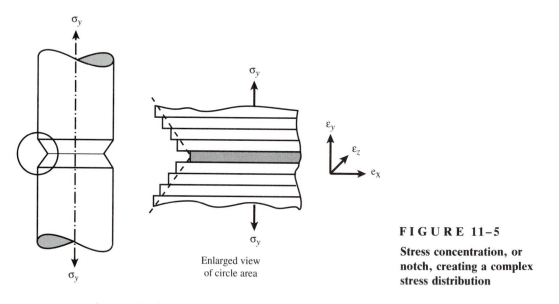

FIGURE 11–5

Stress concentration, or notch, creating a complex stress distribution

the notch, the original y-oriented stress has created stresses in three directions—a triaxial stress distribution.

Assuming that the specimen stays in one piece and doesn't split up into a multitude of thin wafers (a reasonable assumption), the shaded wafer can't move in the direction of ε_x and/or ε_z; i.e., it behaves as if it is brittle, even though it is an otherwise ductile metal.

Since the complex stress distribution at the root of the notch inhibits slip, the stress to make the notched specimen yield is greater than for the unnotched specimen. That is, the yield strength of the notched specimen, based on the reduced area at the notch, is greater than the yield strength of the unnotched specimen. However, the most noticeable effect of the notch in a tensile test is the reduction in elongation; this measure of ductility will be a fraction of the value for an unnotched specimen.

The larger the change in size or geometry in a part or component, the larger the "notch effect," or stress concentration. Thus failures in threaded fasteners often occur at the thread next to the unthreaded portion; the first thread in effect relieves the stress for the rest of the threads.

In machinery and equipment a major consequence of notches or stress concentrations is seen in the role they play in contributing to fatigue failures.

11.6   FATIGUE FAILURES

Metals subjected to cyclical stresses often fail by a process called **fatigue.** Fatigue failures occur after a number of cycles of stresses that are less than the tensile strength of the metal and usually less than its yield strength. Since a cyclical stress occurs on a "per time" basis (cycles per second, cycles per hour, etc.), time is

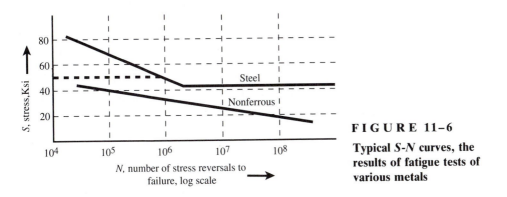

FIGURE 11-6

**Typical *S-N* curves, the
results of fatigue tests of
various metals**

required to cause the failure, but fatigue failures are caused by varying stresses
not by the mere passage of time.

Fatigue Testing. Numerous methods are used to investigate a metal's ability to
withstand cyclical stresses. One of the simplest concepts makes use of the fact
that the maximum stresses in a deflected beam are in its outer fibers and can be
calculated. If a rotating shaft is transversely loaded, a point on the surface goes
through a complete reversal of stress from tension to compression with each
rotation. Laboratory testers are built to count the number of rotations and to shut
off when the specimen fails. Testers are also used that apply cyclical loads in
tension or torsion.

S-N diagrams. The data that result from such tests are used to generate *S-N*
diagrams similar to the one seen in Figure 11-6. An *S-N* diagram plots the applied
cyclical stress versus the number of cycles needed to cause failure at that stress.
Stresses are shown in normal stress units, but the number of cycles is usually
shown on a logarithmic scale because of the large numbers required for some
stresses to cause failure.

Inspection of Figure 11-6 shows that larger stresses cause failure in fewer
cycles, or conversely, in order to have a machine member last through many
cycles the applied stress has to be reduced. Note that the curve for the steel has
two distinct portions; i.e., it has a hockey-stick shape. In the portion with the
steep slope, small reductions in stress cause large increases in the number of
cycles required for failure; then the curve levels off, indicating that if the stress
stays below some limit, fatigue failure will not occur.[†]

Fatigue Strength and Endurance Limit. If the steel of Figure 11-6 is stressed at 50
ksi (dashed line), failure is predicted to occur in about 10^6, or 1 million, stress
reversals. That is, 50,000 psi is the **fatigue strength** at 1 million stress reversals for

[†]The number of cycles to which some equipment is subjected gets quite large. An unexceptional
example is an automobile whose engine turns at about 2000 rpm when driven on a level highway at 50
mph. If the car is driven in this manner for 100,000 miles, the crankshaft goes through about 240×10^6
cycles or stress reversals.

the steel, or fatigue strength is the stress required to cause failure in a certain number of cycles.

However, if the cyclical stress on the steel is kept below about 40 ksi, then fatigue failure should not occur; the **endurance limit** for this steel is 40,000 psi. The endurance limit is the stress below which specimens can withstand an infinite number of stress reversals. Actually, it is the median stress value of the specimens that survived the number of cycles used in the test; i.e., half the specimens were above this value and half were below. Note that for the fatigue strength the number of cycles is included in the specification, while for the endurance limit it is not.

Typically, nonferrous metals have no endurance limit, so their *S-N* diagram is a sloping or gently curving line similar to that shown in Figure 11-6. Because of this lack of an endurance limit, the fatigue strength at a certain number of cycles has to be specified; often this is 10^8 cycles. This is one of the reasons aircraft drive shafts and rotor blades made of aluminum are inspected and changed periodically.

Because notches and stress concentrations significantly reduce fatigue strength, fatigue tests are normally performed using specimens that have polished surfaces. Tests with *polished specimens* have shown that the endurance limit for steel is approximately 0.4 to 0.5 times its ultimate tensile strength. A key point is that this is the best that can be expected. In machine and equipment applications it is probably not good to assume that surfaces can be maintained in an undamaged, polished condition.

The yield strength of some steels, e.g., hot-worked steels, is about one-half the tensile strength, about the same as the endurance limit. This is merely coincidental; there is no relationship between the two values. The point is that when cyclical loads are expected design should be based on the endurance limit, which for many steels is a value much lower than the yield strength.

One alternative, which is what has to be done with nonferrous metals, is to design on the basis of fatigue strength with a specified number of cycles, and then monitor the part closely and remove it from use before it fails.

Patterns of Cyclical Stresses. There are four patterns in which an **alternating,** or **cyclical, stress** can be applied. These relate to the magnitude of the constant stress upon which the cyclical stress is superimposed. Figure 11-7 shows a wave, approximately sinusoidal in form, with an axis marked 0_1 through its median value. In pattern 1 the stress varies from a maximum value in one direction, through zero, to an equal maximum value in the other direction; this pattern has no constant stress, just a varying one. This stress pattern occurs in a transversely loaded rotating shaft, for example.

In pattern 2 a constant stress is superimposed on the fluctuating stress, shifting the axis to the 0_2 position. The stress still changes direction, but its value is greater in one direction than the other. This pattern might occur in the piston rod of a fluid power cylinder required to exert a force in both directions; in one direction the full area of the piston has fluid pressure, but in the reverse direction the piston area is reduced by the area of the rod; i.e., the stresses in the rod are opposite but not equal.

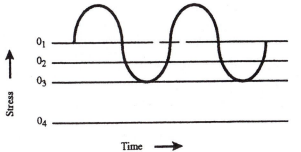

FIGURE 11-7

An alternating stress shown as a wave with four possible patterns or zero-stress axes

When the zero-stress axis is shifted to the extreme of the alternating stress (axis 0_3), it means the cyclical stress is applied in one direction only and then completely removed. This type loading occurs in almost any lifting device.

In pattern 4 the zero axis (0_4) is well removed from the cyclical stress, or the system has a "prestress" with the cyclical stress superimposed on it. This is the stress pattern to which threaded fasteners are subjected. Just tightening the fastener automatically provides prestressing; some moving device, e.g., reciprocating, rotating, etc., then provides the alternating stress. Such a stress pattern also occurs in a spring preloaded to hold something in place and then stretched when a force is applied.

In addition to categorizing stress patterns, the discussion above is significant because as the stress patterns move from 1 to 4, i.e., move from stresses varying in both directions to the prestressed condition, the plot of the *S-N* diagram moves up. That is, *fatigue strength is increased by changing to the prestressed condition.*[‡]

In the previous discussion the stresses were assumed to have a symmetrical shape or contour. It should be realized that the actual form of the alternating stress can be sinusoidal, triangular, square, trapezoidal, or combinations of these shapes, and their amplitudes can be random rather than equal. Some of these shapes are illustrated in Figure 11-8.

Initiation and Propagation of Fatigue Failures. Fatigue failures are initiated by localized shearing stresses that exceed the shear strength. It appears that the alternating stress causes slip to occur in two directions rather than just one. This causes a nonuniform sliding of the slip planes, which in turn causes a small notch. This surface irregularity is the beginning of the crack.[§]

After the formation of an initial crack, the alternating stress proceeds to propagate the crack through and/or around the specimen, depending on whether stress concentrations are present and what type they are. The crack itself is responsible for creating a complex stress distribution so that limited yielding and work-hardening can take place. The crack proceeds in small steps as separating

[‡]Robert E. Reed-Hill and Reza Abbaschian. *Physical metallurgy principles* (3d. ed.). Boston: PWS-Kent, 1992, 752.

[§]Ibid., 756–8.

F I G U R E 11–8 **Various waveforms for cyclical stresses**

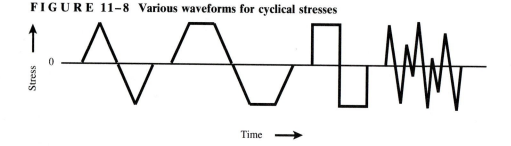

stresses exceed the value needed to cause the crack to move forward; at lower values and during compressive stresses the crack does not propagate.

Figure 11-9 shows that whether an initial crack propagates or not is dependent on the stress level. For a particular metal if the stress is σ_3 the crack does not propagate, but at higher stresses, σ_2 and σ_1, the larger the stress, the fewer the cycles needed to cause failure. (When the curves are parallel to the y-axis it means the crack grows with no additional cycles, i.e., the part fails.)

As the crack grows and spreads, it reduces the area of the metal available to carry the load. Since a large number of cycles are usually required for failure, the crack may propagate for years. When the load-carrying area is reduced enough that the load exceeds it tensile strength, the metal pulls apart, usually very suddenly.

The fractured surface usually has two very distinct areas; the portion formed by the fatigue crack is very smooth, but the portion that failed last is very rough and/or distorted.

Typical Fatigue Failures. Most fatigue failures originate at stress concentrations; these can be internal defects and inclusions, but more often they are caused by a sharp notch of some kind. Machined sharp corners, screw threads, splines, keyways, and tool marks are examples of potential stress concentrations that often lead to eventual failure by fatigue. Many failures begin on the outer surface of parts because this is where the maximum bending and torsional stresses are and where machine marks and surface damage can cause stress concentrations.

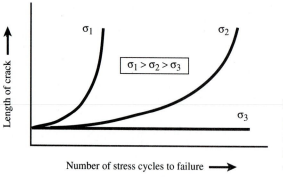

$\sigma_1 > \sigma_2 > \sigma_3$

F I G U R E 11–9

Relationship of crack growth (length) and stress cycles to failure to the relative value of the stress

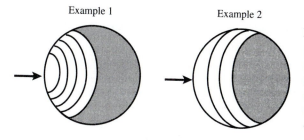

FIGURE 11–10

Illustration of two fatigue failures that progressed from left (beach marks) to final failure on the right (roughened and distorted area)

Fatigue failures usually leave behind telltale signs that enable their origin to be determined. As a fatigue crack progresses through metal, it leaves **beach marks**‖ that mark the limit of the crack at that particular stress cycle. Cycle after cycle these marks are imbedded in the metal so that when the failure surface is finally exposed, the beach marks can often be seen with the naked eye. The failed surfaces shown in Figures A-36, A-37, and A-39 are good examples of beach marks.

Figure 11-10 is a representation of the failure surface of two parts that have failed by fatigue. The fatigue crack of both examples began on the left and moved to the right, but the extent of the stress concentrations present was different. In example 1 the crack initiated at a localized stress concentration; once initiated, its easiest path was through the metal. In example 2 the crack was initiated about where the arrowhead is, but the stress concentration was so significant that the easier path for the crack was to follow it part of the way around the piece before continuing through it. (See Figure A-37.)

The type stress concentrations that might cause a failure like example 1 are a set screw impression, a machining mark, or a burn mark from a cutting torch used to remove a bearing. (See Figure A-38.) Screw threads are a common cause of failures that have the appearance of example 2. (See Figure A-37.)

The cross-sectional area of the metal involved in the final failure is an indication of how much the part was overdesigned on the basis of the load to be carried. Figure A-39 shows the failure surface of a reciprocating hydraulic piston rod designed on the basis of buckling rather than strength; it failed by fatigue in spite of being overdesigned by a factor of 10 or more. The presence of screw threads caused a stress concentration that was just too powerful.

Uniform cyclical stresses generate a smooth failure surface on which beach marks may be difficult to see; these marks are more obvious when the stresses vary, leaving different distances between the marks. (See Figure A-37.)

11.7   STRATEGIES TO AVOID FAILURE

Strategies to avoid failure will be discussed from two standpoints: metallurgical and design. A general concept, admittedly not always possible to accomplish, is to

‖These marks are similar to those left by waves that mark the tide on a beach.

design and select materials so that the load is always in the elastic region and complex stress distributions are avoided. This means the design can be based on a much more predictable ductile failure.

Since any sharp interruption in the surface of a part or component can become a stress concentration, it is important that flaws caused by improper manufacturing techniques (tool marks, heat-treat cracks, etc.) be avoided. Although this is an important area, it is beyond the scope of this text.

Ductile Failures.   Avoiding ductile failures was discussed earlier, in Section 11.3, but some general conclusions are repeated here. From a metal standpoint, if the assumption can be made that multiaxial stresses are not present, then the major strategy is to choose metals with higher yield strengths. From a design standpoint, the strategy is to use members with larger cross-sectional areas and to use load-limiting protective measures, e.g., circuit breakers, clutches, couplings, and shear pins.

Brittle Failures.   If failure is anticipated using brittle metals, e.g., cast iron, then, similar to forestalling ductile failure, the strategy is to increase the size and strength of the member. Usually, however, the concern is to avoid brittle failures in ductile metals. As discussed in Section 11.4, these failures are usually related to a multiaxial stress distribution, so an obvious design strategy is to use pin joints or point loadings, for example, to keep the loading simple (unidirectional) rather than complex. Since notches and other stress concentrations can turn simple stresses into complex ones, they should also be avoided.

If all these measures still leave a brittle failure likely, then, from the metal standpoint, the normal strategy is to use a metal with a high yield strength.

Fatigue Failures.   From the standpoint of the metal, the most obvious strategy to avoid fatigue failures is to choose a metal with a high endurance limit or fatigue strength. For steels, high-carbon content and 100% martensitic structure give higher endurance limits. The decision as to whether to use ferrous or nonferrous metals will be made on the basis of many factors, but to choose a ferrous metal just because it has an endurance limit rather than a fatigue strength may be a mistake. A good design using a nonferrous metal, properly monitored during its expected life, is preferable to a poor design that uses a ferrous metal and trusts the endurance limit will offset the stress concentrations present. Representative mechanical properties for various metals are given in Table XI-1.

The need to avoid stress concentrations also suggests that metallurgical measures be used to strengthen or harden the outer surface of parts and thus improve their resistance to fatigue. The surface-hardening processes discussed in Chapter 7 are helpful in improving the fatigue properties of steel members subjected to cyclical stresses. Hardening the outer surface helps to reduce the initiation of fatigue cracks and helps prevent the creation of notches from surface damage.

Hardening the surface by heat-treating and/or cold-working (peening, shot-peening, or cold-rolling) places the surface under compressive stresses, which gives the surface the prestressed stress pattern discussed earlier and raises the

T A B L E XI–1 Representative mechanical properties, including fatigue properties

Metal	Alloy	Temper	Tensile Strength, ksi	Yield Strength, ksi	Elongation, %	Fatigue Strength or Endurance Limit		Source
						ksi	Cycles	
Aluminum	6061	T6	45	40	12	14	5×10^8	1
Aluminum	7075	T6	83	73	11	23	5×10^8	1
Brass	70–30	Hard	76	63	8	21	1×10^8	2
Bronze	5% P	Hard	82	80	8	33	1×10^8	2
Alloy steel	4340	Q/T	160	140	18	70	End. lim.	2
Stainless steel	302	Annealed	90	37	55	34	End. lim.	2

Sources: 1. *Aluminum standards and data.* Washington, D.C.: Aluminum Association, 1984, table 2.1.
2. *Metals handbook* (8th ed.). Materials Park, OH: ASM International, 1961, vol. 1.

S-N diagram. For example, rolled rather than machined threads are preferred where fatigue is a factor, and many highly stressed aircraft parts are shot-peened.

The use of metal that is as inclusion-free as possible helps to prevent the initiation of fatigue cracks by internal defects.

In the design of parts subject to cyclical stresses much can be done to reduce fatigue failures. Obviously, the need for any stress-concentrations—any shoulder, notch, keyway, reentrant angle, or hole—must be challenged, and if required, must have the stress concentration reduced by generous fillets and/or radii. Parts should be designed so that they can be manufactured with surfaces free of stress concentrations caused by machining marks.

The design should anticipate the energy that must be absorbed during the stress reversal. For example, threaded members can be undercut, i.e., made with a reduced-body diameter, so that more of the member is involved in absorbing the energy of the stress reversal. This is very much dependent on the resilience of the metal. (See Section B.6.)

Earlier, protective load-limiting devices were proposed as methods to help avoid ductile failures. Although such devices cannot stop the propagation of fatigue cracks, they can help avoid the excessive stress that may initiate such a crack, and therefore they may be useful here also.

Inspection Techniques. One of the problems with fatigue failures is that the final failure happens suddenly and unexpectedly and can be a real threat to safety. There are a number of nondestructive inspection techniques that, although they do not prevent the initiation of fatigue failures, can help prevent the final failure from being catastrophic. As might be expected from the previous emphasis on stress concentrations and surface conditions, most of these are methods for in-specting the surface of metals.

- Penetrant dyes are available that can be sprayed on a surface to help locate fine cracks and openings. Fluorescent penetrants are also used that reveal cracks under "black light." These methods can be used on both ferrous and nonferrous metals.
- Eddy currents induced into metal parts are interrupted by defects and can be detected by special instrumentation. This technology can also be used to detect metal and film thickness; it can be used on both ferrous and nonferrous metals.
- Magnetic particle inspection is used to find defects in ferrous parts. The part is placed in a magnetic field; a defect that interrupts the field is outlined by the pattern of iron powder that is sprinkled on it.
- Internal defects can be found by using X rays, but the test has rather limited application.
- Ultrasonic testers measure sound waves reflected by internal defects. The test can be used to detect fatigue failures in critical components, e.g., shafts of overhead cranes, by testing from the end of the shaft.

SUMMARY

1. Quality manufacturing requires that the correct specifications be met throughout the whole manufacturing process.
2. Mechanical failures of components often have multiple causes.
3. Ductile metals deform because of shearing stresses if the loading is not complex; these stresses will cause the metal to move at an angle to the applied stress.
4. In tension, ductile metals fail with a "cup-and-cone" fracture with 45° angles; in torsion, they twist off like taffy.
5. Interruption of that sideways movement, i.e., the canceling out of the shear stresses, means the ductile metal cannot fail in its normal manner; it has been subjected to a complex stress distribution and fails as if it were a brittle metal.
6. Brittle metals fail because of separating or cleavage stresses that pull the metal apart with little distortion or deformation. In tension specimens break with a surface 90° to the stress direction; in torsion the failure occurs with a 45° spiral.
7. Complex, or multiaxial, stress distributions are most easily caused by a notch or stress concentration; these stress distributions limit the yielding and work-hardening of the ductile metal.
8. Alternating or cyclical stresses can also cause stress concentrations or notches by fatigue.
9. Fatigue tests are used to determine the fatigue properties of metals.
10. Ferrous metals have an endurance limit; at this stress value, the metal has some probability, usually 100 or 50%, of withstanding an infinite number of stress reversals.

11. Nonferrous metals generally do not have an endurance limit; their fatigue strength is the stress that can withstand a stated number of cycles, usually 10^8 cycles.

12. In general, endurance limits and fatigue strengths are less than half the tensile strength; see Table XI-1.

13. Alternating stresses occur in four patterns, which differ by whether or not a constant load is also present and how large that load is. The constant load, or prestress, has the effect of raising the fatigue strength.

14. Fatigue cracks are often initiated by stress concentrations; they continue to propagate if separating stresses are large enough.

15. Continued propagation of the crack leads to failure when the load exceeds the strength of the metal that is still intact. The failure surface usually has telltale beach marks that can identify where the crack initiated.

16. Specific strategies can be used to avoid failures:
    a. Use metals with high yield strengths.
    b. Avoid designs that create complex stress distributions; try to keep metals loaded so they can fail in simple and predictable stress modes.
    c. Avoid notches, sharp corners, and surface imperfections in design, manufacture, and use. Design with generous radii and fillets. The largest stress concentrations occur where the section thickness or geometry makes the biggest change.
    d. If cyclical stresses are to be present, choose metals with high endurance limits or fatigue strengths; in steels, the martensite structure and higher carbon content contribute to higher fatigue strengths. In critical applications inspect the part for fatigue-type defects.
    e. Hardening the outer surface by heat-treating or cold working will reduce the potential for notches and place the outer surface in compression; both help to reduce fatigue failures.
    f. If the alternating stress involves large masses, take into account the energy that must be absorbed when the stress reverses, and use metals and designs that maximize resilience.

REVIEW QUESTIONS

1. Sketch typical cyclical stress waveforms with different stress patterns and describe their effect on the S-N diagram.

2. How can shear stresses, or lack of them, determine whether a failure is a ductile or brittle failure? Use a sketch of the stresses in your answer.

3. Describe how reducing the diameter of the nonthreaded portion of a bolt can help reduce the likelihood of a fatigue failure in the threaded portion.

4. If a low-carbon steel has the properties of 100 ksi TS, 55 ksi YS, approximately what would be its *maximum* endurance limit?

5. Sketch the pattern of beach marks that might be generated by the cyclical axial loading of a horizontally mounted rectangular steel bar that has a burn mark (from a cutting torch) on its upper surface.

6. Given Figure 11-6, describe how you might go about trying to estimate the *tensile* strengths for the two metals.

7. Outline some of the strategies you might use to avoid the failure of the motor-boat drive shaft shown in Figure A-36.

FOR FURTHER STUDY

DeGarmo, E. Paul, J. Temple Black, and Ronald A. Kohser. *Materials and processes in manufacturing* (7th ed.). New York: Macmillan, 1988, 53–6.

Dieter, George E. *Mechanical metallurgy* (3d ed.). New York: McGraw-Hill, 1986, chap. 7.

Metals handbook (9th ed.). Materials Park, OH: ASM International, 1986, vol. 11, Failure analysis and prevention.

Reed-Hill, Robert E., and Reza Abbaschian. *Physical metallurgy principles* (3d ed.). Boston: PWS-Kent, 1992, chap. 21.

Wulpi, Donald J. *Understanding how components fail*. Materials Park, OH: ASM International, 1985.

CHAPTER 12

Failure Due to Impact and Low Temperature

OBJECTIVES

After studying this chapter, you should be able to

- describe the relationship between impact strength and temperature for BCC, FCC, and CPH metals
- sketch impact strength vs. temperature curves for a normalized carbon steel compared with a quenched and tempered steel
- describe the effect each of the following factors has on impact strength and transition temperature for BCC steels: carbon content, alloy content, microstructure, grain size, surface condition, deoxidation and removal of other discontinuities from the metal, orientation of the metal during working, and tempering temperatures
- identify those factors that influence the impact strength of nonferrous metals
- make broad comparisons, based on data like those of Table XII-1, of low-temperature tensile properties vs. low-temperature impact properties for metals used in engineering applications

DID YOU KNOW?

In World War II there was a need to ferry large numbers of troops and supplies to Europe. For this purpose thousands of cargo ships and tankers (Liberty ships), many built to very similar designs, were fabricated by welding rather than the previously used riveting method. In a short period of time approximately 200 of these ships failed, sometimes catastrophically, e.g., breaking in two. In some failures the ships bent, with the fore and aft ends going down and the midsection going up; some ships divided into completely separate fore and aft sections.*

Previous failures were known in riveted ships and other structures (pipelines and tanks, for instance), but initially the culprit was thought to be the welding. It was thought that riveted plates would shift to redistribute stresses, whereas the rigid welded structure wouldn't. After much investigation and testing, welding turned out not to be *the* problem, but it did accentuate the real causes—subjecting steel with low impact strength to high impact stresses at low temperatures with stress concentrations present.

A PREVIEW

This chapter discusses the properties of metals at low temperatures. However, most of our emphasis is on impact strength and how it varies with temperature, so some appreciation of how that property is measured is required; if you're not familiar with the impact test, study Appendix F before proceeding.

We begin by discussing the low-temperature notch sensitivity of steels and relate it to the problem of the Liberty ships in World War II. The transition temperature of steels is then described and compared with the low-temperature properties of metals in general. We'll identify and discuss the factors that influence the impact strength and transition temperature of steels. Finally, we'll study low-temperature data for representative metals and relate the data to the principles discussed earlier.

This chapter makes use of plots of impact strength and temperature; using those data, we can visualize the transition temperature as occurring at the temperature midway between the upper-level and lower-level impact strengths. However, it should be remembered that the transition temperature can also be defined according to the fracture surface (see Section F.5).

*Summaries of the investigation into the ship failures are found in *Metals handbook* (8th ed.). Materials Park, OH: ASM International, 1961, vol. 1, 225; and Robert E. Reed-Hill and Reza Abbaschian. *Physical metallurgy principles* (3d ed.). Boston: PWS-Kent, 1992, 830. For more complete information, see: Brittle fracture and structural failure of the Liberty ships during WW II. *The Welding Journal,* American Welding Society, July 1947, or M. E. Shank. Brittle failure of steel structures—A brief history. *Metals progress,* American Society for Metals, September 1954, 83–88.

12.1   LOW-TEMPERATURE NOTCH SENSITIVITY IN STEELS

Stresses.   A large oceangoing ship, such as the Liberty ships mentioned above, can be thought of as a fabricated beam with its ends closed. In calm water the ship would be supported by hydrostatic pressure, following Archimedes principle, and the stresses would be fairly equal from one end to the other. However, in the rough waters of the North Atlantic, and especially with high waves some distance apart, the support would vary from end to end. The ship could go from being a simply supported beam between the crests of two waves at one moment, to being supported at its center like a seesaw the next. Under such conditions the magnitude and direction of the stresses would vary from moment to moment.

Notches and Low Temperatures.   Cargo vessels require easy access to their interiors, which means there must be large openings or hatches in their decks, i.e., the top side of the beam. These hatches weaken the ships, act as notches or stress concentrations, and create complex stress distributions. In the case of the Liberty ships, protection from submarines was also a consideration; the ships traveled along the coastline so they could be covered by land-based aircraft. This took them into the cold waters of the North Atlantic, which added to the specific problem we'll explore in detail below.

Catastrophic Failures.   High seas associated with winter storms in the North Atlantic strained the welded plates of the Liberty ships described in the opening paragraphs of this chapter. Because of the rigid construction, stresses were not redistributed, but instead small cracks were formed. Because the steels used were notch-sensitive, the cracks increased in length until one last strain united them in a catastrophic failure across the deck and sometimes down the sides. Some failures occurred in heavy seas, but others happened in the calm waters of rivers and at dockside. The latter failures were accounted for by the difference in expansion between a hull in cool water and a deck warmed by the sun.

The immediate solution to this serious wartime problem was to provide stress relief at the corners of the hatches and place riveted sections in the welded structure to act as "crack arresters."

12.2   TRANSITION TEMPERATURES IN BCC METALS

Variations with Temperature.   The more basic cause of the Liberty ship failures was that notch sensitivity or impact strength of BCC metals varies with temperature. The Charpy impact test shows that these metals fail in a ductile mode at higher temperatures, in a brittle mode at lower temperatures, and go through a transition at intermediate temperatures. (See Section F.5.) Figure 12-1 illustrates a curve typical of low-carbon steels.

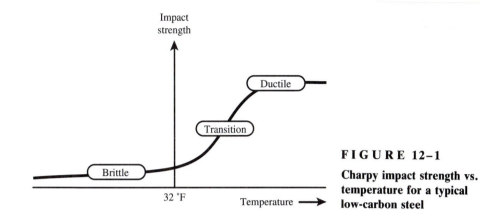

FIGURE 12–1

Charpy impact strength vs. temperature for a typical low-carbon steel

After performing many Charpy impact tests on the steels specified for the Liberty ships, researchers reached the conclusion that the steels were too notch-sensitive, i.e., too sensitive to stress concentrations to avoid cracking at the cold temperatures to which they were subjected in the North Atlantic. At higher temperatures they failed in a ductile mode, but at temperatures closer to 32°F (0°C), and under the loading experienced, they were too brittle.

Specifically, the Charpy test showed that the steels went through their ductile-to-brittle transition at temperatures near to those being experienced by the ships in the North Atlantic. Although the strain rate and stress concentration of the Charpy test did not duplicate actual conditions, the test did identify the temperature of the steel's transition from ductile to brittle behavior *under standard conditions*. Obviously, the impact strength of the steels being used was too low since the ships were failing.

Standardized Testing. It is stated in Appendix F that the data that result from the impact test can be used only to compare, select, or specify metals; they cannot be used for design purposes. The major reasons for this are that *the standardized conditions of notch size, notch shape, and strain rate rarely simulate the actual conditions,* and the results of the test depend on these standardized parameters.

Comparison of Variables. To demonstrate this, the author tested specimens of the same steel at various temperatures but with sawed notches of 1/32 and 3/16 in.[†] The results of the tests are shown in Figure 12-2. As might be expected, the shallower notch had a higher impact strength and a lower transition temperature, i.e., it failed in ductile mode at lower temperatures, than the specimens with a notch closer to standard.

As a generality, the upper-level strength is increased (the curve is raised), and the transition temperature is lowered (the curve is moved to the left) by

• reducing the stress concentration of the notch (making it less sharp or shallower) or

[†]At all temperatures unnotched specimens only bent, they did not fracture.

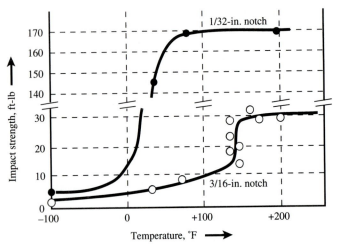

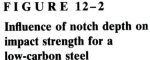

F I G U R E 12–2

Influence of notch depth on impact strength for a low-carbon steel

- reducing the strain rate (making the hammer travel more slowly, e.g., dropping it from a lower height)

So, in the case of the Liberty ships there is no direct or causal correlation between the strain rates, notches, or temperatures experienced in the ships and those used in the tests. The operating temperatures experienced were close to the transition temperatures as defined by the test, but they were not exactly the same. The test is merely a standardized way to define and study the ductile-brittle transition.

12.3   LOW-TEMPERATURE IMPACT PROPERTIES OF METALS

Metals with other atom lattices do not go through the dramatic change in ductility or impact strength experienced by the BCC ferrous metals. The basic relationships are shown in Figure 12-3.

FCC Metals. FCC metals, e.g., nonferrous metals like copper and aluminum alloys and austenitic steels, generate impact strengths that are very close to being constant, regardless of temperature, as shown in Figure 12-3. In general, the impact strengths of these metals go up slightly as temperature goes down.

BCC Metals. BCC metals are a different story; the impact strength of a low-carbon steel may decrease by a factor of 10 between test temperatures of 100 and $-100°F$ (38 to $-73°C$). Alloy steels, on the other hand, particularly those containing nickel, retain their ductility until far lower temperatures. Cooling quenched-and-tempered (martensitic) AISI 4340 from room temperature to $-50°F$ only reduces its impact strength about 10%; its transition temperature is about $-150°F$.

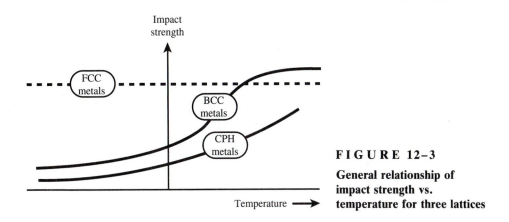

FIGURE 12–3

General relationship of impact strength vs. temperature for three lattices

On the other hand, if the same steel is normalized (pearlitic), its impact properties are about the same as a low-carbon steel.

A number of variables influence the low-temperature impact strength of BCC ferrous metals and they are investigated in the next section.

Other BCC metals also undergo the ductile-brittle transition. For example, the impact test identifies a transition temperature of 35°C (95°F) for a cast 75% tungsten–25% rhenium alloy.[‡]

CPH Metals. Metals and alloys with the CPH lattice do not go through a transition, but their impact strengths gradually decrease as temperature decreases. Data for titanium,[§] CPH at room temperature and below, and zinc[‖] show this behavior.

12.4 FACTORS AFFECTING THE IMPACT STRENGTH OF BCC STEELS

Stress concentrations and high strain rates are *mechanical* variables that influence the impact properties of metals in general. For steels there are also a number of chemical-composition and physical-structure variables that influence impact strengths and/or transition temperatures. Since composition and physical structure are interrelated, they are discussed together, but some introductory generalities will be helpful as you follow the flow of the discussion:

- Lowering the carbon content improves impact strength.
- **Tempered martensite** has the best impact properties.
- Small-grained steels have better impact properties than large-grained steels.

[‡]*Alloy digest.* Upper Monclair, NJ: Engineering Alloys Digest, January 1970, filing code W-15.
[§]Brandes, Eric A. (ed.). *Smithells metals reference book* (6th ed.). London: Butterworth, 1983, table 22–40.
[‖]*Alloy digest.* Upper Monclair, NJ: Engineering Alloys Digest, May 1969, filing codes Zn-12, and October 1978, Zn-34.

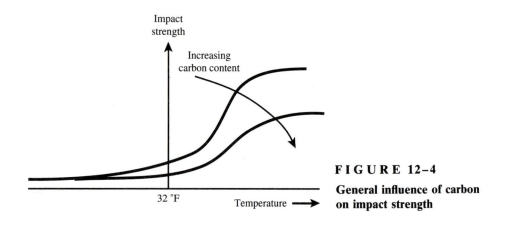

F I G U R E 12–4

General influence of carbon on impact strength

- Discontinuities in the steel tend to lower impact strength and raise the transition temperature.

Carbon Content. Carbon is detrimental to impact properties. As illustrated in Figure 12-4, increasing the carbon content has the effect of rotating the upper part of the curve to the right, lowering the upper-level impact strength and raising the transition temperature. In general, for the same heat-treated condition, e.g., normalized, this holds true for alloy and carbon steels: The higher the carbon content, the lower the upper-level impact strength and the higher the transition temperature, or the lower the carbon content, the higher the upper-level impact strength and the lower the transition temperature. Thus for engineering applications involving low temperatures and/or impact, it is preferable to use a steel with as low a carbon content as possible. Note that this is in the direction of lower strength, which is not good for fatigue life.

Manganese Content. In slow-cooled and normalized steels, manganese in amounts up to 1.5% *reduces* the transition temperature; the curve is moved to the left but changes shape very little (Figure 12-5). The reduced transition tempera-

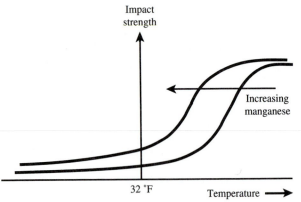

F I G U R E 12–5

Influence of small amounts (< 1.5%) of manganese on impact strength vs. temperature curves for slow-cooled steels

ture comes about probably because the manganese reduces the spacing of the pearlite laminations, which reduces the thickness of the weak ferrite plates. In martensitic steels, manganese tends to *promote* brittleness and reduces their impact properties.

Boron Content. Reducing carbon content improves impact strength, but this also reduces tensile strength. In lower-carbon quenched-and-tempered steels, boron (as little as 0.0005%) has the ability to *raise* tensile properties without reducing impact strength. Thus a steel with less carbon can have the same tensile properties as a higher-carbon steel, but the lower carbon improves its impact properties.

Just the reverse of manganese, boron is only beneficial to quenched-and-tempered steels; it *reduces* the impact strength of slower-cooled steels.

Nickel Content. In general *nickel lowers the transition temperature* while reducing upper-level impact properties. When used in moderate amounts (1 to 3%), it reduces the transition temperature enough to put it below normal ambient temperatures without reducing the upper-level properties too adversely. Used in larger amounts (7% and more), as in austenitic stainless steels, it eliminates the transition altogether by converting the steel to the gamma-FCC phase.

Microstructure. Microstructures that have low impact strengths at room temperature tend also to have high transition temperatures. Since soft ferrite is easily cleaved, it has the poorest impact strength and highest transition temperature among steel microstructures. Pearlite is slightly better, but its ferrite plates are paths of easy cleavage. Bainite is better than pearlite, but tempered martensite has the best impact strength and lowest transition temperature. Or, comparing the impact properties for a given steel with its microstructures,

- Tempered martensite is better than lower bainite.
- Lower bainite is better than upper bainite.
- Upper bainite is better than pearlite.

Significant amounts of austenite retained in martensite increase the impact strength. Apparently the austenite either blocks the propagation of the crack or consumes energy as it goes through a complex transformation to martensite triggered by the strain of the crack.

Generally, for a given steel, normalizing or quench-and-temper processes that give *lower* strengths and hardnesses at room temperature produce higher impact strengths[#] and lower transition temperatures. That is, a martensitic steel will have a higher impact strength and lower transition temperature if tempered at a high temperature; i.e., it will have lower room-temperature hardness and strength than if tempered at a lower temperature. Similarly, a lower-carbon martensitic steel will have better impact properties than a martensitic steel with higher carbon.

[#]FCC metals are similarly affected. Annealed aluminum, copper, and austenitic stainless steel have higher impact strengths than when heat-treated or cold-worked. See *ASM Metals handbook* (1948 ed.). Materials Park, OH: ASM International, 207–15. See also Eric A. Brandes (ed.). *Smithells metals reference book* (6th ed.). London: Butterworth, 1983, table 22–10.

High-carbon, 100% martensitic steels were identified in Chapter 11 as having the best fatigue resistance. From the previous discussion it is clear that the same steel composition and/or the same microstructure will not provide the *best* resistance to fatigue *and* impact failures. Compromises based on the most likely type of failure have to be made.

Spheroidizing treatments of pearlitic steels improve impact properties by changing the brittle cementite plates into spheres that represent less of a stress concentration and by removing the easily cleaved ferrite plates. Annealed metals in general have higher impact strengths than cold-worked metals.**

Hardenability.   Other alloying elements, e.g., molybdenum and chromium, are important in improving impact properties by improving the hardenability of the steel and its ability to become completely tempered martensite, the best structure for low-temperature impact resistance in steels.

Surface Conditions.   Surface-hardening techniques in general reduce impact strength. Carburizing, nitriding, and surface peening all improve fatigue resistance, as discussed in Chapter 11, but in doing so, they create a brittle layer that contributes to crack propagation under impact. The reverse is also true; a soft ductile layer on the outside surface improves impact strength.

Decarburizing the surface of a steel, e.g., heating the steel in air,†† softens the surface, increases impact strength, and lowers the transition temperature. It is unlikely that this would be done purposely, however, since the decarburized outer layer reduces fatigue resistance.

Grain Size.   Grain boundaries raise the energy required to propagate the failure crack. So, for any given steel and microstructure, the smaller the grain size (and thus the more grain boundaries), the higher the upper-level impact strength or toughness‡‡ and the lower the transition temperature. This makes important any alloying additions or processing methods that reduce grain size. For example, alloying additions of vanadium, niobium, titanium, or aluminum (see Section 8.7) are used to control grain size and thus are helpful in improving impact properties.

Thermal practices and processes that avoid grain growth and promote fine-grained steels contribute to better impact properties. Hot-working performed and completed at as low a temperature as possible facilitates this. Normalizing at lower temperatures helps to prevent the ferrite plates of the pearlite from getting thicker, thus improving the impact properties.

The difference in the grain size possible between slowly cooled large sections and more rapidly cooled thin sections is one of the reasons that thinner steel sections tend to have better impact properties than thicker sections.

**See *ASM Metals handbook* (1948 ed.). Materials Park, OH: ASM International, 207–15.
††That is, heating the steel in air as opposed to in a controlled-atmosphere furnace.
‡‡*Impact toughness* is the ability to withstand a load at high strain rate with a stress concentration present that prevents ductile behavior; it is improved by *small* grains. The *modulus of toughness* (see Section B.6) is dependent on a metal's ability to undergo slip, that is, to be ductile; it is facilitated by *large* grains. These are two different measures of toughness.

Discontinuities.   In general, structures that interrupt the normal atom lattice in steels tend to reduce impact properties. Interstitial elements such as carbon, oxygen, nitrogen, and hydrogen and sulfur inclusions or stringers such as manganese sulfide all reduce impact strength. So those practices and alloying additions that produce cleaner, gas-free steels have higher upper-level impact strengths and lower transition temperatures. Aluminum ($\approx$0.075%) and silicon (<0.30%) are often used to deoxidize or "kill" steels; aluminum also combines with nitrogen and helps control grain size. Manganese reduces the sulfur content by forming sulfides that come off in the slag during the original melting process.

Tempering.   Tempering steels at certain temperatures can promote brittle behavior. **Blue brittleness** occurs when medium- and high-carbon steels are tempered in the range of 450 to 700°F (230 to 370°C); this condition should be avoided. **Temper embrittlement** can occur if medium- and higher-carbon steels containing chromium, manganese, phosphorus, tin, arsenic, or antimony are tempered at higher temperatures (700 to 1100°F, or 375 to 575°C) or cooled slowly through this range from a higher tempering temperature; thus cooling rates and alloy composition must be controlled.

Orientation in Working.   The products of commercial working processes such as drawing, rolling, forging, or extruding have properties that vary with the direction of working because the metal grains tend to elongate in the direction of working. Since grain size and grain boundaries influence mechanical properties, impact tests and other tests are often performed on specimens with various orientations to the direction of working.

Figure 12-6 illustrates three orientations for impact specimens in a rolled plate. The long axes of the two labeled "*L*" are parallel to the working direction, but their notches are perpendicular to the direction of working and at 90° to each other. The notch of the "*LS*" specimen lies in a plane parallel to the surface of the plate; that of the "*LT*" is in the thickness of the plate. These two specimens are oriented such that the elongated grains lie across their notches. The third speci-

F I G U R E 1 2 – 6 **Variation in impact strength with working orientation**

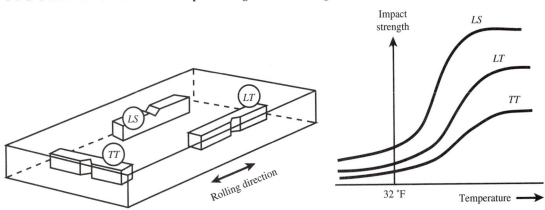

men, "*TT*," is oriented transverse to the direction of working, with its notch lying in the thickness of the plate.

The curves of Figure 12-6 show that the upper-level impact strength and lower transition temperatures are in the sequence just presented. The longitudinal *L* specimens have the better properties; the specimen with its notch parallel to the surface, *LS,* has the best. The transverse specimen, *TT,* has the poorest low-temperature impact properties.

12.5 LOW-TEMPERATURE PROPERTIES OF VARIOUS METALS

Table XII-1 is a compilation of actual data from tensile and impact tests at room temperature and below for various metals, it demonstrates some of the principles discussed in this chapter. The metals are grouped by atom lattice—FCC at the top, CPH in the center, and BCC at the bottom.

Figure 12-7 shows the general relationships of impact properties typical of some of the metals of Table XII-1. These properties are discussed below.

Tensile Properties. Examine the tensile strength column of Table XII-1; note that in every case the tensile strength *goes up* as the temperature goes down, and in most cases the elongation increases also. Figures for the modulus of elasticity are not shown in Table XII-1, but in general this value increases slightly as the temperature goes down. So, as measured by the tensile test most metals do *not* get more brittle at low temperatures, though they do get a little stiffer.

Impact Strength. A comparison of the impact data (Table XII-1) shows that the FCC alloys of aluminum, copper, and nickel, at the top of the table, retain their impact strength at low temperatures. The CPH titanium and magnesium alloys reduce in impact strength somewhat, but do not go through the ductile-brittle transition of the BCC metals. Both the carbon and alloy steels listed go through decided transitions. The data for the carbon steel at +160°F (85 ft-lb) is included to show the extent of the reduction that occurs.

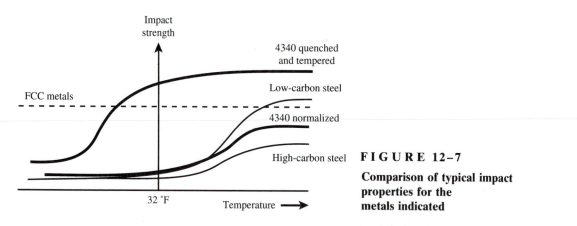

FIGURE 12–7

Comparison of typical impact properties for the metals indicated

T A B L E XII-1 Comparative low-temperature properties for representative metals

Alloy Metal & Lattice	Condition*	Temperature, °F	Tensile Strength, ksi	Elongation, %	Impact Strength ft-lb	Test*	Reference
2024 Aluminum, FCC	T3 Heat-treated & naturally aged	70 −112	65.5 74.1	23.3 25.5	12 12	CV	1
Beryllium-copper, FCC	Solution-heat-treated rod	70 −112 −292	68 78 100	36 38 41	41 40 40	I	2
	Solution-heat-treated precipitation-hardened rod	70 −112 −292	166 180 192	2.6 0.4 3.0	2 3 3	I	2
Cartridge brass, 70% Cu-30% Zn, FCC	Annealed	70 −100 −150 −300	50 56 60 75	48 60 56 75	66 68 70 79	I	3
Monel, 71% Ni-29% Cu, FCC	Annealed	70 −112 −292	70.8 85.3 113.0	41 40 51	90 90 97	I	1
304 stainless steel, FCC	Annealed	70 −60 −113 −150 −300	87.2 143.7 164 225	69.0 66.0 54.0 44.0	119 117 113 119	I	1
	Annealed & cold-drawn	70 −58 −110 −300			66 67 68 65	I	1
IMI 680, titanium, CPH	Quenched & aged rod	70 −108 −148 −320	146 179 224	15.0 10.0 8.5	8 6.5 6	CV	2
AZ 61-magnesium CPH	Extruded	70 −110	46.8 50.0	12.7 12.2	6.9 2.9	CK	1
Carbon steel, 0.22% C, BCC	As-rolled	160 70 −114 & −120	 62.6 69.0	 35.5 35.5	85 55 5	CV	1 & 3
4340 alloy steel, BCC	Normalized @ 1630 °F	70 0 −100			13 7 5	CV	1
	Oil-quenched from 1525°F Tempered @ 1100°F (app. 32 HRC)	70 0 −100			82 82 77		1
	Oil-quenched from 1575°F Tempered @ 450°F (app. 55 HRC)	70 −100 −320	270 280 320	9.5 10.0 10.0			4

*CV = Charpy vee notch; CK = Charpy keyhole notch; I = Izod.

Sources: 1. *ASM Metals handbook*. Materials Park, OH: ASM International, 1948.

2. Eric A. Brandes (ed.). *Smithells metal reference book* (6th ed.). London: Butterworth, 1983.

3. *Metals handbook* (8th ed.). Materials Park, OH: ASM International, 1961, vol. 1.

4. *Metals handbook* (9th ed.). Materials Park, OH: ASM International, 1978, vol. 1.

Tensile Properties vs. Impact Strength. A comparison of the data for beryllium copper and 304 stainless steel shows that strengthening processes reduce impact strengths. The solution-heat-treated/precipitation-hardened beryllium copper has much higher tensile strengths than the same metal heat-treated only, but it has much lower impact strengths at all temperatures. Although their difference is not quite as great, the same principle is true for cold-drawn stainless steel compared with the softer annealed microstructure.

Steel Microstructure. The AISI 4340 steel has three groups of data (Table XII-1). The first two compare the impact strengths of this alloy steel at different temperatures in the normalized (pearlitic) and quenched-and-tempered (martensitic) conditions. These data clearly show the superiority of martensite at low temperatures. Other data[§§] indicate that the transition temperature for martensitic AISI 4340 is about $-150°F$ ($-100°C$); at $-300°F$ ($-185°C$) its impact strength drops to 20 ft-lb. Thus although the alloy steel retains its impact strength to a far lower temperature, it still eventually goes through the ductile-brittle transition.

The third group of AISI 4340 data are for a steel tempered to a much higher hardness (55 versus 32 HRC). Note that the tensile strength goes up as the temperature goes down, as happens with the other metals in the table.

S U M M A R Y

1. The failure of cargo ships during WW II brought to light in a very significant way the problem of low-temperature notch sensitivity in carbon steels.

2. The problem was investigated using the impact test. Results showed that the steels in use were too notch-sensitive for the temperature and loading to which the ships were subjected by the North Atlantic.

3. BCC metals go through a transition from ductile- to brittle-type failure as temperature decreases. The impact properties of FCC metals change very little with temperature. CPH metals loose some impact strength as the temperature decreases, but the change is much more gradual than for the BCC metals.

4. The impact strength and transition temperature of BCC steels can be influenced by factors such as the following:
   • The *lower* the carbon, the higher the upper-level impact strength and the lower the transition temperature.
   • In slow-cooled steels, manganese, up to 1.5%, shifts the impact curve to the left, i.e., to lower temperatures; it does not change upper-level impact strength much.
   • Nickel lowers the transition temperature. Large amounts lower the upper-level impact strength, but in the amounts used in austenitic stainless steels, the BCC lattice is replaced by the FCC and no transition temperature occurs.

[§§]H. J. French. *Trans. Metall. Soc. AIME, 206,* 1956, 770 (cited in George Dieter. *Mechanical metallurgy* (3d ed.). New York: McGraw-Hill, 1986, 482, fig. 14-9).

- Tempered martensite has the best impact properties of the steel microstructures. Alloying elements that improve hardenability thus indirectly improve impact properties.
- Hard, brittle outer surfaces tend to reduce impact properties; soft, ductile outer surfaces give higher impact properties.
- Smaller grains improve impact strength and reduce the transition temperature. Alloying additions such as aluminum, vanadium, niobium, and titanium, used to control grain size, indirectly improve low-temperature properties. Temperatures at which the steel is worked should be controlled to prevent grain growth.
- Impact properties of steels are improved by keeping inclusions and interstitial particles to a minimum.
- Tempering practices and direction of working also influence the impact strength of steels.

5. The tensile strength of metals improves as the temperature is lowered; most metals also have higher elongations at lower temperatures.

REVIEW QUESTIONS

1. Explain how the impact test could help define the problem with the WW II ships but not actually test the steels under the same conditions as the North Atlantic.

2. Identify engineering applications where low temperatures would be important in the selection of metals.

3. For use at low temperatures would you specify that a steel have a high transition temperature or a low transition temperature? Explain.

4. Discuss the desired composition and microstructure of a pearlitic steel intended for use at low temperatures or with impact loads.

5. Discuss the preferred composition and microstructure of a martensitic steel used at low temperatures with impact-type loads.

6. Are room-temperature stress-strain data helpful in selecting metals for use at low temperatures? Are low-temperature stress-strain data helpful in selecting metals for use at low temperatures? Discuss.

7. List and explain the factors that can improve the impact properties of steels. Do the same for nonferrous metals.

FOR FURTHER STUDY

Avner, Sidney H. *Introduction to physical metallurgy* (2d ed.). New York: McGraw-Hill, 1974, secs. 1.33 and 17.3.

Dieter, George E. *Mechanical metallurgy* (3d ed.). New York: McGraw-Hill, 1986, chap. 14.

Metals handbook (8th ed.). Materials Park, OH: ASM International, 1961, vol. 1, 225–43.

Metals handbook (9th ed.). Materials Park, OH: ASM International, 1978, vol. 1, 689–709.

Metals handbook (10th ed.). Materials Park, OH: ASM International, 1990, vol. 1, 737–54.

Reed-Hill, Robert E., and Reza Abbaschian. *Physical metallurgy principles* (3d ed.). Boston: PWS-Kent, 1992, 827–33.

Failure Due to High Temperatures

OBJECTIVES

After studying this chapter, you should be able to

- describe the effect of high temperatures on the tensile properties of metals
- define creep, describe a creep test, sketch the results of a creep test, and indicate the three stages of creep on the sketch
- describe the creep-rupture test and the data obtained
- differentiate between the creep and stress-rupture tests and between creep strength and stress-rupture strength
- identify and describe the metallurgical factors that improve high-temperature properties of metals
- describe the mechanism by which the superalloys are strengthened
- state the approximate temperature limits for applications of carbon steels, stainless steels, and superalloys

In Chapter 11 we saw that ductile metals fail in a brittle mode when they are subjected to stress concentrations that interrupt or cancel the shear stresses. At higher temperatures, brittle-type failures are less of a problem because ductility is encouraged by raising the temperature. At higher temperatures failure is far more likely because of slip than because of separating-type stresses. Thus those strategies that improve resistance to slip become important.

However, at higher temperatures some of the normal techniques don't work. Cold work raises strength, but if a metal is used at or above its recrystallization temperature the metal softens, grains coarsen, and the effects of the cold work are lost. Similarly, phases that strengthen by precipitating at low temperatures, e.g., age-hardened aluminum alloys, overage at higher temperatures and redissolve if held near the solvus line. Heat-treated steels are essentially retempered if used at temperatures above their tempering temperature; at higher temperatures they are annealed or spheroidized and their advantageous properties are lost.

In addition to these problems, when metals are subjected to stresses that they are normally capable of withstanding at lower temperatures, they exhibit a phenomenon called *creep*—a slow stretching. Creep is the major subject of this chapter.

A PREVIEW

We explore first the effect of high temperatures on tensile test results and compare some data to illustrate the effect. Then we look at the phenomenon of creep and the results of creep tests; we define the three stages of creep. We'll emphasize that the second stage of creep is the basis of creep strength and describe another measure of a metal's ability to resist high temperatures, stress-rupture strength. We'll investigate the factors that influence a metal's resistance to high temperatures. Alloy content is one of these factors, and we'll explore it by comparing the composition of metals in industrial use. Finally, we'll describe the strengthening mechanisms used by the superalloys and relate those mechanisms to the compositions and properties of three alloys.

13.1   EFFECT OF HIGH TEMPERATURE ON TENSILE PROPERTIES

At higher temperatures atoms are more mobile, dislocations and vacancies can move through lattices more easily, and diffusion occurs more readily. The result of all these factors is that metals lose their strength properties at higher temperatures; i.e., metals become weaker, less stiff, and more ductile. That is, the modulus of elasticity and tensile and yield strengths go down, and elongation goes up.

High-Temperature Tensile Tests.   Tensile tests conducted with specimens held at higher temperatures have a rather characteristic result. As the temperature increases, the strength values go down, only slightly at first, but as the annealing temperature is approached, the data go down abruptly.

Table XIII-1 shows tensile-test data for various alloys at different temperatures. Note that the aluminum alloys retain some amount of their strength until about 300 or 400°F (150 or 200°C) and then drop off rapidly. The carbon and low-alloy steels hold their strengths reasonably well until 800 to 1000°F (425 to 540°C) and then fall quickly. If data for 1400 and 1600°F (760 and 870°C), which are above A_1, were available, they would show a very large decline.

The titanium alloy goes through similar changes and maintains some semblance of strength up to about the same temperatures as the two steels.

The data for the stainless steels shows that they are stronger at high temperatures than regular steels. The ferritic-type 410 is slightly stronger at room temperature than the austenitic 316, but it does not retain that advantage as the temperature increases. Note that when the temperature is high enough to convert the ferrite to austenite, about 1400°F, the strength of 410 goes up slightly.

Hot-working.   High-temperature tensile tests are essentially hot-working operations (see Section 3.6). Since the tests are done at a slow rate, the metal does not work-harden when the temperature is near that required for recrystallization. Thus the strength values determined from tensile tests conducted at annealing temperatures are far lower than the values given for as-annealed strengths. The Table XIII-1 data for the carbon steel and the titanium alloy came from specimens that were annealed before reheating, so the room-temperature data are their annealed values. For comparison, the annealed tensile strengths for 6061 and 7075 are 18 and 33 ksi, respectively, also far higher than their strengths at the higher temperatures.

13.2   CREEP

The rate of loading used in the tensile test is low compared with that used in the impact test, but it still causes failure in a much shorter time than experienced by metals that undergo a phenomenon called **creep.** Creep is plastic flow that occurs under a constant stress; as its name suggests, it occurs *slowly*. The stress causing

TABLE XIII–1 Tensile data at high temperatures for various metals

Temperature, °F	6061-T6 Aluminum[1]		7075-T6 Aluminum[1]		0.15% C Steel Killed, Annealed[2]		Cr-Ni-Mo* HSLA, Q/T[2]		IMI 318 Titanium[3]		410 S.S. 12 Cr[2]		316 S.S. 18-8 Mo[2]	
	Tensile Strength, ksi	Elongation, %	Tensile Strength, ksi	Elongation, %	Tensile Strength, ksi	Elongation, %	Tensile Strength, ksi	Elongation, %	Tensile Strength, ksi	Elongation, %	Tensile Strength, ksi	Elongation, %	Tensile Strength, ksi	Elongation, %
70	45	17	83	11	70	30	120	10	135	15	89		82.5	
300	34	20	31	30	70	30								
400	19	28	16	55	70	28	115	10	111	18				
600	4.6	85	8	70					97	18				
700					64	35	110	10						
800									78	26	66		71.5	
900					35	45	90	10						
1000							85	11			44.5		67.5	
1100					20	57	65	13	34	58	22		56.5	
1200							50	19	17	127				
1300														
1400											9		35	
1600											9.5		22	

*0.14% C, 0.49% Cr, 0.90% Ni, 0.52% Mo.

Sources: 1. *Aluminum standards and data* (8th ed.). Washington, D.C.: Aluminum Association, 1984.

2. *Metals handbook* (1948 ed.). Materials Park, OH: ASM International, 491, 564.

3. Eric A. Brandes (ed.). *Smithells metal reference book* (6th ed.). London: Butterworth, 1983, table 22.36.

the creep is usually less than the yield strength of the metal at the temperature used in the test. Metals exhibit significant creep at temperatures that are about 0.5 of their absolute melting temperature.*

Most metals only creep when stressed at an elevated temperature; metals that exhibit significant creep at ambient temperatures are obviously poor choices as structural materials. However, metals like tin, lead, and zinc, which have very low recrystallization temperatures, will creep at room temperature. (See Table III-1.)

In comparison with other mechanical tests and mechanical properties, the unusual thing about creep is that *time is one of the variables*. The combination of stress and temperature are important—i.e., increasing temperature and stress increases the amount of deformation (creep)—but the total amount of deformation depends on the length of time the load is applied.

13.3   THE CREEP TEST

The creep test is conducted very simply by subjecting a tensile-type specimen to a stress while it is in a heated chamber. The stress is usually generated by hanging a weight on the end of the specimen. The variables controlled are the stress, determined by the hanging weight and the area of the specimen, and the temperature. The variables measured during the test are the linear extension, or stretch, of the specimen (strain) related to the elapsed time of the test.

Although the basic principles of the test are simple, the equipment is not. The heating chamber must be capable of maintaining the temperature of the specimen within a few degrees of the nominal test temperature. The length, or change in length, of the specimen must be measured accurately and correlated with the elaspsed time.

ASTM Specification E139, vol. 03.01, covers the creep and similar tests, including specimen and equipment requirements, but the test is not as standardized as other tests discussed in the appendixes of this text. The concluding section of the standard perhaps explains why: "The property variation of the materials tested have a greater effect on the measured results than the inaccuracies in the test method."[†]

13.4   THE STAGES OF CREEP

Three Stages of Creep. Plotting the values of strain and time measured during a creep test results in a curve such as seen in Figure 13-1. When the load is initially

---

*If the melting temperature of a steel is 2600°F (3059°R), then the steel might be expected to undergo significant creep at a temperature of 1529°R, or about 1070°F.

[†]*ASTM Standards* (1992). Philadelphia: ASTM, vol. 0301, 332.

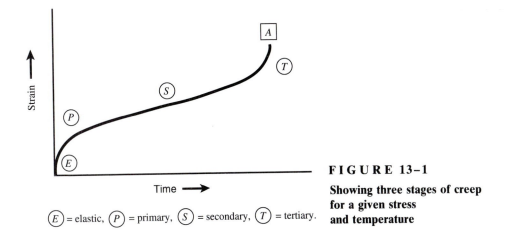

FIGURE 13–1

**Showing three stages of creep
for a given stress
and temperature**

put on the heated specimen, there is an immediate stretching or elastic strain; in the figure this is indicated by E. There then begins a period where some work-hardening occurs so that the metal begins to resist the strain. That is, the rate of creep (amount of strain per time or the slope of the tangent to the curve) begins to slow. This is the **primary stage** of creep and is indicated by P.

As time continues, the rate of deformation and the rate of work-hardening come into balance and a line of almost constant slope results; this is the **secondary stage** of creep and is indicated by S in Figure 13-1. As the test continues, and for the conditions assumed for Figure 13-1, the third stage, or **tertiary stage,** of creep occurs; this is shown as T. In this stage the rate of creep, i.e., the slope of the line, increases and the specimen pulls in two.

Two Stages of Creep. The loading and temperature assumed for Figure 13-1 are such that the specimen goes through all three stages of creep, but there are two other scenarios possible. Figure 13-2 shows the results of a creep test where the combination of load and temperature preclude the third stage. The specimen goes through the elastic and primary stages and then settles into a long second stage of creep.

From a practical standpoint, keeping metal parts or components loaded so they stay in the second stage of creep at their operating temperature is very useful. Since this creep is a steady-state situation, it can be anticipated and allowed for in the original design and successful applications can result. Failure in this case might not be in separation, but rather in the stretching of a part or component that prevents continued operation.

In the other scenario the combination of loading and temperature is such that there is no secondary stage of creep. That is, the specimen is, in effect, over-loaded and goes right from primary to tertiary creep and failure. This is shown in Figure 13-3. Obviously, this type of loading is not desirable in machine applications, although it is useful for testing.

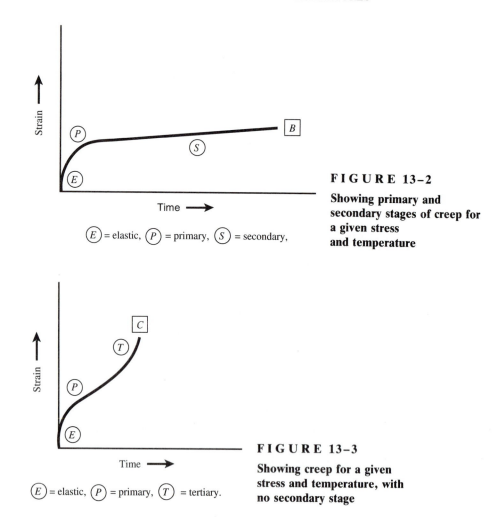

FIGURE 13–2

Showing primary and secondary stages of creep for a given stress and temperature

E = elastic, P = primary, S = secondary,

FIGURE 13–3

Showing creep for a given stress and temperature, with no secondary stage

E = elastic, P = primary, T = tertiary.

13.5   CREEP STRENGTH

From the description of the creep curves of Figures 13-1 and 13-2, you can see that the slope of the line[‡] in the secondary stage of creep represents the minimum creep rate experienced during the test. If a structure or part is made of the same metal and subjected to the same load and temperature as used in the test, then an estimate can be made, based on the slope of the second stage, as to how long it will take for the part either to break or to be stretched beyond its useful length.

The stress required to cause a specified rate of creep at a given temperature is called the **creep strength**; it has units of pounds per square inch (psi) or megapas-

[‡]The slope units are strain, as a percent, per time, usually 1000 hours, or %/1000 h.

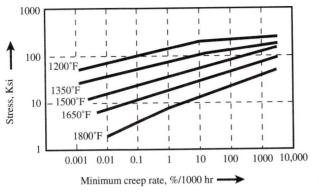

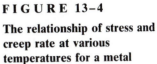

FIGURE 13-4

The relationship of stress and creep rate at various temperatures for a metal

cals (MPa). The rate of creep specified can be of any value, but two rates frequently used are 0.1% per 1000 h and 0.01% per 1000 h. These are sometimes written "per hour" and become 0.0001% and 0.00001% per hour, respectively.

For more generalized information, a series of creep tests can be run for a metal at predetermined temperatures and stress levels and the creep rates determined. From the temperature, stress, and creep-rate data, curves such as those illustrated in Figure 13-4 can be developed. Thus, for a specified temperature and stress a designer can estimate the rate at which a part will creep; or, for a specified operating temperature, the design stress can be chosen to limit the amount of deformation over an anticipated operating life.

The log-log plots of the data are basically straight lines with occasional changes of slope due to structural changes. For example, if a transformation occurs, or a precipitate dissolves, or the mode of failure goes from being transgranular (through or across grains) to being intergranular (between grains or along grain boundaries), the lines can change slope. Note that if these changes can be prevented, i.e., the slope at the lower stress values continues, the properties will be higher. In general, transgranular deformation is preferable to intergranular.

13.6   STRESS-RUPTURE STRENGTH

One disadvantage of the creep test is the time it takes to run it; it is not uncommon for the tests to run for 10,000 h, or almost 14 months. And to get enough data to construct a set of curves like that shown in Figure 13-4 requires hundreds of tests. To reduce this problem, the stress-rupture test is used. As the name suggests, this test loads the specimen so that it ruptures (e.g., Figure 13-3). It is similar to the creep test in that a heated specimen is subjected to a constant load; it is different in two ways: The load-temperature combination is selected so that the specimen will break in a reasonable length of time and only the length of time for rupture is recorded. So the variables controlled are the stress and temperature, and the variable measured is the time required for failure.

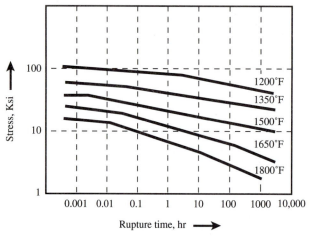

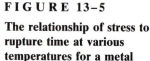

FIGURE 13–5

The relationship of stress to rupture time at various temperatures for a metal

The **stress-rupture strength** *is the amount of stress required to cause rupture in a measured time at a given temperature.* When a number of tests are run at a whole series of temperatures, curves like those of Figure 13-5 can be prepared. The stress-rupture strength, or the results of the stress-rupture test, can be used to estimate the time required for a part to rupture at temperature, but they cannot be used to design for a certain rate of creep. Although the test measures properties over a shorter time period, there is some correlation with the creep test; i.e., *metals that have good stress-rupture strength tend to have good creep strength.*

13.7   FACTORS INFLUENCING HIGH-TEMPERATURE PERFORMANCE

The major strategies for resisting slip and the movement of dislocations and vacancies at high temperatures rely on solid-solution strengthening and generation and retention of second phases that remain stable at operating temperatures.

Carbon Steels.   For steels used at low temperatures where their structures are not affected by sustained high temperatures, increasing carbon content improves creep resistance. However, at temperatures closer to A_1, say 900 to 1000°F (480 to 540°C), the pearlite plates tend to spheroidize, so increasing carbon content is of little value; thus *lower carbon contents are preferred at higher temperatures.* Generally, the normalized condition provides the best long-term results.

Grain Size.   In previous chapters we discussed the advantages of small grains and alloying elements that promoted small grains in improving strength properties at room temperature and below. At higher temperatures this is not the case. At higher temperatures *intergranular* (between grain) rather than *transgranular* (across grain) deformation begins to occur; i.e., slip occurs along grain bound-

T A B L E XIII–2 A comparison of creep properties for a variety of metals

Alloy (%) & Condition	Temperature, °F	Tensile Strength, ksi	Stress Rupture @ 1000 h, ksi	Creep Strength, 0.1%/1000 h, ksi
0.15 C steel	70	60		
	800	50	20*	20
	1000	30	11	5
	1200	18	3	0.8
	1400	8	1	
4140 Q/T 0.95 Ni- 0.20 Mo	70	177		
	800	130	55*	45
	900	120	40*	18
	1000	75		6
0.14 C, Q/T 0.49 Cr- 0.90 Ni- 0.52 Mo	70	120		
	700	110	100	100
	900	90	60	45
	1000	85	22	10
	1100	65	10	2.5
410 stainless steel, Q/T 12 Cr	70	105		
	900	70	42	20
	1000	60	25	10
	1100	50	14	4
	1200	35	6	1.3
316 stainless steel, annealed 17 Cr-12 Ni- 2.5 Mo	70	95		
	1000	70		30
	1100	65	35	18
	1200	55	25	12
	1350	40	12	6
	1500	30	6	3
A-186 15 Cr-26 Ni- 1.25 Mo-2 Ti- 0.2 Al- 55.2 Fe- 0.04 C	70	150		
	1000	133	85	80
	1100			70
	1200	110	46	28
	1350	77	22	
	1500	36	8	
S-816 42 Co-20 Cr- 20 Ni-4 Mo- 4 W-4 Nb- 4 Fe-0.38 C	70	140		
	1000	122		
	1200	111	50	30
	1350	94*	30	20
	1500		17	10
	1600	52	10	5.5
D-979, Q/T 45 Ni-15 Cr- 4 Mo-4 W- 3 Ti-1 Al- 0.05 C- 0.01 B-27 Fe	70	204		
	1200	160	78	65
	1350		43	32
	1400	104	35	
	1500		22	≈10
	1600	50		

*Data are for temperature 50°F higher.

Source: *Metals handbook* (8th ed.). Materials Park, OH: ASM International, 1961, vol. 1, 491–521.

aries.[§] The additional grain boundaries provided by small grains encourages inter-granular deformation, and, conversely, large grains tend to inhibit it. So, *for strength at high temperatures large grains are preferred*. The removal of grain boundaries is such an important factor in improving creep resistance for some critical applications that components like gas-turbine blades are cast as single crystals.

Alloying Additions. Table XIII-2 compares high-temperature properties for a variety of alloys, going from the simple carbon steels, to alloy and stainless steels, to the superalloys at the bottom. Among those metals at the top, note that a slight amount of alloying, especially with chromium and nickel, can contribute substantially to improving properties at higher temperatures.

13.8   SUPERALLOYS

Alloys based on nickel, nickel and iron, and cobalt and used at temperatures above approximately 1000°F (540°C) are termed **superalloys.** Although the metallurgical principles important to strengthening these alloys are similar to those already discussed, they are unusual enough that they deserve separate treatment. Also, the ability of superalloys to withstand high-temperature creep and high-temperature corrosion put them in a completely separate class from ordinary steels and stainless steels.

The three metals used as the basis of superalloys occupy consecutive positions (Fe-Co-Ni) in the periodic table. At room temperature, their atom lattices are BCC, CPH, and FCC, respectively (see Table I-1). However, at the high temperatures where superalloys are normally used they are all FCC.[‖]

Strengthening Mechanisms. The strengthening mechanisms used in these alloys are solid-solution strengthening and the precipitation of stable second phases; principal among the latter are carbides of a variety of compositions and other intermetallic compounds.

The carbides are quite complex and are usually identified as forming between an alloying element or elements, M, and carbon, C. For example, the simplest carbides are MC; these form between hafnium (Hf), tantalum (Ta), niobium (Nb), titanium (Ti), or zirconium (Zr) plus carbon; which carbide forms depends on the amounts of the elements present, the relative chemical activity, and the temperature, not unlike the factors discussed in Section 8.2.

Other carbides are $M_{23}C_6$, or $(Cr, Mo, Fe, Ni)_{23}C_6$, and M_6C, or $(W, Mo)_6C$. These carbides can form between one element and carbon or as combinations of

---

[§]This change in mode of deformation is one of the things that causes the ''kinks'' in the curves of Figures 13-4 and 13-5.

[‖]Iron transforms from BCC to FCC at 1666°F (908°C), cobalt transforms from CPH to FCC at 842°F (450°C), and nickel is FCC at all temperatures.

the metals; that is, $M_{23}C_6$ could be $Cr_{23}C_6$ or $(Cr, Mo)_{23}C_6$ or $(Mo, Fe, Ni)_{23}C_6$; M_6C would be W_6C or $(W, Mo)_6C$; and so forth.

Many of these alloying elements—hafnium, tungsten, and titanium, for example—have large atomic diameters compared with iron, cobalt, and nickel and therefore also contribute to strengthening when in solid solution.

Besides carbides, the other major second phase that contributes to high-temperature strength is known as **gamma-prime,** or γ'. Gamma-prime (γ') is a compound that forms between nickel and aluminum and/or titanium, i.e., Ni_3Al, Ni_3Ti, or $Ni_3(Al, Ti)$ Gamma-prime (γ') has an FCC lattice with a lattice parameter different from the FCC lattice of nickel. This mismatch in lattice size serves to reduce the movement of dislocations and enourage precipitation-hardening, with the result that the strength properties increase.

Phases at Different Temperatures. One of the unusual things about the second phases used in the superalloys is that they form at different temperatures. Thus the compositions of the alloys are designed to cover certain temperature ranges. Therefore just because an alloy has better properties at a high temperature does not mean it is the best choice for an application at a lower temperature; a superalloy designed to form the stable phases at the lower temperature will probably give better results.

Alloy Comparisons. At the bottom of Table XIII-2 are three examples of superalloys, with their compositions given. All three of these alloys contain nickel; nickel's FCC lattice and solubility for a variety of elements makes it an ideal metal upon which to base superalloys. Note that chromium is also present in rather large amounts; chromium is very helpful in combating corrosion at high temperatures.

The A-186 and D-979 alloys contain iron in large amounts, and both alloys have titanium and aluminum to form gamma-prime (γ') with the nickel available. The S-816 alloy is a high cobalt alloy and contains very little iron; it does not rely on gamma-prime (γ'), but instead will form carbides with molybdenum, tungsten, niobium, and iron and/or use the larger-diameter atoms for solid-solution strengthening.

Unfortunately, the temperature ranges for the data of Table XIII-2 are not the same, but an appreciation of the improvement possible by alloying can be gained by comparing the data at various temperatures. For the steels and stainless steels at the top of the table, note that as alloying level increases the high-temperature properties improve. The creep strength of stainless steel 316 at 1000°F makes it clearly superior to any of the steels above it in the table. Compared with the 0.15% C steel at 1200°F, the 316 stainless steel is almost 10 times better in stress rupture and 15 times better in creep strength.

When the superalloys are compared with the 316 stainless, you can see that they are vastly superior; at 1200°F the creep strengths of these superalloys are 2 to 5 times higher.

Also note two things relative to a comparison of the tensile strengths shown in Table XIII-2:

- Alloying improves the *high-temperature* tensile strength, but not as dramatically as it improves creep properties.
- The more highly alloyed metals do not necessarily have the highest *room-temperature* tensile strengths.

The point is that for good high-temperature performance, the proper alloys must be used and that room-temperature tensile properties cannot be used to indentify metals that will have good high-temperature properties.

Temperature Limits.   The data of Table XIII-2 suggest some temperature limits for the applications of these metals. These limits are

- for carbon steels, about 800°F (425°C)
- for alloy steels containing nickel and chromium, about 900 to 1000°F (480–540°C)
- for 410 stainless steel, about 1000°F (425°C), though other types can go a little higher
- for 316 stainless steels, 1000 to 1200°F (540 to 650°C), though other types can go a little higher
- for superalloys, in general, over 1000°F (540°C) and as high as 1800°F (980°C), but there are many types of superalloys and they cover a wide range of properties; some heat-resistant types are used at over 2000°F (1090°C)

Disadvantages of Superalloys.   The major disadvantage of the superalloys is their cost; nickel sells for $3 to $3.60 per pound, chromium for about $4.60, cobalt for $12.50 to $25, tungsten for $9.50 to $10.[#] Superalloys carry a superprice. Another problem with superalloys is that they are difficult to fabricate. As ways are found to improve their performance, which would require that the movement of dislocations at high temperatures become more limited, superalloys will become even more difficult to form, which means that hot-working will not be an option. Casting, which can be done, is one of the ways around this problem with superalloys.

S U M M A R Y

1. Higher temperatures increase the mobility of atoms and make it easier for dislocations and vacancies to move through lattices.

2. As temperatures go up, the modulus of elasticity and tensile and yield strengths go down, and, in general, elongations go up.

3. Tensile tests done at high temperatures are hot-working operations done very slowly, so no cold work can be done during the test: strength values obtained during these tests will be much lower than the annealed values.

[#]Prices determined from a survey of market prices listed in 1992 issues of *Iron Age* (Hitchcock Publishing Co., Carol Stream, IL 60188).

4. Creep is slow plastic flow that occurs under a constant temperature and stress; it normally occurs at temperatures that are about 0.5 of the absolute melting temperature, or higher.

5. Time is one of the variables of creep; the amount of creep is dependent on the amount of time the load is applied at temperature.

6. In a creep test a weight is hung on a specimen kept at a constant temperature; the amount of strain is then measured relative to the time.

7. After the initial elastic reaction to creep, there are three major stages—primary, secondary, tertiary. The creep rate in the primary and tertiary stages is high; the minimum creep rate occurs in the secondary stage.

8. This almost constant slope of strain versus time in the second stage can be used to establish a creep strength, i.e., the stress required to cause a specified rate of creep at a given temperature.

9. In the loading of parts or components it is advantageous to keep the loading within the second stage.

10. The stress-rupture test measures the time required for a given stress to rupture the specimen.

11. The stress-rupture strength is the stress required to cause the specimen to rupture in a specified time at a given temperature.

12. Resistance to high-temperature creep is aided by having large grains, even to the extent of casting metals as a single grain.

13. Resistance to high-temperature creep in steels can be improved by keeping the cementite lamellar (using high carbon contents at low temperatures and low carbon at high temperatures), by using alloying elements such as chromium and nickel, and by having large grains.

14. Superalloys are based on nickel, nickel and iron, and cobalt; they are typically used in the 1000-to-1800°F (540-to-980°C) temperature region.

15. Superalloys are strengthened by solid-solution strengthening (by alloying with large-diameter atoms such as molybdenum and titanium) and by the formation of stable second phases.

16. The stable second phases are carbides (various compositions formed by carbon and other alloying elements, e.g., tungsten and chromium) and gamma-prime [γ', or $Ni_3(Al, Ti)$].

17. Second phases have their strengthening effect at different temperatures.

18. The high-temperature properties (creep strength and stress-rupture strength) of the superalloys are many times better than those of ordinary steels and stainless steels.

19. For best performance, alloys specifically designed for the required temperature range should be used. The performance of metals at high temperatures cannot be evaluated by testing done at room temperature.

REVIEW QUESTIONS

1. In general, how are the tensile properties of metals affected when they are tested above room temperature?

2. If the metal of Figure 13-4 is used for a component 10 in. long operated at 1500°F and loaded to 50 ksi, what length will the component be after one year?

3. If the metal used to develop the data for Figure 13-5 is stressed at 10 ksi while at 1650°F, approximately how long will it take it to rupture?

4. Contrast creep strength and stress-rupture strength. Where are they similar? Where are they different? What can each one be used for?

5. Compare the strategies used to strengthen metals at room temperatures and at higher temperatures. Describe how superalloys are strengthened.

6. Contrast the most advantageous grain size for good high-temperature properties with that for good impact properties. Explain the difference.

FOR FURTHER STUDY

Betteridge, W. *Nickel and its alloys*. New York: Halsted (Wiley), 1984.

Dieter, George E. *Mechanical metallurgy* (3d ed.). New York: McGraw-Hill, 1986, chap. 13.

Metals handbook (1948 ed.). Materials Park, OH: ASM International, 115–8.

Metals handbook (9th ed.). Materials Park, OH: ASM International, 1987, vol. 13 (Corrosion), 641–3.

Metals handbook (9th ed.). Materials Park, OH: ASM international, 1987, vol. 15 (Casting), 811–24.

Superalloys source book. Materials Park, OH: ASDM International, 1984.

C H A P T E R **14**

Failure Due to Corrosion

OBJECTIVES

After studying this chapter, you should be able to

- explain the difference between the electromotive and galvanic series and the significance of that difference
- describe the conditions required for electrochemical corrosion to occur and explain how the electrochemical corrosion cell operates
- explain how the relative areas of the anode and cathode, polarization of the cathode, and cathodic protection influence corrosion
- describe the various methods of combating electrochemical corrosion and apply them to solve simple corrosion problems

Corrosion means "rust" to most people; it's what happens to the unpainted steel shed or the tool that is left outside. To those who live in certain areas of the United States, corrosion is something that happens to their cars.

The economic importance of the corrosion of metals is tremendous; consider the cost of protecting metals, repeated polishing of bright surfaces, replacement of weakened or roughened equipment, the use of expensive corrosion-resistant metals or of thick sections to give a greater factor of safety, plus the cost of the metal that is never reclaimed. In addition, corrosion causes contamination and discoloration of foods, textiles, paper, and chemicals; continual and elaborate research and testing programs are devoted to combating it. Perhaps the most serious problem is the hazard implied in failure by corrosion; often the damage due to corrosion goes undetected until it is too late to prevent a catastrophic failure.

The system by which the world works seldom seems to allow solutions that have no negative consequences. If, somehow, all corrosion were eliminated, many failures could be circumvented and money saved, but we would have no batteries to start automobiles and we would not be able to etch metallurgical specimens.

A PREVIEW

In devising strategies to combat metallic corrosion, it helps to see that corrosion is the action of the environment tending to cause metals to revert to their natural state, i.e., to revert to their lowest energy state, the ores from which they were refined. This view also suggests that corrosion is never prevented or eliminated completely; at best it is controlled or "managed."

The environments in which corrosion takes place can be divided into gases, aqueous solutions, and molten salts and liquid metals. Electrochemical corrosion in aqueous solutions is the most frequently encountered form and is the major topic of this chapter, though the dry gaseous corrosion that occurs at ambient and relatively low temperatures and is referred to as *oxidation* will be discussed first. Corrosion due to high-temperature gases and to liquid salts and molten metals is of a more specialized nature and we will not discuss it here. The forms of corrosion that are assisted by mechanical effects, e.g., cavitation and fretting will be dealt with briefly.

14.1   OXIDATION

At ordinary temperatures aluminum and chromium, for example, react with atmospheric oxygen, oxidize,* and spontaneously form invisible oxide films. These films are nonporous, adhere well to the metal, and act as a barrier to retard additional attack. Such protective action is called **passivation.** The thin, adherent black oxide finishes produced on steel by various methods offer similar, **passive** protection.

Dry corrosion at high temperatures, as occurs in hot-working, annealing, or heat treatment, can cause a layer of oxide, or **scale,** to form on the metal surface. If the oxide is not porous and adheres to the metal, the metal can be protected from further oxidation through passive protection. However, if the oxide layer and the metal have different coefficients of thermal expansion, as happens in most steels, the stresses generated at the metal-oxide interface cause the oxide to **spall,** leaving the metal subjected to further attack.

Long exposure of structural steels to moisture at atmospheric temperatures causes similar problems. The oxide layer that is generated has a volume greater than the metal it replaces and does not adhere well to the metal. When the stresses at the metal-oxide interface become too great, the oxide layer spalls. A somewhat different problem is known to have been caused in at least one instance by this volume increase: Steel columns, encased in brick that leaked rainwater, corroded and expanded enough in volume to force the bricks to break loose and fall to the floor of a mill building.

Protective passivating films are formed in some cases of chemical corrosion. For example, lead in dilute sulfuric acid becomes coated with a film of insoluble lead sulfate; steel in concentrated sulfuric acid is passivated by ferrous sulfate, which is insoluble unless the acid is dilute.

Metal under conditions where it is not passive is said to be **active.**

14.2   ELECTROCHEMICAL CORROSION IN AQUEOUS SOLUTIONS

Formation of Metal Ions.   For corrosion to proceed at a serious rate at ordinary temperatures, moisture is usually necessary; if metal is kept completely dry, little attack will occur. A moderately humid atmosphere greatly accelerates rusting of steel, particularly if salts from fingerprints or imperfect washing after **pickling** or plating are present; these salts and their corrosion products absorb water from the atmosphere and an invisible film of moisture develops on the metal surface.

---

*The term *oxidation* is used here in its chemical sense, regardless of whether the gas is oxygen or not; that is, oxidation is an increase in valence (loss of electrons) of the metal, as opposed to a reduction in valence (gain of electrons).

Gases that are otherwise corrosive often are harmless to metals if water vapor is absent, or if the metal is moderately warm so sweating from condensation cannot occur.[†]

Wet corrosion, which includes specimens in humid atmospheres, requires that the water, and any solutes present in it, be considered an **electrolyte**[‡] and involves **electrochemical** principles. A single metal **M** in contact with water, the electrolyte, tends to dissolve in the water according to the following reaction:

$$\mathbf{M}_{(metal)} \leftrightarrow \mathbf{M}^+_{(ion)} + \mathbf{e}^-_{(electron)}$$

The metal ions formed in the electrolyte are metal atoms that have lost their negatively charged valence electrons, so the ions become positively charged.

Equilibrium. When the metal dissolves in water, positive metal ions are formed and electrons are set free following the reaction shown above. The positive ions go into solution in the electrolyte, but the electrons remain in the electron cloud of the metal's lattice. The metal thus becomes negatively charged and attracts its positive ions back to itself so the reaction is in equilibrium; i.e., it comes to a standstill. The reaction cannot proceed further in either direction unless this balance is somehow disturbed.

Corrosion. Disturbing this balance and *moving the equation to the right* is the basis of electrochemical, or **galvanic,** corrosion. If electrons are removed from the right side of the equation, the reaction will continue moving to the right, generating metallic ions, which are, in fact, the metal being corroded.[§]

Electromotive Force Series. This tendency to ionize varies with the metal; it is greater for the active, or easily corroded, metals such as magnesium and zinc, and less for the less corrodible metals, such as silver and gold.[‖] Metals and alloys can be arranged in the order of their tendency to form these metallic ions when placed in specific solutions. This **electromotive force (emf) series** is shown in Table XIV-1.

As the name suggests, this series is based on the relative value of the electromotive force, or voltage, of the metals. In Table XIV-1 the more corrodible, or **anodic,** metals are listed at the top, and the less corrodible, or **cathodic,** metals at the bottom. Hydrogen is in the list because it is the standard against which the

---

[†]A metal can be cooled by reduced ambient temperatures overnight, or during shipment in an airplane or truck, and then "sweat" when later exposed to warm humid air; this corrosion often proceeds without detection.

[‡]An electrolyte is a chemical substance, usually a liquid solution, that will ionize, allowing the transfer of ions and electrons.

[§]The reaction is reversible. Moving the reaction to the left, say by providing electrons from a battery, will convert the metal ions in the solution to the metal itself; i.e., they will plate out of the solution. This is the basis of electroplating.

[‖]The less corrodible metals, e.g., gold, are also termed *noble;* the more corrodible or active metals, e.g., zinc, are said to be *base* in the sense of being low grade.

TABLE XIV–1 Electromotive force series of metals

More corrodible **(or more active or anodic)** **end of series**	Cobalt Nickel Tin Lead
Potassium Calcium Sodium Magnesium Aluminum Zinc Chromium Iron Cadmium Titanium	*Hydrogen* Copper Mercury Silver Platinum Gold
	Less corrodible **(or more protected** **or more noble or cathodic)** **end of series**

voltage of each metal is compared when tested in an electric cell whose *electrolyte is a salt of the metal under test.*# Since the sequence of the metals in the list is dependent on the test electrolyte, the relative position of the metals will change if the electrolyte is changed. Also, some metals will ionize to different ions, which have different potentials, so some listings show metals in more than one position.

14.3 GALVANIC CELL

In many applications, two different, **dissimilar metals** are in contact in the presence of an electrolyte. The metals are thus **coupled** together by being in electrical contact and form a **cell,** called a **galvanic cell,** capable of generating an electric current that flows from one metal to the other. Figure 14-1 is a schematic representation of the operation of a galvanic cell using copper and zinc as the dissimilar metals. In such a cell the metals are referred to as **electrodes.**

Each of the metals (electrodes) has a tendency to ionize in accordance with the general reaction $M \leftrightarrow M^+ + e^-$. In the previous discussion there was only one metal dissolving in the electrolyte. Now, with two metals in an electrolyte, and an external circuit between them, the electrons do not have to stay in the electron cloud of the metal lattice; they can be removed so the equation does not have to come to equilibrium and one of the two metals will be corroded.

The galvanic cell is used schematically to study electrochemical corrosion. Actual applications probably won't look like Figure 14-1; more practical examples are seen in the weathering of a tin-coated steel can, a steel bolt in an aluminum

---

#The other metal used in the standard cell is usually platinum, a very corrosion-resistant metal. The solution is a salt of the metal being tested that is saturated with hydrogen.

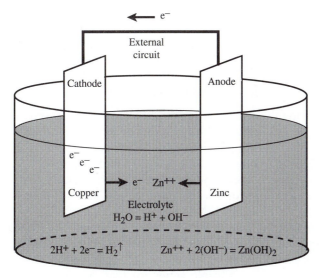

FIGURE 14–1

Galvanic cell showing couple action

Source: Clyde W. Mason. *Introductory physical metallurgy*. Materials Park, OH: ASM International, 1947, figs. 74 and 75, 118 and 125. Used by permission.

lighting pole exposed to rainwater containing salts, copper piping connected to pumps and fittings of other metals all carrying a chemical, or a stainless steel motorboat drive shaft in a brass coupling. A galvanic cell can occur in any situation that allows the coupling of dissimilar metals in the presence of an electrolyte.

The Galvanic Cell and Electrochemical Corrosion. In the galvanic couple of Figure 14-1, zinc, the more anodic metal, has more of a tendency to ionize (put more metal ions into the electrolyte) than does the copper; thus it releases electrons into the external circuit to the cathode. At the cathode these electrons attract any copper ions that might tend to corrode (go into the electrolyte). Thus the zinc is corroded, while the copper is protected. At the cathode any excess electrons unite with the hydrogen ions of the water and go off as a gas. The zinc ions from the anode unite with the OH^- radical and the electrolyte becomes basic as zinc hydroxide is generated.

Electron flow. As shown in Figure 14-1, the electrons flow in the external circuit *from* the anodic, or more corrodible, metal, *to* the cathodic, or less corrodible, metal. Removing the electrons from the anode permits the equation to go to the right at the anode, and the anode is corroded. Just the reverse happens at the cathode; the supply of electrons forces the equation to the left at the cathode and prevents corrosion.

For corrosion to continue, the reaction must continue; that is, the electrons must continue to flow in the external circuit. If the external circuit is interrupted, or if the electrons flow to the cathode and stop there—i.e., they are not removed—the cell will stop functioning. The first condition is one of the obvious ways to reduce galvanic corrosion of dissimilar metals, e.g., by placing electrical insulators between them. The second condition, accumulation of electrons on the cathode, is termed **polarization.**

Polarization. Polarization is usually forestalled by the presence of hydrogen ions in the electrolyte. The hydrogen ions remove the electrons from the cathode by forming hydrogen gas, which escapes, and thus keeps the reaction going. The reactions given in Figure 14-1 show this process. Although polarization doesn't occur normally, it can be encouraged; this is one of the strategies for reducing electrochemical corrosion that will be discussed later.

Galvanic series. In the galvanic cell (Figure 14-1), different metals are *electrically connected while in the same electrolyte.* This is a different set of circumstances than used in the electromotive series (Table XIV-1), where each metal is tested in its own electrolyte and compared with a standard electrode.

Since most common corrosion problems occur when dissimilar metals are used in the same electrolyte or environment, the emf series is not as helpful as the **galvanic series** shown in Table XIV-2.

In the galvanic series, metals are compared with each other when in the same electrolyte, usually seawater. Since salt is found in much of our water and air the galvanic series is a helpful ranking. However, if another electrolyte is responsible for the corrosion, then the galvanic series for that electrolyte must be used to study its effect on different metals. If data are not available, tests should be run.

Voltage and current. The voltage developed by a galvanic cell depends on the difference between the ionization tendencies (or potentials) of the two metals or electrodes. The farther apart the two metals are in the galvanic series (Table XIV-2), the larger the voltage (emf) of the cell.

The larger the area of electrodes, the more metal is available to undergo the ionization process. Thus the current capacity or amperage of the cell is dependent on the areas of the electrodes.

T A B L E XIV–2 Galvanic series of metals and alloys (in seawater)

More corrodible (or more active or anodic) end of series	Aluminum bronze Red brass Copper Silicon bronzes
Magnesium	Copper-nickel alloys
Zinc	Nickel (passive)
Aluminum	Inconel (passive)
Cadmium	Monel
Aluminum alloys	Silver
Ordinary steel	Stainless steel (passive)
Alloy steel	Titanium
Cast iron	Gold
Ferritic stainless steel (active)	Platinum
Austenitic stainless steel (active)	
Tin	**Less corrodible**
Lead	**(or more protected**
Nickel (active)	**or more noble or cathodic)**
Yellow brass	**end of series**

Summary. The key factors in the operation of a galvanic cell are the following:

1. There are two (at least) dissimilar metals.
2. The metals are in electrical contact.
3. The metals are in the presence of an electrolyte.
4. The electrons that accumulate on the cathode are capable of being removed. (If they aren't, the cell will become polarized and cease operating.)
5. The farther apart the two metals are in the galvanic series, the higher the emf (voltage) the cell will generate.
6. The larger the areas of the electrodes, the larger the quantity of electrons (amperage) involved in the cell.

Of the two metals in the cell, the metal that is lower in the galvanic series is the cathode and is not attacked; i.e., it is protected. The metal that is higher in the galvanic series is the anode and is the metal corroded.

14.4 CORROSION IN A SINGLE PIECE OF METAL

So far we have discussed electrochemical corrosion in the presence of two separate pieces of dissimilar metal. The same concepts, however, can be used to explain the corrosion of a single piece of metal, e.g., bare steel exposed to rain and snow. Most metals, especially those used in structures, are mixture alloys. Low-carbon steel, for example, is a mixture of the ferrite and cementite phases; it has a microstructure containing ferrite and pearlite.

The phases that make up these mixtures have different ionization potentials or emf's. Thus, within the same piece of metal, one portion of the surface will be more anodic or cathodic than another; the portion that is more anodic will corrode "at the expense of" that portion that is cathodic.

Residual stresses set up similar differences in potential; if a piece of metal is stress-free in one area, but has residual stresses elsewhere, that portion that has residual stresses will be anodic (more easily corroded) than the stress-free portion. (These subjects will be dealt with again in Section 14.6.)

14.5 COMBATING ELECTROCHEMICAL CORROSION

To combat electrochemical corrosion a basic strategy is to interrupt the operation of the cell; this involves offsetting one or more of the key factors listed in the summary of Section 14.3. The design and the environment in which a metal is used determine how many of these factors have to be considered.

For electrochemical corrosion to proceed, there must be (1) dissimilar metals (2) in contact with each other (3) in an electrolyte, and (4) the electrons must be removed from the cathode. Strategies are found by challenging each of these requirements. That is, don't use dissimilar metals, or, if dissimilar metals are required, insulate them from each other; make sure any fluids that may be electro-lytes are excluded, i.e., coat the metals or keep them dry, attempt to prevent the electrons from being removed from the cathode, etc.

Dissimilar Metals. If dissimilar metals must be used, have the cell generate as low a voltage as possible by using metals that are as close together in the series as possible. The lower the voltage, the lower the rate of corrosion. Since mixture alloys will have dissimilar potentials within their own structures, use purer metals if possible. Single-phase alloys (solid solutions) should perform better than mix-ture alloys.

Current Flow. If the dissimilar metals are in intimate contact, e.g., a steel shaft in a bronze coupling, it may be difficult to separate them electrically, but often in structural applications nonconductive washers and sleeves can be used to isolate one metal from the other. If the metals must be in contact, then efforts should be made to shield the junction from the electrolyte and prevent the circuit from being completed elsewhere.

Relative Area. The quantity of electrons flowing, amperage, also affects the rate of corrosion. If electrode areas can be kept small, the rate of corrosion can be kept low, but the area of the cathode *relative* to the anode is also important.

As the cell operates, the electrons flow through the external circuit to the cathode and are removed by hydrogen ions in the electrolyte (see Figure 14-1). If the area of the cathode is *small* compared with the anode, it will be more difficult to maintain the flow of electrons into the electrolyte. That is, there is less relative area on the cathode to get rid of the electrons generated by the larger anode, and the rate of corrosion can be reduced.

Removal of Electrons from the Cathode. In addition to keeping the cathode area small, there are other strategies that can aid in reducing the ability of the electrons to be removed from the cathode: shielding the cathode, reducing the flow of the electrolyte, and reducing the oxygen content of the electrolyte.

- If the cathode can be shielded from the electrolyte, the flow of electrons will be inhibited. This can be done using various plastic coatings or paints. Since the cathode is the more corrosion-resistant of the two metals and usually has a pleasing appearance, it is often left unpainted or uncoated. Apparently the thinking is that if it's not going to corrode why paint it? In fact, the overall corrosion can be reduced by shielding the cathode.
- A moving electrolyte makes a good source of new hydrogen ions to complete the reaction. Slowing the flow of the electrolyte discourages the availability of the hydrogen ions and slows the corrosion rate. In recirculating systems, e.g., cooling systems, commercially available inhibitors can also be added to reduce the corrosion rate.
- The hydrogen generated at the cathode will not necessarily go off as a gas. If oxygen is available in the electrolyte, it can combine with the hydrogen to form

H_2O, which can increase the rate of corrosion. Remember that oxygen can be entrained, for instance, in turbulent streams, in pipelines that aspirate air, and in mixing equipment. Corrosion can be reduced if the oxygen content of the electrolyte can be lowered.

A combination of keeping the cathode area small plus the above strategies may cause the cell to polarize and cease functioning at all.

An Illustration. Consider a situation where two metals, say steel and copper, are used to fabricate a trough to handle seawater. If the trough is formed from copper sheet and fastened with steel rivets, the corrosion of the rivets will be very rapid. The moving seawater will easily remove the electrons from the large copper cathode, and corrosion of the anodic steel rivets will proceed at a high rate. If, on the other hand, the trough is made from steel and fastened with copper rivets, the corrosion of the rivets will be negligible. The small area presented by the cathodic copper will make a large electron flow between the dissimilar metals difficult.

Note that the recommendation to minimize the area of the cathode is opposite to the "commonsense" approach of using as much of the corrosion-resistant metal as possible. Of course, for the problem presented, the best solution would be to use copper for the rivets and the trough. The seawater would corrode the bare steel very rapidly, regardless of what kind of rivets were used.

The removal of the electrons from the cathode in this example is facilitated by the fact that the electrolyte is *flowing* seawater, a good source of hydrogen ions to complete the reaction. Slowing this flow would reduce the availability of hydrogens ions and slow the corrosion rate. Also, the flowing water is a potential source of oxygen to form H_2O with the hydrogen at the cathode.

Galvanic Protection. In some applications, corrosion of important members can be avoided by using the galvanic series to give them **cathodic protection.** In the galvanic cell the anode is the metal that corrodes. If there is equipment that needs corrosion protection, e.g., a steel tank, a cast-iron pipeline, or a copper water heater, connecting that equipment to a more anodic metal will convert the equipment to a cathode and protect it from corrosion. That is, *the metal to be protected is made the cathode* by linking it with a metal that is above it in the series. The anode then is **sacrificed** for the cathode.

Coating sheet steel with zinc, or **galvanizing,**** serves the same purpose. The zinc, near the top of the galvanic series, is anodic to the steel. In the presence of an electrolyte, the anodic zinc corrodes and protects the cathodic steel.[††] In a similar method aluminum sheet and tubing are protected with an **alclad** coating of a zinc-rich aluminum alloy that is anodic to the base alloy to be protected.

**Note the root "galvan" in "galvanizing."
[††]Note that "tin-plated steel" is not protected galvanically; tin is below steel, or cathodic to steel in the galvanic series, so the steel corrodes rather than the tin. The tin in this case merely serves as a good corrosion-resistant coating.

The cathodic-protection method is often used to protect tanks and pipelines buried in the ground. Large anodes are buried near the metal to be protected and connected electrically to it. Anodes used for such protection should be well up on the galvanic series; usually magnesium or zinc is used.

Similar protection can be obtained by using a direct current, or **impressed current,** to supply electrons to the metal to be protected so that it becomes a cathode; i.e., the impressed current drives the previously discussed equation to the *left.* An impressed current can be visualized in Figure 14-1 as a battery inserted into the external circuit to supply electrons to the *anode.*

14.6   **TYPES OF CORROSION**

Pitting Corrosion.   The structure of most metals is not homogeneous enough for a **uniform corrosion** of the whole surface, so most forms of corrosion are of a localized nature. **Pitting corrosion** is one type that results from this lack of homogeneity. Because of coring, mixtures of phases, and residual stresses, differences of electrical potential occur at the metal surface. In the presence of an electrolyte, a galvanic cell is established and the more anodic portion of the surface is attacked at localized sites, causing pits.

Intergranular Corrosion.   This form of corrosion, similar to pitting corrosion, is due to a difference in electrical potential within the metal itself. In this case the electrical potential difference is between the grain boundaries and the grain, causing a higher corrosion rate at the boundaries. This potential occurs because of differences in composition between the two areas. Precipitation usually begins at the grain boundaries, which raises the percent composition of that phase in the grain boundary; but it may also deprive the grain of a phase needed for corrosion resistance. For example, improperly heat-treated stainless steels may have excess chromium in the boundaries and not enough chromium within the grain. The solution to intergranular corrosion is to avoid the segregation that causes the problem, i.e., to use proper heat treatments, fabricating practices, and compositions. These practices vary with the type of metal. Also, certain corrosive environments should be avoided for specific metals, e.g., brass in ammonia.

Crevice Corrosion.   The term *crevice corrosion* is used to describe two modes of localized corrosion that are somewhat related. The first usage is for those physical situations where the electrolyte can be trapped or retained in the the metal structure and not drain away, thus allowing the galvanic cell to continue to function, e.g., rainwater that cannot drain out of crevices in an automobile body.

The second usage is more complicated and relates to differences in the concentration of ions in the electrolyte. That is, the crevice or narrow gap in the metal prevents the electrolyte from circulating and becoming uniform in ion concentration. At the metal's open surface the ions of the electrolyte can freely exchange, but within the narrow crevice they can become concentrated.

For example, at the open surface, not in the crevice, oxygen entrained in the electrolyte can remove hydrogen. Since removal of hydrogen normally occurs at the cathode, this surface functions like a cathode. In the crevice, on the other hand, oxygen consumed in corrosion cannot be replaced, the removal of hydrogen is hindered, and the crevice becomes anodic, and thus corrodible, compared with the metal outside the crevice.

By a similar but reverse phenomenon, metallic ions (i.e., + ions) can concentrate in the crevice and make the crevice positive or cathodic to the area outside the crevice. In such conditions the area outside the crevice becomes anodic and corrodes rather than the crevice. The metal and electrolyte determine which of these two effects occurs.

Stress Corrosion. This form of corrosion occurs *when a metal containing residual stresses is subjected to a corrosive environment*. The areas of the metal containing the stress in effect become anodic to those areas that are unstressed. Thus the corrosion occurs locally. This is an important type of corrosion because of the frequency of its occurrence. Solutions are (1) to use recovery-type anneals and protective coatings to shield the metal from the hostile environment and (2) to avoid certain combinations of metals and corrosive materials, e.g., austenitic stainless steels and chloride environments.

Mechanically Aided Corrosion. The types of corrosion that fall in this category involve both chemical and mechanical (wear or fatigue) components, which are also discussed in Chapter 15.

Fretting corrosion occurs between mating surfaces that have very little relative motion, e.g., splined or keyed shafts or wheels press-fitted to axles. Although there is very little relative motion, the adhesive forces between the two surfaces cause welding and then breaking away of small particles; these particles then react with the environment, to become, for example, oxides that form a powder between the two surfaces. Some of the solutions to fretting corrosion are to separate the surfaces with a resilient material, to use tighter fits to reduce the relative motion, or to use harder surfaces.

Cavitation corrosion occurs when a metal vibrates against a liquid. Vaporous bubbles in the liquid collapse against the metal surface and cause metal particles to be removed. Cavitation may appear as pockmarks, pits, or deep depressions. Most solutions to cavitation involve using a harder material surface to resist metal removal or a resilient coating to absorb the impact of the collapsing bubbles.

SUMMARY

1. Oxidation of some metals—for example, aluminum and chromium—provides passive protection of their surfaces and inhibits further corrosion.
2. Other metals, like steel, form oxides that do not adhere well; i.e., the oxides spall off and the metal's surface is *not* passively protected from further corrosion.

3. Metallic ions form in the presence of an electrolyte.

4. Electrochemical or galvanic corrosion occurs when the electrons generated during the ionization are removed so more metallic ions form.

5. Metals are ranked by this tendency to ionize. The electromotive force (emf) series is based on tests done in a standard cell.

6. The galvanic series is done comparing metals in a common electrolyte, usually seawater, and is usually of more use in selecting metals for engineering applications.

7. The galvanic cell, i.e., electrochemical corrosion, requires:
   a. dissimilar metals (the more easily corroded metal is the anode, the less corrodible metal is the cathode)
   b. an electrolyte
   c. electron flow between the metal electrodes in an external circuit
   d. electron flow into the electrolyte

8. To reduce electrochemical corrosion you can adopt the following practices:
   a. Use pure metals where possible, or use other special alloys like stainless steels and monel.
   b. Avoid using dissimilar metals in electrical contact; if this is not feasible, place insulation between the metals.
   c. Avoid moisture and humid environments when dissimilar metals are in contact; if this is not feasible, protect the metals with coatings and drain the liquid away so the metals do not have to stay in it any longer than necessary; if the electrolyte is water, use inhibitors.
   d. If dissimilar metals must be used, keep them as close on the galvanic series as possible to reduce the voltage capacity of the cell.
   e. Keep the area of the cathode, the more noble metal, as small as possible compared with the area of the anode to reduce the ability of the ions to be removed and to encourage the polarization of the cell.
   f. If possible, use the more corrodible, or anodic, metal in places where it can easily be inspected and replaced.
   g. Take advantage of cathodic protection by using a sacrificial anode or an impressed current to protect the important part in the design.
   h. Use stress-relief anneals to avoid anodic areas.
   i. Avoid agitating or aspirating oxygen into the electrolyte.

9. A single metal can corrode (pitting corrosion) because of surface areas that are dissimilar electrochemically because of mixtures of phases, segregation of alloying elements, or residual stresses.

10. Intergranular corrosion occurs when grains and grain boundaries have different electrochemical potentials because of improper distribution of alloying elements; proper heat treatment is usually the solution.

11. Crevices concentrate corrosion in very localized areas; if the electrolyte is drained away, corrosion will not occur.

12. Areas of metal with residual stresses are more anodic and thus more subject

to corrosion; problems can be avoided by using improved fabricating practices and recovery anneals.

13. Fretting corrosion occurs when metallic surfaces are subjected to small amplitude vibrations in the presence of an electrolyte. Cavitation corrosion is due to the collapsing of vaporous bubbles against metallic surfaces.

REVIEW QUESTIONS

1. Identify the conditions necessary for electrochemical corrosion.

2. Describe the operation of the corrosion or galvanic cell. Use a sketch of the cell to identify the conditions of Question 1.

3. What is the limitation on the application of the galvanic series (Table XIV-2)? The emf series (Table XIV-1)? Explain.

4. Explain how cathode protection functions and how it can be used. Give an example.

5. Will corrosion occur more rapidly in a running stream or a quiet lake? Explain your answer.

6. Describe the process of polarization and give two practical ways it can be used to reduce corrosion.

7. In electrochemical corrosion what factors influence the rate of corrosion?

8. Identify and discuss the similarities and differences among the types of corrosion discussed in Section 14.6.

9. Write your own strategy for reducing electrochemical corrosion; check it against item 8 of the summary section above.

10. For a marine application plans specify that a brass coupling be used with a stainless-steel drive shaft. What types of corrosion are likely? What metal should be used for the key to connect the shaft and coupling? Explain your answer.

FOR FURTHER STUDY

Avner, Sidney H. *Introduction to physical metallurgy* (2d ed.). New York: McGraw-Hill, 1974, chap. 15.

Corrosion source book. Materials Park, OH: ASM International, 1984.

Jones, Denny A. *Principles and prevention of corrosion.* New York: Macmillan, 1992.

Mason, Clyde W. *Introductory physical metallurgy.* Materials Park, OH: ASM International, 1947, chap. VIII.

Metals handbook (9th ed.). Materials Park, OH: ASM International, 1987, vol. 13.

Wulpi, Donald J. *Understanding how components fail.* Materials Park, OH: ASM International, 1985, chap. 13.

CHAPTER 15

Failure Due to Wear

OBJECTIVES

After studying this chapter, you should be able to

- describe the parts of the tribological system
- identify the motions possible in systems subject to wear
- describe the primary and secondary mechanisms by which wear occurs, describe the deterioration these mechanisms cause in metal parts, and identify possible solutions
- explain how lubrication works, identify the major parameters required to form a lubricating film, and discuss the effects of lubrication on wear mechanisms

DID YOU KNOW?

Wear is somewhat like corrosion in that without it much money could be saved. But, without the frictional forces that cause wear, we would have a very slippery world indeed, a world with no brakes.

Wear is the loss of material from one surface because of relative motion with another surface. For many years the solution to the problem of mechanical wear between mating surfaces was to make one surface hard and the other surface (preferably replaceable) soft, and to lubricate as much as possible. It worked. And in many applications today it still works. But we now realize that, except in the simplest of mechanisms or machines, wear involves more than just two surfaces rubbing together.

It wasn't until the 1960s that the study of wear was first approached on a systems basis. That study, which has been given the name *tribology,* * is the study of all aspects of the wear process—the surfaces on which the wear occurs, the motions that the surfaces undergo, the mechanisms by which the surfaces wear, and the influence of lubrication.

There are a number of mechanisms (or ways) by which wear occurs; since each mechanism has different consequences, it is important that each mechanism be identified so that when remedies are conceived they are appropriate to the total problem.

A PREVIEW

The chapter begins by discussing the tribological system and its parts. The general system in which wear is studied has four parts:

- the surface of a solid body that is being worn
- a countersurface (that may not always be present as a solid)
- the element(s) that may exist between these two surfaces
- the environment in which this all takes place

Since wear is dependent on relative motion, we'll discuss the motions possible in most systems. These motions lead to forms of wear which are organized into three primary mechanisms and a number of secondary mechanisms. (This organizational structure is shown in the table that appears in the summary section at the end of this chapter.) Each of the primary and secondary mechanisms is described in detail with an emphasis on how or why each happens. Then we explore "solutions" that help avoid or reduce each form of wear.

*The word comes from Greek "tribo," to rub or wear down, and "ology," a branch of learning.

15.1   THE WEAR SYSTEM

The tribological system has four parts: the surface that is being worn, the counter-surface with which it mates, various intersurface elements, and the environment.

Surface and Countersurface.   In straightforward mechanical systems the surface is worn by a mating **countersurface;** e.g., a door hinge and its pin, the race and ball of a ball bearing, the tool head and ways of a shaper.

In some cases the countersurface is less obvious; e.g., even though a chute (the surface) will be worn by sand or concrete sliding down it, the countersurface is not solid but the sliding material. Fluids traveling through a pipe can cause wear to the inside surface by a number of mechanisms, but the countersurface is not a solid.

Intersurface Elements.   In many mechanisms and machines it is intended that there be a lubricant between the two surfaces; that is, the lubricant is an **intersurface element.** Other examples of intersurface elements are particles from one or both surfaces generated when they rub, the oxide layer that is generated on the surface(s) of some metals when they rub under light loading, and sand or other particles, present with or without a lubricant, that may be foreign to both surfaces.

Environment.   For the mechanisms and machines used in the home or office, the typical environment could be described as ambient temperature, atmospheric air pressure, and dry air. Obviously the parts inside an aircraft or automobile engine exist in a different environment; likewise those deep in the hull of a seagoing ship, or in Antarctica, or 100 miles out in space, or in a pump used to convey a hot acid, or on off-the-road vehicles or farm equipment. In extreme conditions the environment can be decision-controlling in the design and selection of materials used for parts subjected to wear.

15.2   TYPES OF RELATIVE MOTION

In the four-part tribological system it is possible for the surface and countersurface to have a number of different relative motions; these are illustrated schematically in Figure 15-1.

a. *Sliding,* the motion most often thought of when considering wear, occurs as two bodies rub or slide against each other. Consider the piston rings and cylinder walls of an internal combustion engine; although a shaft rotates in a journal bearing the relative motion is really sliding.

b. *Rolling,* the action exemplified by the movement of tires on roadways, roller bearings in their races, and rolls of a rolling mill.

c. *Vibrating,* or small amplitude oscillating, occurs, for example, when an out-of-balance rotating shaft causes relative motion between a coupling and the shaft.

FIGURE 15–1 Possible relative motions related to wear

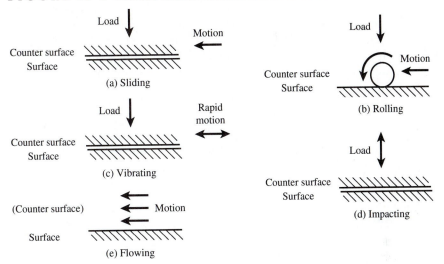

(a) Sliding

(b) Rolling

(c) Vibrating

(d) Impacting

(e) Flowing

d. *Impacting* involves the rapid application and removal of a load, as is typical of some metalworking machines.

e. *Flowing* is the action seen in a stream of fluid—liquid or gas—which constitutes the countersurface and which may or may not have particles in it.

Even some of these "basic" motions can be combined. When rolling and sliding occur together, the term used by car drivers is *skidding*.

15.3   MECHANISMS OF WEAR

Wear proceeds by a number of ways or mechanisms. Organizing these into a concise outline is complicated by the fact that the mechanisms overlap the relative motions discussed above. That is, one type of motion, say sliding, can be involved in more than one mechanism of wear.

The outline used here identifies three primary mechanisms—abrasion, contact, and erosion. We devised these categories because the words describe approximately what is going on during the wear process; they cannot be confused with the words used to describe the relative motions. The three primary mechanisms are introduced and then the contributing factors (secondary mechanisms) are discussed in more detail.

Abrasive wear is caused by moving particles, i.e., intersurface elements that remove metal from the surface(s) involved. The particles may be trapped between the two surfaces (high-stress abrasive wear) or loosely packed (low-stress abrasive wear); they may be present by themselves or with liquids, gases, lubricants, and such. Wear by abrasion is probably the most damaging and costly of the wear mechanisms.

Contact wear occurs when the surface and countersurface are forced to-gether and moved relative to each other. Under certain conditions of temperature and load, mating surfaces may oxidize and control the amount of wear. Under higher loads and without the protection of oxides, the two surfaces may adhere; this is referred to as **galling,** and in the extreme case results in the parts seizing or welding together. If the loading is such that subsurface shear stresses are large enough and the load is cyclic in nature, the part may fail by subsurface contact fatigue. If the load has a vibratory component, fretting wear occurs.

Erosive wear is characterized by flow, especially fluid flow. The fluid may contain particles that may be solid, liquid, or gas bubbles. The fluid may also be a thick slurry containing solid particles. In some respects erosive wear is similar to low-stress abrasive wear, but it can also involve corrosion.

15.4   ABRASIVE WEAR

In **abrasive wear,** particles can be thought of as cutting tools driven by the counter-surface. The type and amount of damage done are dependent on the hardness, size, and shape of the particles and on whether the particles are packed or trapped in the intersurface gap.

High-Stress Abrasion.   If the particles are large with respect to the gap between the surfaces (see Figure 15-2), the movement of the surfaces can trap them be-tween the two surfaces and high-stress abrasion occurs. Under this condition particles may be fractured, creating new sharp edges.

Low-Stress Abrasion.   If the intersurface gap is large enough that the particles can pack into the gap, they are in a low-stress situation and are much more likely to stay intact.

Grooving Abrasion.   Particles that are forced into the surface abrade by a process of **grooving.** Sharp angular and/or imbedded particles will do more damage than rounded particles that can freely move along a surface, e.g., dirt sliding down a chute.

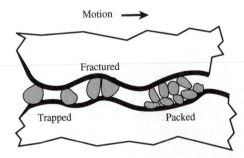

F I G U R E 15–2

Abrasive particles shown trapped and packed between wear surfaces

Abrasion by Cutting, Plowing, and Chipping. Depending on particle shape and its hardness relative to the metal, the particle will abrade by cutting, plowing, or chipping. In **cutting** abrasion, a sharp particle acts like a shaper tool and removes a chip of metal from the surface. In **plowing,** a more rounded particle makes a groove in the surface but leaves a furrow rather than removing metal; i.e., the metal is "piled" alongside the groove. **Chipping** occurs when the metal is brittle enough that a particle causes a fracture along its path and the metal surface chips out.

Inspection of the damaged surface under low-power magnification usually enables the type of abrasion to be identified.

Abrasion can be **three-body,** i.e., the surface, countersurface, and the particle(s), or **two-body,** where a hard tip or projection of the countersurface acts like a particle and abrades the surface.

Particle-to-Surface Hardness Relationship. In general, harder surfaces are helpful in resisting abrasion, but the relative hardness of particle and surface, the shape or angularity of the particle, and whether the particle is trapped or can move along the surface are all important variables.

When particles are softer than the metal surface, wear rates are low and relatively constant, regardless of particle hardness. As the ratio of particle hardness to metal-surface hardness approaches 0.9 to 1.0, the wear rate begins to increase; it continues to increase until the hardness ratio is about 1.2 to 1.5, where it almost levels off and becomes somewhat insensitive to particle hardness.

That is, wear rates are on two plateaus—a lower level of wear when particles are softer than the metal and a higher level when the particles are harder than the metal. *The big increase in wear rate occurs between a hardness ratio of 0.9 to 1.0 and 1.2 to 1.5.*

At the lower hardness ratios the wear rates for two-phase metals—e.g., steel = ferrite matrix + carbides—are less than those for single-phase metals *of the same surface hardness*. The reverse is true for the hardness ratios above 1.2 to 1.5, where the wear rate is lower for homogeneous metals. (You should realize that although carbides are quite hard the ferrite matrix is quite soft, and the metal performs well against softer particles but less well against the harder.) In the transition region between ratios of 1.0 and 1.2 to 1.5, there is a switch between which is the more resistant microstructure. This makes it difficult to select metals for use in this region.

The evidence of abrasive wear is imbedded particles, grooves, valleys, and/or cracks in the surface. These can usually be seen under low-power magnification.

15.5   CONTACT WEAR

Metal surfaces, even if finely ground, are not perfectly smooth but are a series of peaks; these projections are termed **asperities.**[†] When two metal surfaces are

---

[†]"Asperity" means roughness or harshness.

brought together, the presence of these asperities hides the true area that is in contact. That is, the area actually under load is smaller than it appears; in **contact wear,** the stresses applied to the asperities are large enough to make them deform. Relative motion may cause the asperities to break off and become intersurface elements; if the broken asperities are hard, they can become abrasive intersurface particles.

Oxides. When the loading on the surfaces is relatively light, and especially if the environment is dry, oxides of the metals form. This is encouraged by heat, either from the rubbing of the surfaces or the environment. At low loads the oxides protect the surfaces and wear is retarded. At higher loads the oxides are stripped away and become intersurface debris.

Galling. If the load between the surfaces is increased, their asperities will begin to interlock and adhere to each other; this is termed *galling*, or adhesive wear. If the load and movement continue, and especially in the absence of lubricant, the adhesion of the metals can increase to the point where the members are "frozen" together, or "welded," or "seized."

Since this wear mechanism is based on surface deformation, softer metals should adhere more readily than harder metals, and vice versa. This is correct for many metals, but metals with the CPH lattice tend to have poor adhesion that varies little with hardness. This is thought to be related to the fact that the CPH lattice has so few slip planes. When the other lattices deform they do so on a number of different planes that can intersect; these intersecting slip planes can stop the dislocations from moving at the wear surface and the metals adhere rather than slip. Because the CPH lattice has few intersecting slip planes, the dislocations have no impediment and slip can take place easily at the surface.

Galling is recognized by the transfer of one metal to the surface of another.

Contact Fatigue. When large loads are concentrated on a surface, seemingly to compress it, *transverse shear stresses* are generated below the surface. In Figure 15-3 a load is being transferred by a cylinder to a horizontal surface. As described in Chapter 11, loads can't force atoms closer together (or farther apart), so if the horizontal surface is going to deform it has to move sideways by the slipping of the atoms on lattice planes. Subsurface shear stresses are thus generated.

If the cylinder is rolling, as in a roller bearing, the load is applied along the surface in a cyclic fashion; i.e., the surface is subjected to fatigue stresses. If these loads continue for enough cycles and at high enough stresses the surface will fail by fatigue. This type loading occurs on the race of roller and ball bearings and near the pitch line of gear teeth. Failures usually begin with small surface cracks that slope down and in the direction of movement (Figure 15-3). Eventually these cracks return to the surface and a small crater forms; i.e., the metal **spalls.**

As these cracks form, their propagation may be aided by lubricants that enter at the surface; the added hydraulic pressure created by the passing load helps to propagate the crack. Failures can also be initiated at preexisting subsurface defects. If the countersurface is large, as compared with the theoretical line contact

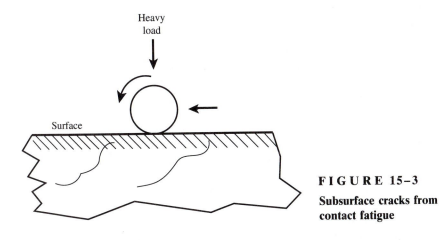

FIGURE 15-3

Subsurface cracks from contact fatigue

of a roller or gear tooth, the cracks created will tend to be longer and be parallel to the surface.

The signs of failure by contact fatigue are pits, spalls, or cracks in heavily loaded surfaces. Metals selected for heavy-load applications should have good fatigue properties.

Fretting. Fretting normally occurs between surfaces that are not intended to have relative motion, e.g., a shaft in a coupling or a gear mounted on a shaft. When these surfaces are subjected to a *vibratory motion* of small amplitude, fretting occurs. The damage that results is of a pitting nature, and it can be accelerated if an electrolyte or corrosive chemical is also present, in which case it is termed *fretting corrosion*. Figures A-36 and A-41, two views of the same motor-boat drive shaft, are examples of this. The dark pitted areas show fretting corrosion caused by operating a stainless-steel shaft in a brass coupling in a wet environment.

15.6   EROSIVE WEAR

Erosive wear is divided into four categories—solid-particle erosion, slurry erosion, cavitation, and liquid-droplet erosion.

Solid-Particle Erosion. *Solid particles carried by a gas* will erode the surface of the metal they strike. The rate of erosion is dependent on the velocity of the particles, their size and shape, and the angle at which they strike. The higher the velocity, the more rapid the erosion; up to a point the larger the particle, the more damage it does, with more angular particles causing more damage than smoother ones. For some softer metals the most severe deterioration occurs when particles strike at angles of 20° to 30° to the surface. The surface of brittle metals deteriorates more rapidly as the impingement angle approaches 90°.

Solid-particle erosion resistance is improved by hardness, ductility, and work-hardening ability.

Slurry Erosion. **Slurry erosion** occurs when *liquid streams carry particles against metal surfaces.* The flow of these particles along the metal surface makes this mechanism much like low-stress abrasion, except that the liquid medium adds a corrosion component. This becomes a special problem since the eroding action of the particles may remove any passivating films that would form on the surface of some metals. So, for the softer, corrosion-resistant-type metals, if corrosive liquids are present erosion resistance will largely depend on forming and maintaining a passive film. In noncorrosive liquids the erosion will be more dependent on the maximum cold-worked hardness, which is also the criterion in low-stress abrasion.

For multiphase metals, e.g., steels, the erosion mechanism is similar to that for solid-particle erosion and is related to the angle at which the slurry strikes the surface. Erosion resistance in these cases seems related to ductility rather than hardness or strength.

Cavitation. When a *metal vibrates against a liquid* it sends shock waves through the liquid. In the low-pressure troughs of these waves vaporous bubbles of the liquid can form. As the vibrations cause the pressure to change these bubbles collapse or implode at the liquid-metal interface, sending shock waves against the metal, in effect subjecting it to impact loads that cause small pits or cavities in the surface. These pits tend to cluster in the areas of low pressure so attacks tend to be localized. Since they are localized, they can penetrate quite deeply into a metal part. Pits may grow to a large size on the low-pressure side of devices like propellers and impellers and may completely penetrate thin members.

Since cavitation occurs in the presence of liquids, the deterioration is frequently accelerated by corrosion. The cyclical and impact nature of the loading suggests that resistance to cavitation is improved by good fatigue strength and resilience.

Note that the cause of cavitation stems from vaporous *bubbles of the liquid itself,* not from entrained air or other gases. These entrained gases are high-pressure vaporous bubbles that can reduce the load-carrying capacity of lubricants and encourage corrosion, but they do not collapse rapidly enough to cause cavitation.

Liquid-Droplet Erosion. Examples of this type of erosion occur in steam turbines and aircraft flying at high speed through rainstorms. The process involves *liquid droplets striking with an impact force sufficient to produce plastic depressions* in the surface of the metal. The cyclical nature of the load results in fatigue failures and pitted and spalled areas form on the surface. Since liquids are involved the possibility of accompanying corrosion is very high.

In some aspects this failure mechanism is the same as for cavitation, and fatigue strength and resilience are important in the selection of metals subjected to this form of erosion.

15.7 SOLUTIONS: ABRASIVE WEAR

It has been pointed out that wear is a systems problem, so it is virtually impossible to separate worthwhile corrective actions into categories of metallurgy, design, operation, etc.; each one affects the other. Rather, we'll discuss the possible solutions under the mechanism to which they apply rather than according to the category of the solution itself.

Closed Systems. In closed systems, e.g., internal combustion engines, the abrasive particles are foreign materials; the proper solution is to exclude the particles and/or filter them out. This is the purpose of oil and fuel filters, sumps, and shaft seals. If particles are small enough, lubrication may reduce deterioration by keeping surfaces cool and separated. If, through design or maintenance procedures, the abrasive particle size can be kept small enough that the particle is in a low-stress situation, i.e., packed rather than trapped, the rate of wear will be less.

Open Systems. In open systems, such as those used to handle abrasive materials, or where particles can enter from the environment, such as farm equipment, the ratio of particle-to-surface hardness is important. If possible, materials should be selected to keep this ratio less than 1.0. In general, harder metals fare better, although for single-phase metals used in low-stress applications, maximum cold-worked hardness is a better criterion for wear resistance. In any case, unless a great amount of data are available on the hardness of the abrasive(s) to be encountered and the metal(s) to be used, comparative trials under actual conditions will probably be necessary to choose the optimum material.

Resistance to low-stress abrasion is aided by second phases, whereas under high-stress conditions they are less helpful. In single-phase metals subjected to low-stress abrasion, it appears that maximum cold-worked hardness is the important determiner of resistance.

The best solutions to abrasive wear come about when its existence is anticipated and the system designed for it. Hard surfaces can be formed by cold-working, anodizing, plating, and hard-surface welding plus all the alloying and selective heat-treating methods for steel described in Chapter 7.

In that regard, it is not necessarily beneficial to "harden everything." For example, if a V-shaped digging tooth for a back-hoe bucket is hardened on both sides of the V, when the point wears through, the soft interior is exposed and the tooth becomes very dull. If instead the tooth is hardened on only one side, the softer side will wear away, keeping the tooth sharp.[‡]

15.8 SOLUTIONS: CONTACT WEAR

Oxide Layers. Maintaining a wear system so that stable oxides separate wear surfaces is desirable, but it requires low loads and a favorable environment. If this is not possible, then some transfer of metal, i.e., adhesion or galling, is likely.

[‡]This is how the teeth of rodents stay sharp.

Galling. Since galling occurs because of metal-to-metal contact, the most obvious solution is to keep surfaces separated by a layer, and a lubricant is most often used for this purpose. (Because lubrication is a somewhat involved subject, we'll discuss it separately below.) Some other steps that can be taken to avoid or reduce wear by galling include:

- Keep surfaces smooth so that projections (asperities) don't interfere with the other surface.
- Use metals for the surface and countersurface that don't naturally weld.
- Design, operate, and maintain the systems so that wear surfaces stay cool and don't overheat.
- Use hard surfaces opposite softer ones, preferably making the softer surface replaceable.

Lubrication. Lubricants serve a number of important functions: They cool metal surfaces to reduce their tendency to gall or weld together, physically separate the metal surfaces, and reduce the frictional forces. Some lubricants form oxide coatings that can help protect surfaces from wear. In closed systems contaminated by abrasives, lubricants can help to convey the particles away to be removed by filtration or settling in a sump.

For purposes of explanation, assume that two stationary surfaces are separated by a film of lubricant and can carry a certain load such that the lubricant is not forced from between them. If these surfaces are now moved relative to each other, they can carry a much larger load and still maintain their separation. The key parameters in forming this hydrodynamic film, or **slipper effect,** are the *load, lubricant viscosity,* and *relative speed* of the surfaces.

We can speak of "lubrication" in three different conditions:

1. There is no lubricant; i.e., the system is "dry," or lubrication has broken down, so there is continuous metal-to-metal contact.
2. Lubrication is a boundary condition, where there is thick-film separation part of the time and metal-to-metal contact part of the time.
3. Lubrication provides thick-film separation, or the slipper effect just described, where the surfaces are separated by an oil film and wear is at a minimum.

Figure 15-4 illustrates the formation of a hydrodynamic or lubrication slipper in a **journal bearing.** Before start-up the shaft is in the bottom of the bearing and there is metal-to-metal contact. Because of the "dry" condition, when the shaft begins to rotate it climbs the bearing surface to the left. However, as the shaft turns, lubricant adhering to the surface, plus lubricant supplied by a pump if any, accumulates under the shaft, forming the slipper. For a given load and lubricant viscosity, as long as the shaft operates at high enough rpm the slipper is maintained; if the shaft slows too much, the slipper is lost and metal-to-metal contact occurs. The lubricants visualized here are liquids, but gases, e.g., air, can do the same thing. Shafts are operated at very high rpm's using "air bearings."

The slipper effect occurs on flat surfaces as well as round and in rectilinear as well as rotary motion. In reciprocating motion the slipper is difficult to maintain since the velocity of the relative motion goes to zero at the ends of the stroke.

F I G U R E 15–4 Start-up of a shaft in a journal bearing and the formation of a hydrodynamic slipper

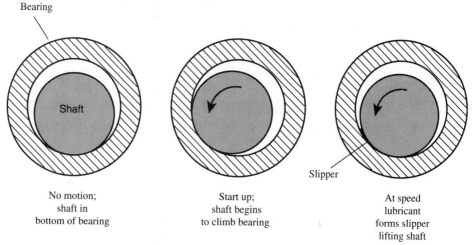

| No motion;
shaft in
bottom of bearing | Start up;
shaft begins
to climb bearing | At speed
lubricant
forms slipper
lifting shaft |

Lubricant viscosity can be a difficult parameter to optimize since it often varies with temperature. Higher-viscosity lubricants help support loads better but may not flow well enough, especially at low temperatures, to develop the slipper before severe damage occurs. However, over some ranges of operating conditions, the speed and viscosity can be self-regulating. For example, as speed goes up, system temperature would tend to go up, which would lower the viscosity of most oils, which improves their ability to extract heat, which would lower the temperature, etc.

Some journal bearings use grooves to distribute lubricant within the bearing surfaces. These grooves reduce the area of the bearing carrying the load so they should be used sparingly and especially be kept out of the area where the slipper forms. Many times lubricants are supplied to bearings under pressure. In journal-bearing systems these connections must also be kept out of the slipper area since the hydrodynamic pressure generated by the slipper effect is much greater than the pressure supplied by the lube pump.[§]

A caution about lubrication: Many machines have systems to distribute lubricant to bearing surfaces. Although these systems help reduce galling wear, the openings and holes required for the distribution system can be stress concentrations that lead to other types of failures.

Contact Fatigue. The solutions to avoiding contact fatigue are to use metals of high fatigue strength, which generally means high strength-high hardness, and to avoid overloading. Lubrication can be of help in cooling the system and reducing friction, but high overloads of, say, a roller bearing, will eventually lead to failure.

[§]The approximately 40-psi oil pressure supplied by the oil pump of automobile engines times the area of the main bearings is hardly enough force to lift the crankshaft. Most of the force has to come from the hydrodynamic slipper.

The subsurface shear stresses cause spalling, and the gear box and lube system will become full of small chips of expensive steel.

Fretting. Since this failure is due to unintended vibratory motion between surfaces, the best solution is to eliminate the vibration. Diagnostic devices are now available that make this a viable solution. Avoiding the use of dissimilar metals in wet environments can help reduce the tendency for corrosion to accompany the fretting. More secure fastening of parts is also a possibility but may lead to stress concentrations that can cause other problems.

15.9   **SOLUTIONS: EROSIVE WEAR**

Good solutions to combat solid-particle, slurry, and liquid-droplet erosion begin with the recognition that they are going to occur because of the application, and design and metal selection should be done accordingly.

Solid-Particle Erosion. The resistance of metals to this form of erosion is dependent on the angle of the particle stream, particle velocity, and particle size. At low angles of impingement, solid-particle erosion is similar to grooving abrasive wear, and resistance is improved by high surface hardness.

At higher angles of impingement, the ability to absorb the impact of the particles becomes important. Although many properties might have to be considered for components subjected to this erosion, ductility and work-hardening ability are important.

Slurry Erosion. Wear resistance in these applications is dependent on the particle and the angle at which it strikes the surface. At low angles of impingement, the wear is similar to low-stress abrasion plus corrosion. In general, hardness and ductility seem the desirable characteristics.

Cavitation. The cyclical nature of the loading causing cavitation suggests that resistance is improved by good fatigue strength, i.e., high surface hardness and strength. The need to absorb impact loads means the resilience of the part is important. And since the problem also involves vibration, a change of the vibrational characteristics of the system may reduce the deterioration; e.g., stiffening the surface with ribs. Changing the characteristics of the liquid and streamlining its flow are also steps that can be taken.

Liquid-Droplet Erosion. In this form of erosion the load is cyclic in frequency and impact-like, so fatigue and resilience properties are important. The metal selection here would be similar to that for resistance to cavitation.

SUMMARY

1. The following table summarizes the parameters relevant to wear.

Wear parameters

Tribological System	Motion	Mechanism	Lubrication Condition
Surface Countersurface Intersurface elements Environment	Sliding Rolling Vibrating Impacting Flowing	Abrasion • High-stress • Low-stress • Cutting, Plowing, Chipping Contact • Oxide • Galling • Contact fatigue • Fretting Erosion • Solid-particle • Slurry • Cavitation • Liquid-droplet	None Boundary Thick film

2. Abrasive wear is caused by hard particles. In general, surface hardness is important in combating abrasive wear, but the relationship of particle to surface hardness is important in selecting metals.

3. In contact wear, the mechanisms can be separated by the size and concentration of the load: Under lightly loaded and dry conditions, oxides form; under heavier loads and with no lubrication, galling occurs; heavy, concentrated, and moving loads cause contact fatigue; and a vibrating load brings fretting.

4. In machine applications galling is often encountered. Lubrication is the principal method of avoiding galling, although metal selection is also important.

5. Thick-film lubrication requires the proper relationship between surface speed, viscosity, and load in order to generate and maintain a hydrodynamic slipper. If speed, for example, is not high enough boundary conditions will exist and metal-to-metal contact will eventually occur.

6. Erosion is caused by moving particles or bubbles and in general requires metals with surface hardness and resilience.

REVIEW QUESTIONS

1. What is tribology? Describe a tribological system.
2. Hard foreign particles are responsible for what wear mechanism?
3. What effect does the angle of impingement have on the mechanisms of erosion and how do those mechanisms relate to low-stress abrasion?
4. What type or mechanism of wear occurs when metallic surfaces actually touch or rub each other?
5. Lubrication is most often used to reduce which type of wear? Why?
6. How might galling contribute to abrasive wear?
7. When people "fret" what do they do? How is this similar to what metals do when they "fret"?
8. How does cavitation differ from erosion by a flowing liquid full of air bubbles?
9. What are the primary mechanisms for wear? What are the secondary mechanisms?

FOR FURTHER STUDY

ASM handbook. Materials Park, OH: ASM International, 1992, vol. 18 (Friction, lubrication and wear technology).

Crook, Paul. Materials performance. *Practical guide to wear for corrosion engineers,* vol. 30, no. 2 (February 1991), 64–66.

Czichos, Horst. *Tribology* (Tribology Series, 1). Amsterdam: Elsevier Science, 1978.

Jones, Marvin H., and Douglas Scott. *Industrial tribology* (Tribology Series, 8). Amsterdam: Elsevier Science, 1983.

Metals handbook (9th ed.). Materials Park, OH: ASM International, 1987, vol. 13 (Corrosion), 663.

Sarkar, A. D. *Wear of metals*. Oxford: Pergamon, 1976.

Zum Gahr, Carl-Heinz. *Microstructure and wear of materials* (Tribology Series, 10). Amsterdam: Elsevier Science, 1987.

APPENDIX A

Photomicrographs and Photographs

A.1   METALLOGRAPHIC EXAMINATION

The surfaces of metallic specimens are often examined using metallurgical microscopes, which are similar to the optical microscopes used in biology laboratories, except that they are equipped to light opaque surfaces. The photomicrographs of Appendix A were taken using optical metallographs, metallurgical microscopes equipped to take pictures of opaque surfaces.

A.2   SPECIMENS

To examine metals metallographically, specimens are first selected to represent the metal under investigation; they are then cut to sizes appropriate for viewing with the microscope. The specimens are cut with a metal saw or abrasive wheel, depending on their hardness. They are then rough-smoothed by grinding or filing. Depending on their size, specimens may be mounted in plastic or held in a clamp so they can be easily gripped for further preparation and examination.

A.3   METALLOGRAPHIC POLISHING

The objective of preparing a specimen for examination under the optical microscope is to generate a flat surface with a mirrorlike surface. This is usually accomplished by smoothing the surface (in one direction) with an abrasive belt followed by hand sanding with a series of abrasive papers of finer and finer grits. Between each grit size, the specimen is turned 90° so it can be seen that all lines from the previous grit were removed. After sanding with the finest paper, the specimen is

334

polished on a rotating cloth lap; a suspension of a fine oxide is used as the abrasive. Depending on the extent of polishing required, more than one lap and fineness of abrasive may be used.

A.4 ETCHING

The final product is a small piece of metal with a very highly polished surface. Sometimes this is all that is required for microscopic examination, but more often the observer wishes to see other characteristics that require the specimen be etched. Various acids and bases are used to selectively attack the surfaces of metals to reveal grain boundaries, twins, second phases, precipitates, inclusions, and variations in composition.*

A.5 INTERPRETATION

In interpreting photomicrographs or views through a microscope, remember that:

- The view seen is two-dimensional but the metal is three-dimensional. So, while grains are seen as bounded by two-dimensional boundaries, they are actually volumetric and the grain boundaries are like the skin of an irregularly shaped potato. Depending on where the potato is sliced, the boundary may appear roughly round or oblong, but if the slice is tangent to the skin (i.e., boundary) it will have a very different appearance.
- The structure seen in the microscope may be a small part of a larger one that exists just above or below the surface. Dendrites, for example, have a tree-like appearance, and at times these structures will be seen complete with trunk and limbs. But often the evidence of dendritic structure can be seen in regular spacings of just the limbs, as if a giant chain saw (the surface of the specimen) trimmed all the tree limbs at an angle to the main trunk.
- In interpreting photomicrographs it is helpful to consult the phase diagram and to have some knowledge of the composition and thermal history of the specimen. If both resources are used, phases can often be identified and the sequence in which they formed can be traced.
- With experience, the above process can be reversed; that is, a metallographic examination can determine the thermal history of a specimen; for example, it can determine if a failed part was overheated.

*A good source of information on metallography is the *Metals handbook*, Materials Park, OH: ASM International. Volume 8 (8th ed., 1973) and volume 9 (9th ed., 1985) contain details of specimen preparation, etching, and examination for both optical and electron microscopes.

The magnifying power of a microscope is limited by the relationship of the wavelength of the light source to the size of the details to be resolved or seen. In optical microscopes this limit is about 2000 times, and even then special techniques must be used. High-power electrons, however, have wavelengths significantly smaller than those of light and permit magnifications of 10 million times or more.

The photomicrographs in this appendix were done with optical microscopy and range from magnifications of 25× (see Figure A-1) to 3000× (see Figure A-21). The photographs at the end of the appendix demonstrate that useful information can also often be obtained with the naked eye.

F I G U R E A–1

Cavity in cast-monel metal (33% Cu-67% Ni), × 25. Metal is partially frozen and then molten metal poured out, leaving the center starved for metal. Dendrite arms and trunks can be seen reaching out into cavity. The variation in composition (coring) of the dendrites has caused variations in the etching, thus exposing the dendritic structure.

Source: Clyde W. Mason. *Introductory physical metallurgy*. Materials Park, OH: ASM International, 1947, fig. 6, 12. Used by permission.

FIGURE A–2

85% Cu-15% Ni, chill-cast, × 50. Small cored dendrites of alpha (α) solid solution. Etchant has selectivity etched the copper-rich spaces between the nickel-rich dendrite arms, thus exposing them. This should be a homogeneous structure, but because of the coring it is not.

Source: R. Brick and A. Phillips. *Structures and properties of alloys*. New York: McGraw-Hill, 1949, micro 3.1, 62. Used by permission.

FIGURE A–3

85% Cu-15% Ni, chill-cast and reheated to 750°C for three hours, × 50. Reheating has caused diffusion of the copper and nickel, reducing the composition differences somewhat, but there is still evidence of segregation.

Source: R. Brick and A. Phillips. *Structures and properties of alloys*. New York: McGraw-Hill, 1949, micro 3.2, 62. Used by permission.

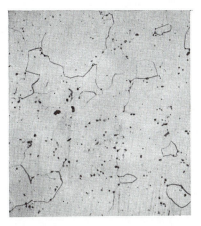

FIGURE A–4

85% Cu-15% Ni, chill-cast and reheated to 950°C for nine hours, × 50. The long high-temperature treatment has homogenized the structure and composition; grain boundaries can be seen.

Source: R. Brick and A. Phillips. *Structures and properties of alloys*. New York: McGraw-Hill, 1949, micro 3.3, 63. Used by permission.

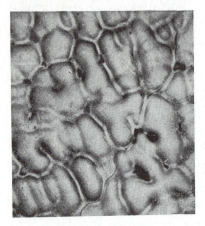

FIGURE A–5

85% Cu-15% Ni, cast in a hot mold and slowly solidified, × 50. The dendritic structure is much coarser than that of the chill-cast specimen of Figure A-2. The raised portions are nickel-rich dendrite arms; the valleys between are the copper-rich metal that froze later.

Source: R. Brick and A. Phillips. *Structures and properties of alloys*. New York: McGraw-Hill, 1949, micro 3.4, 63. Used by permission.

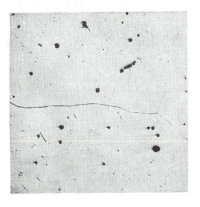

FIGURE A–6

85% Cu-15% Ni, cast in a hot mold and slowly solidified, then reheated to 950°C for 15 hours, × 50. Although this structure is homogenized, longer time was required; the composition differences across the much larger dendrites mean that atoms had to diffuse longer distances.

Source: R. Brick and A. Phillips. *Structures and properties of alloys*. New York: McGraw-Hill, 1949, micro 3.5, 64. Used by permission.

F I G U R E   A – 7   Brass (70-30) as-cold-worked, 33%. Large original grains with slip lines revealed by etching, × 75.

Source: Used through the courtesy of Joseph E. Burke, Burnt Hills, NY.

F I G U R E A–8 Cold-worked brass of Figure A-7 after heating three seconds at 580°C (1075°F). Recrystallization is beginning; notice that nuclei forming preferentially at slip lines are much smaller than original grains, × 75.

Source: Used through the courtesy of Joseph E. Burke, Burnt Hills, NY.

F I G U R E A–9 **Cold-worked brass of Figure A-7 after heating eight seconds at 580°C (1075°F). Recrystallization is complete with grain size much smaller than originally, × 75.**

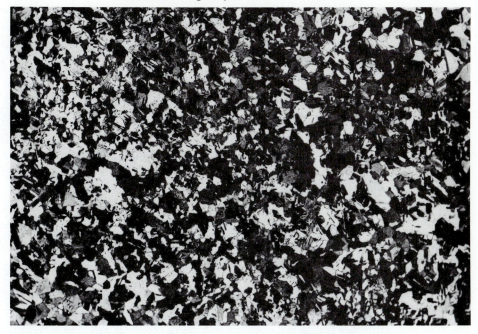

Source: Used through the courtesy of Joseph E. Burke, Burnt Hills, NY.

F I G U R E A – 10   **Cold-worked brass of Figure A-7 after heating 15 minutes at 580°C (1075°F). Grains are experiencing grain growth by being held at temperature for a longer time.**

Source: Used through the courtesy of Joseph E. Burke, Burnt Hills, NY.

F I G U R E A–11 **Cold-worked brass of Figure A-7 after heating 10 minutes at 700°C (1290°F). At the higher temperature grains recrystallize and grow to a larger size than those of Figure A-10.**

Source: Used through the courtesy of Joseph E. Burke, Burnt Hills, NY.

F I G U R E A – 12   Cold-worked brass of Figure A-7 after additional heating at 700°C (1290°F). Continued heating at the higher temperature produces very large grains.

Source: Used through the courtesy of Joseph E. Burke, Burnt Hills, NY.

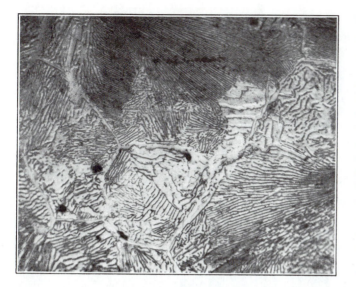

FIGURE A–13

Proeutectoid cementite and pearlite in 1.3% carbon steel, × 400. The proeutectoid cementite formed at the boundaries of the austenite grains, which transformed at a lower temperature to pearlite. The variation in the closeness of the pearlite laminations is due to their being sectioned at different angles in different grains.

Source: Clyde W. Mason. *Introductory physical metallurgy*. Materials Park, OH: ASM International, 1947, fig. 51, 88. Used by permission.

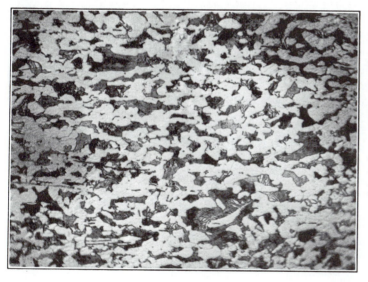

FIGURE A–14

Ferrite and pearlite in hot-rolled 0.35% carbon steel, × 500. The light ferrite grains have been elongated by the rolling. The dark pearlite transformed from the austenite between them.

Source: Clyde W. Mason. *Introductory physical metallurgy*. Materials Park, OH: ASM International, 1947, fig. 52, 90. Used by permission.

F I G U R E   A–15

Ferrite and pearlite in cast 0.4% carbon steel, × 200. Blades of (light) ferrite have precipitated, or transformed, along crystal planes of the original large grains of austenite; the remainder of the austenite transformed to (dark) pearlite.

Source: Clyde W. Mason. *Introductory physical metallurgy*. Materials Park, OH: ASM International, 1947, fig. 53, 90.

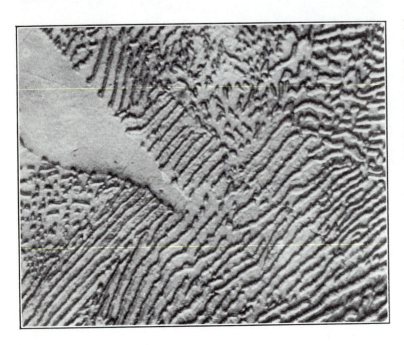

F I G U R E   A–16

Pearlite and ferrite in 0.4% carbon steel, × 2000. Enlarged view of Figure A-15. The tip of one of the ferrite blades is shown. The cementite of the pearlite is shown in relief because the ferrite has been attacked by the etchant.

Source: Clyde W. Mason. *Introductory physical metallurgy*. Materials Park, OH: ASM International, 1947, fig. 54, 91. Used by permission of J. E. Burke.

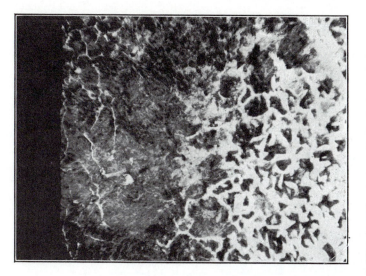

FIGURE A–17

Section through case of
carburized steel, × 50.
Carbon, supplied to the
surface from solid or gaseous
carburizing agent at high
temperature, diffuses inward
(left to right) and raises the
carbon content of the left
portion. On cooling from the
austenitic range, cementite
separates at the austenite
grain boundaries of the
hypereutectoid zone on the
left and decreases through a
pearlitic area to the
hypoeutectoid, uncarburized,
interior to the right. With
heat treatment, this steel will
have a hard martensitic case
with a softer and
tougher core.

Source: Clyde W. Mason.
*Introductory physical
metallurgy*. Materials Park, OH:
ASM International, 1947, fig. 55,
92. Used by permission.

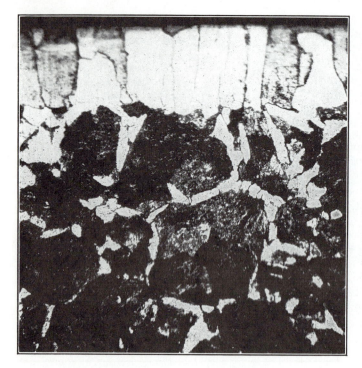

FIGURE A–18

Surface decarburization of 0.6% carbon steel, × 100. By exposure at high temperature to an oxidizing atmosphere, carbon is lost from the exterior surface of the steel (top). On cooling through the austenite range, ferrite (light) grows inward from the low-carbon zone. In the interior, proeutectoid ferrite forms at the boundaries of the grains that finally transform to pearlite (dark). The soft skin cannot be hardened by heat treatment and is undesirable.

Source: Clyde W. Mason. *Introductory physical metallurgy.* Materials Park, OH: ASM International, 1947, fig. 56, 93. Used by permission.

FIGURE A–19

Plate and grain boundary network of cementite in 1.3% carbon steel, × 200. The steel was air-cooled (normalized) from above A_{cm}. The proeutectoid cementite formed at the austenite grain boundaries. The austenite later transformed to a matrix of fine pearlite with "sheets" of additional cementite forming on crystal planes of the austenite.

Source: Clyde W. Mason. *Introductory physical metallurgy.* Materials Park, OH: ASM International, 1947, fig. 57, 94. Used by permission.

FIGURE A–20

Incomplete formation of cementite and fine pearlite, in 1% carbon steel, × 500. Quenching from the austenite range was not rapid enough to prevent some austenite from transforming to softer products. A thin band of cementite can be seen at the old austenite grain boundaries, where (dark) pearlite later formed as cooling continued. With falling temperature, the untransformed austenite was converted to the martensite that can be seen between the rounded pearlite areas.

Source: Clyde W. Mason. *Introductory physical metallurgy*. Materials Park, OH: ASM International, 1947, fig. 58, 95. Used by permission.

F I G U R E A–21 **Martensite in 1.3% carbon steel, × 3000. Light and dark "needles" formed at different temperatures during quenching. It is typically difficult to get martensite needles in focus.**

Source: Clyde W. Mason. *Introductory physical metallurgy*. Materials Park, OH: ASM International, 1947, fig. 59, 96. Used by permission.

F I G U R E   A–22   **Three transformation products of austenite, × 500. As a 0.9% carbon steel was quenched, fine pearlite (dark, rounded) began to form at the austenite grain boundaries. As the temperature dropped, the pearlite stopped growing, and some bainite (dark, feathery) began to spread from the grain boundaries along parallel crystal planes of the austenite. Continued cooling prevented this transformation from being completed, and with further cooling the remaining austenite transformed to martensite (light).**

Source: Clyde W. Mason. *Introductory physical metallurgy.* Materials Park, OH: ASM International, 1947, fig. 60, 97. Used by permission.

F I G U R E   A–23   **Ferrite and martensite in 0.4% carbon steel, × 1000. Cast steel, with the coarse ferrite and pearlite structure shown in Figure A-24, was heated briefly above A_1 but below A_3. The pearlite transformed to austenite, which became martensite (dark) on quenching; the ferrite (light) remained untransformed. The relative hardness of the soft ferrite and the hard martensite is indicated by the scratch of a razor blade on the surface.**

Source: Clyde W. Mason. *Introductory physical metallurgy*. Materials Park, OH: ASM International, 1947, fig. 63, 101. Used by permission.

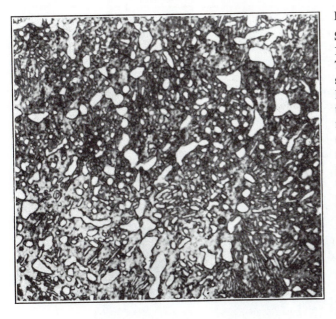

F I G U R E A–24

Spheroidized cementite in 1.3% carbon steel, × 1000. The original structure was similar to Figure A-13. Holding the steel at a temperature near A_1 has rounded and coarsened the cementite. The large particles are the remains of the grain-boundary network, and the fine particles are from the pearlite laminations. High-carbon steels are advantageously machined in this condition, but the spheroidized cementite must be dissolved by sufficient soaking above A_{cm} before the steel is quench-hardened.

Source: Clyde W. Mason. *Introductory physical metallurgy*. Materials Park, OH: ASM International, 1947, fig. 64, 103. Used by permission.

FIGURE A–25 **Austenite-graphite eutectic freezing in cast iron, 3.5% carbon, × 20. Quenched from 2025°F (1107°C) when partly frozen. Dendrites of austenite (gray) in the light background represent the melt. Eutectic froze in spreading patches between dendrites, as a mixture of graphite flakes (black) and austenite (gray). Austenite subsequently transformed to martensite as the temperature dropped. The remaining melt froze as fine austenite-cementite eutectic (white).**

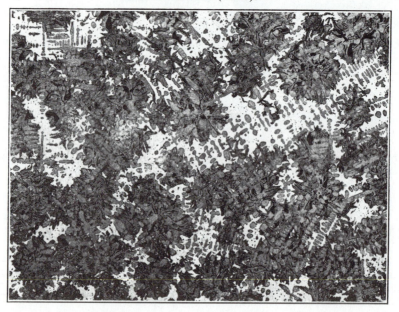

Source: Alfred Boyles. *The structures of cast iron*. Materials Park, OH: ASM International, 1947, fig. 31, 31. Used by permission.

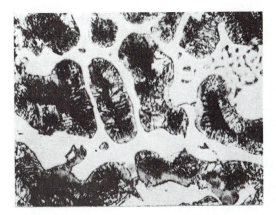

F I G U R E   A–26

Hypoeutectic white cast iron, 2.5% C, × 400. Rounded areas of pearlite (dark) from branches of austenite dendrites. The duplex nature of the eutectic is apparent only in the larger areas, as cementite (light) and pearlite (dark) from austenite.

Source: Clyde W. Mason. *Introductory physical metallurgy*. Materials Park, OH: ASM International, 1947, fig. 66, 106. Used by permission.

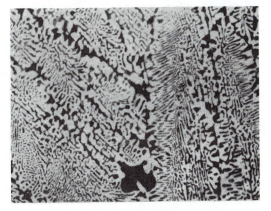

F I G U R E   A–27

Eutectic white cast iron, ledeburite, 4.3% C, × 75. Largely eutectic, with a few small dendrites of austenite (now converted to pearlite) formed because of rapid cooling.

Source: Clyde W. Mason. *Introductory physical metallurgy*. Materials Park, OH: ASM International, 1947, fig. 67, 106. Used by permission.

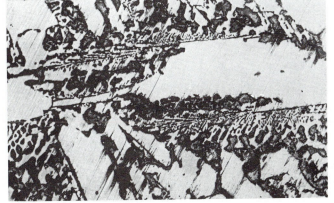

F I G U R E   A–28

Hypereutectic white cast iron, × 75. Very coarse primary cementite with eutectic between.

Source: Clyde W. Mason. *Introductory physical metallurgy*. Materials Park, OH: ASM International, 1947, fig. 68, 106. Used by permission.

FIGURE A–29

Gray cast iron, × 500. Flakes of graphite in matrix of fine pearlite. Small patches (white) are phosphide eutectic.

Source: Clyde W. Mason. *Introductory physical metallurgy*. Materials Park, OH: ASM International, 1947, fig. 73, 112. Used by permission.

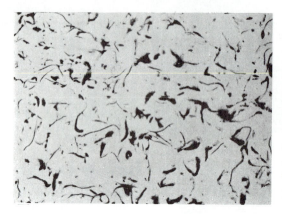

FIGURE A–30

Graphite flakes in gray cast iron, unetched, × 75. A high-strength iron with small amounts of interdendritic graphite. Larger dark areas are flakes sectioned obliquely.

Source: Clyde W. Mason. *Introductory physical metallurgy*. Materials Park, OH: ASM International, 1947, fig. 72, 111. Used by permission.

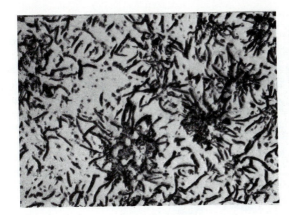

FIGURE A–31

Graphite flakes in gray cast iron, unetched, × 75. A lower-strength iron with more graphite, showing eutectic origin of flakes. (Poor preparation technique makes flakes look thicker than they are.)

Source: Clyde W. Mason. *Introductory physical metallurgy*. Materials Park, OH: ASM International, 1947, fig. 72, 111. Used by permission.

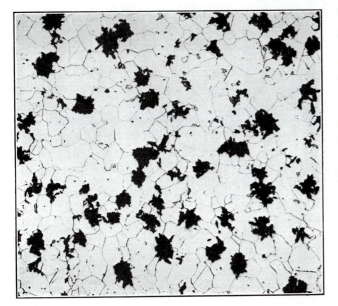

FIGURE A–32

Ferritic malleable cast iron, × 100. Clumps of "temper" carbon or graphite in a matrix of ferrite, etched to show its grain boundaries.

Source: Clyde W. Mason. *Introductory physical metallurgy*. Materials Park, OH: ASM International, 1947, fig. 69, 107. Used by permission.

FIGURE A–33

Aluminum bronze with 10.3% aluminum, solution-heat-treated at 1650°F (899°C) for two hours, water-quenched, tempered at 1200°F (649°C) for two hours, water-quenched, × 200. See Figure 10-1 for phase diagram. Quenching caused martensitic-type reaction. The white needles are alpha; matrix is beta.

Source: *Metals handbook* (8th ed.). Materials Park, OH: ASM International, 1972, vol. 7, micro. 2456, 296. Used by permission.

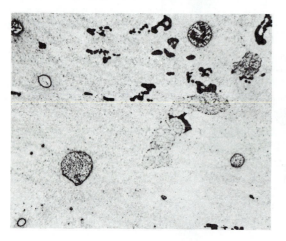

FIGURE A–34

Rounded rosettes formed by incipient melting of eutectic in aluminum alloy 2014 due to heating at too high a temperature during solution heat treatment, × 500.

Source: *Metals handbook* (8th ed.). Materials Park, OH: ASM International, 1972, vol. 7, micro. 2018, 245. Used by permission.

F I G U R E A–35 Splined shaft tested under high torque at high speed. Test intentionally stopped just before failure to show partial failure at 45° to axis from separating stresses in a hardened (brittle) steel.

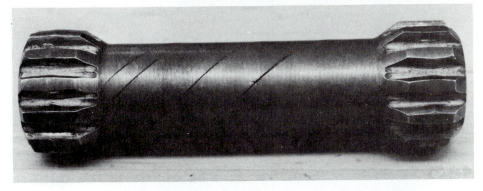

F I G U R E A–36 1-in.-diameter, inboard motorboat drive shaft, stainless steel. Driven in forward direction almost exclusively; when driven in reverse, shaft and propeller unscrewed out the back of boat into the lake. The shaft was driven in a broken condition for a long period, thus the dark fretted corrosion seen. Broken keyway, A′, allowed shaft to come in contact with set-screw recesses, especially at A. Beach marks start there and progress across to the distorted area, the final failure, at B. Purpose of set screws was to hold shaft axially in brass coupling.

FIGURE A–37

Reciprocating water-hydraulic steel piston rod, 6-in. diameter. Stress concentration of thread at A caused fatigue failure to proceed toward center. Use of equipment probably changed; note change in shape of beach marks. Crack proceeded to area near B; final failure area very distorted with surfaces at 45°.

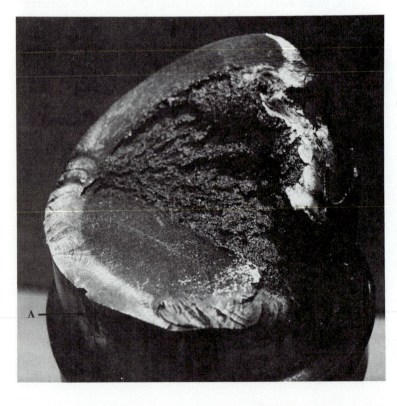

FIGURE A–38

Farm-tractor axle, 3½-in.-diameter steel, with hardened case on outer surface. A indicates burn from cutting torch where frozen bearing was removed in previous "repair." Faint beach marks indicate that the fatigue began near the burn mark, and the fatigue crack proceeded through the shaft until the final, sudden failure.

FIGURE A–39

Reciprocating water-hydraulic steel piston rod, 5¼-in. diameter. Fatigue crack followed stress concentration of threads, A, completely around outer surface to the extent that the mark at 11 o'clock is where the crack had to resolve the separation between the threads. Final failure was at B.

F I G U R E A – 40 The fracture surface of sawed-notch Charpy specimens fractured at following approximate temperatures: (a) −100°F, all brittle fracture; (b) 35°F, gray "picture frame" indicates ductile fracture; (c) room temperature, percent of ductile fracture increases; (d) +100°F, percent ductile fracture increases; (e) +140°F, percent ductile fracture increases; (f) +200°F, 100% ductile fracture.

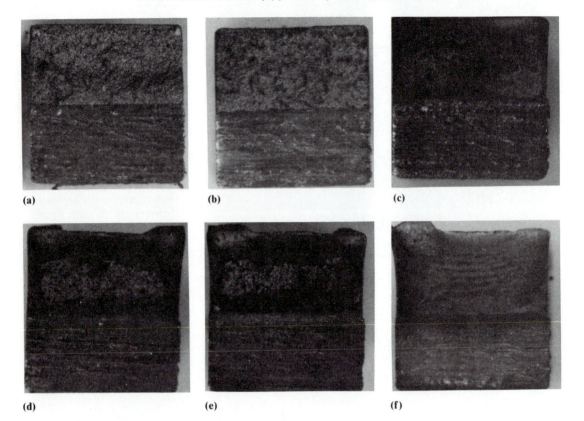

(a)   (b)   (c)

(d)   (e)   (f)

F I G U R E A – 41

Side view of the motorboat drive shaft shown in Figure A-36; the boat was operated in freshwater. The discolored and pitted area to the right is fretting corrosion. The stainless-steel shaft was keyed and set-screwed to a brass coupling.

A P P E N D I X B

Tensile Testing

OBJECTIVES

After studying this appendix and portions of Chapter 2 and, it is hoped, participating in a lab exercise, you should be able to perform a tensile test, and be able to define and determine

- a. Modulus of elasticity
- b. Yield strength
- c. Offset
- d. Tensile strength
- e. Elongation
- f. Reduction in area
- g. Rupture or fracture
- h. Stiffness
- i. Elastic vs. plastic strain
- j. Uniform vs. localized strain
- k. Ductile vs. brittle failures
- l. Modulus of resilience
- m. Modulus of toughness

DID YOU KNOW?

The design and use of metals are usually based on their mechanical or physical properties. This appendix deals with tensile testing, the method used to obtain some of the most important mechanical properties of metals. The tensile strengths of metals vary from 300,000+ psi for the stronger steels, to 70,000 psi for some aluminum alloys, to less than 10,000 psi for lead and tin. You should review the mechanical properties of metals to have a general idea of the strength of metals. Of particular use are the values of the modulus of elasticity for steel, brass, and aluminum; these metals are in such common usage that their approximate values, 30, 17, and 10 million psi, respectively, should be memorized.

A PREVIEW

After introducing and reviewing the basic terms and units used in tensile testing, we'll look at a simulated test and then develop the formulas to calculate pertinent mechanical properties. It's preferable that you participate in or observe an actual tensile test as you study this appendix.

B.1   DEFINITIONS AND TERMINOLOGY

In a tension test known loads are applied to a metal specimen of standard dimensions* and the "stretch" or extension of the specimen is measured. This is usually done using a calibrated "load cell," "strain gage," and "X-Y plotter." From such data the stress (σ) and strain (ε) at any point can be calculated.

Stress. **Stress** (σ) is defined as the *load divided by the area over which the load is applied.* The common practice is to use the original area, and the result is termed simply *stress* or *engineering stress.* When the actual area under load is used in the calculation, the stress is termed **true stress;** we'll deal with it later.

In the English system, used here, the usual units of stress are pounds per square inch (lb/sq. in., lb/in.^{2}, or psi) or thousands of pounds per square inch (ksi). The metric system uses newtons per square meter (N/m^2), pascals (Pa), or megapascals (MPa).†

Strain. **Strain** (ε) is defined as the *change in length, due to the load, divided by the length.* When the length used is the original test length, usually 2 in., then the strain is simply *strain* or *engineering strain.* When the incremental change in length is divided by the immediately preceding length, the strain is termed **true strain;** we'll deal with it later.

The units of strain are dimensionless since it is length divided by length. It is usually expressed in decimal terms for small values, e.g., inch/inch, or as a percentage for larger values.

Modulus of Elasticity. Robert Hooke (1635–1703) is given credit for first stating the relationship that today is known as the *modulus of elasticity*: Within a range of stresses and strains the two have a linear relationship. That is, the modulus of elasticity E is equal to stress/strain or $E = \sigma/\varepsilon$.

It is important to emphasize that *this ratio only holds in the elastic range;* if the metal is loaded such that it is strained beyond the elastic range, the relationship $E = \sigma/\varepsilon$ no longer holds and cannot be used to calculate loads or deflections. Conceptually the modulus of elasticity is the same as the spring constant usually associated with Hooke's name. Just as the spring constant cannot be used to correctly calculate the extension of a spring that has been overstretched, the modulus of elasticity cannot be used to relate stress and strain beyond the elastic region.

For most engineering metals, the value of E in the English system is in the millions and is often written as an integer $\times$ 10^6 psi.

---

* ASTM Standard E8 (Standard Methods of Tension Testing Metallic Materials) should be referred to for more information on this test. Basic information is found in volume 03.01. Information on tension-testing specific metals or products is found in other volumes under the same standard number. Specimens can have a number of sizes and configurations; two common ones are sheet and round specimens with a nominal ½-in. dimension for width or diameter, respectively.
†To convert between the two systems, MPa = ksi $\times$ 6.9.

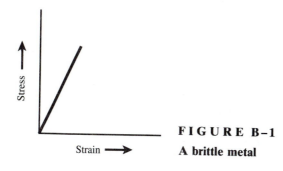

FIGURE B–1

A brittle metal

Brittle vs. Ductile.   A tensile test of a brittle metal will generate the type of curve seen in Figure B-1. Note there is only an elastic portion to the curve. *A completely brittle metal*[‡] *will have no plastic deformation;* it will carry the load to the limit of its elastic region and fracture. Since this line goes through the zero stress-zero strain point, dividing a stress by the corresponding value of strain will equal E, the modulus of elasticity. Or, E is equal to the slope of the line, $\Delta\sigma/\Delta\varepsilon$, which can also be considered $d\sigma/d\varepsilon$; these expressions help make the point that the modulus of elasticity can be interpreted as the "change of stress with strain" in the elastic region.

The tensile test of a *ductile metal*[§] (Figure B-2) *shows a large increase in the strain when the stress goes beyond the elastic region.* The metal goes through plastic deformation before it fractures.

Using plastic deformation to differentiate whether a failure is ductile or brittle is useful in understanding more complex modes of failure, such as fatigue and impact, and is utilized in other parts of this book.

Proportional and Elastic Limits.   There are two points in the elastic portion of the tensile-test curve which are usually discussed in texts but are of little practical value because they are hard to determine and are usually not normal handbook

[‡]White and gray cast irons and tungsten carbide are examples of completely brittle metals.
[§]Brass, aluminum, and most steels are examples of ductile metals.

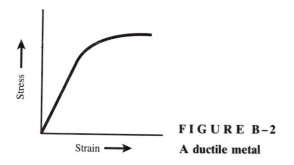

FIGURE B–2

A ductile metal

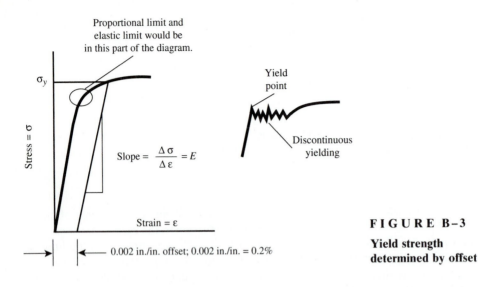

F I G U R E B–3

Yield strength determined by offset

data. The **proportional limit** is the stress at the exact point where Hooke's law ceases; that is, the stress where the σ-ε curve stops being linear. The **elastic limit** is the maximum stress that can be applied and still have the specimen return to its original length when the load or stress is removed. The determination of both these points is influenced by the accuracy of the equipment used to measure very small variations of strain.

Yield Strength. Locating the exact boundary between elastic and plastic deformation in a tension test of metals such as brass, aluminum, and heat-treated or cold-worked steel is difficult, and for engineering purposes is defined in arbitrary terms. The **yield strength** (σ_y) is derived from the stress-strain curve by the **offset** method.[‖] From the point on the strain axis equal to .002 in. per in., or 0.2%, a straight line is drawn at the same slope as the modulus of elasticity. The stress at which this line intersects the stress-strain curve is the 0.2% yield strength of the specimen.[#] This construction is shown in Figure B-3.

Yield Point. Low-carbon and high strength-low alloy (HSLA) steels, especially if in a hot-worked condition, typically undergo discontinuous yielding at the end of their elastic region; this is shown on the right side of Figure B-3. These steels do not change smoothly from elastic to plastic deformation; at the end of their elastic range they abruptly reduce in strength and then increase and decrease strength rather rapidly until the curve smooths out and the strength of the steel increases to the tensile strength. For steels of this type the **yield point** is used in place of the yield strength for design purposes: the yield point is given in handbooks.

[‖] This method would not be used for hot-worked low-carbon steels; they yield upon reaching a *yield point* and strain extensively before stress increases. The stress at that yielding point would be used as the yield strength.

[#] There is also a standard offset of 0.1% or 0.001 in./in. but it is seldom used. Larger offsets are used with metals, e.g., lead, that yield readily.

B.2   THE TENSION TEST AND STRESS-STRAIN (σ-ϵ) CURVE

Before, we calculate stress and strain we'll simulate a tension test, analyzing what happens to the specimen as the load P is applied. Figure B-4 describes the relationship of stress and strain that occurs during a tensile test. In order to describe what is taking place, we'll trace the volume of the specimen.

If the original length and area in the test portion of the specimen are L_0 and A_0, the volume will be $L_0 \times A_0$. When first loaded, the specimen deforms elastically; its length increases and its cross-sectional area decreases. If the length and area at point 1 in the elastic region are L_1 and A_1, the equation

$$L_0 \times A_0 = L_1 \times A_1$$

can be written.** That is, the volume of the metal is a constant, and the deformation is assumed uniform along the test length. In the elastic region this is termed *elastic strain*, or **uniform elastic deformation.**

As loading continues to point 2 (Figure B-4), the specimen deforms plastically because the elastic limit has been exceeded. If unloaded at this point (which would not be done in a regular test), it will not return to its original length, but will return

**The equation is valid for the accuracy of typical measuring instruments.

F I G U R E B–4 **Describing the deformation of a tensile specimen. The specimen is stressed through the elastic range to beyond the original yield strength, unloaded at point 2, reloaded to its new yield strength at point 2 (exhibiting a hysteresis loss), reaches maximum load at point 3, and fractures at point 4.**

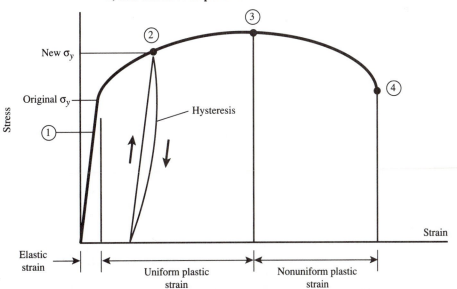

to zero stress along a path *approximating* the slope of the modulus of elasticity.[††] Upon reloading it returns to the stress and strain values it had before unloading, following a line with a slope equal to E; note that the specimen now has a new, higher yield strength. Although the specimen has deformed plastically, the deformation is still assumed to be uniform along the test length, and the volume equation,

$$L_0 \times A_0 = L_2 \times A_2$$

is still valid.

When loading increases to point 3, the maximum load is reached, and when the load is divided by the original area gives the **ultimate tensile strength,** or **tensile strength.** This is the highest load the specimen can carry, and it is the highest strain where the volume equation is valid; that is,

$$L_0 \times A_0 = L_3 \times A_3$$

Between the elastic region and point 3 the strain is termed **uniform plastic deformation.**

As loading continues, the specimen begins to deform locally, or "neck down," and there is less area to carry the load. Since the constant original area, rather than the actual reduced area, is used in the denominator of the stress calculation the value must go down as the load goes down. Thus the stress decreases to point 4, where fracture occurs. *Between points 3 and 4 the volume equation is no longer valid* since the cross-sectional area can vary significantly within the test length. The strain between points 3 and 4 in Figure B-4 is termed **nonuniform plastic deformation,** or localized plastic deformation.

Because the volume formula holds true from the origin of the curve to point 3, the point of maximum load, a relationship between length and area can be written to any point x as long as the deformation is uniform. That is,

$$L_0 \times A_0 = L_x \times A_x$$

Later it will be useful to rewrite this as

$$\frac{A_0}{A_x} = \frac{L_x}{L_0}$$

B.3   **CALCULATIONS FROM THE TENSION TEST**

Yield and Tensile Strength.   The stress σ_x at any point x is found by dividing the load at that point, P_x, by the original area A_0, or

$$\sigma_x = \frac{P_x}{A_0}$$

[††]The difference between the unloading and loading lines generates a hysteresis loop representing the internal energy lost during deformation; it is termed *anelastic behavior*. See W. O. Alexander, et al. *Essential metallurgy for engineers*. Wokingham, U.K.: Van Nostrand Reinhold, 1985, 83.

The yield strength σ_y is found by dividing the load P_y, determined as shown in Figure B-3, by the original area A_0. In like fashion, the tensile strength σ_u is found by dividing the maximum load, the load at point 3 in Figure B-4, by the original area A_0, or

$$\sigma_y = \frac{P_y}{A_0} \quad \text{and} \quad \sigma_u = \frac{P_u}{A_0}$$

Strain. Ordinary **strain** ε has been defined as the *change in length ΔL divided by the original test length L_0*; or the stain at any point x is

$$\varepsilon = \frac{\Delta L}{L_0} = \frac{L_x - L_0}{L_0} = \frac{L_x}{L_0} - 1$$

Modulus of Elasticity. The modulus of elasticity E is found by measuring the *slope of the stress-strain curve in the elastic region.* If a plotter is used, this is done by constructing a right triangle on the plot, measuring the lengths of the sides and converting the lengths, appropriately, to stress and strain, and dividing them to get E, or

$$E = \frac{\Delta\sigma}{\Delta\varepsilon}$$

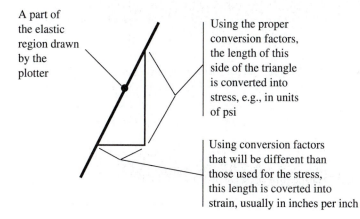

A part of the elastic region drawn by the plotter

Using the proper conversion factors, the length of this side of the triangle is converted into stress, e.g., in units of psi

Using conversion factors that will be different than those used for the stress, this length is coverted into strain, usually in inches per inch

If a plotter is not used and separate stress and strain readings are taken, they must be plotted and the procedure is then the same as described above.

It is not safe to use one test value of stress and divide it by one test value of strain to determine the modulus of elasticity. Such a method would require that all equipment be exactly at zero when the test began and this is not a good assumption. It is much better practice to *determine the modulus of elasticity from the slope of the curve.*[‡‡]

[‡‡]See ASTM Standard E111, vol. 03.01, for further information.

Elongation.   The maximum strain, or strain at rupture, is termed **elongation** and is used as a *measure of ductility*. The larger the elongation, the more ductile the metal is said to be, i.e., the more it is able to stretch under load. Using the letter f to signify the "final length" of the specimen the formula for elongation, ε_L, becomes

$$\varepsilon_L = \frac{L_f}{L_0} - 1$$

This is usually expressed as a percent.

Reduction of Area.   Another measure of ductility is the **reduction of area (R of A)** of the specimen at failure. In Figure B-4 this is the reduction in area from the original area A_0 to the area at point 4, or A_4, or

$$\text{R of A} = \frac{A_0 - A_4}{A_0}$$

or for the more general case,

$$\text{R of A} = \frac{A_0 - A_f}{A_0}$$

A_f is the final area of the specimen after fracture. This number is usually expressed as a percentage.

As with the strain calculation, the reduction in area can be calculated at any point x in the deformation process, or

$$\text{R of A} = \frac{A_0 - A_x}{A_0} = 1 - \frac{A_x}{A_0}$$

If the point x occurs where the deformation is uniform, then the volume formula holds true ($L_0 \times A_0 = L_x \times A_x$), and as mentioned in the prior section, can be rewritten as

$$\frac{A_0}{A_x} = \frac{L_x}{L_0} \qquad \text{or} \qquad \frac{A_x}{A_0} = \frac{L_0}{L_x}$$

In the R of A formula above, the ratio L_0/L_x can be substituted for the area ratio; thus

$$\text{R of A} = 1 - \frac{A_x}{A_0} = 1 - \frac{L_0}{L_f}$$

So reduction of area can be found from lengths; this will be important later when *percent reduction of area will be redefined as percent cold work.* Note that this formula is only applicable where the deformation is uniform; it does not apply after nonuniform deformation starts. Also note that because the areas and lengths are related by their products, the subscripts are inverted when changing from the area to the length calculation.

B.4   TRUE STRESS-STRAIN

In Figure B-4 the engineering stress increases to a maximum at the ultimate tensile strength; when local deformation begins the curve starts to decrease and continues to decrease until fracture occurs. This seems to suggest that the metal gets weaker as it necks. This is true from the standpoint of the strength of the total specimen, but *within the area where the metal is necking, it is actually getting stronger*.

If the *true stress* σ_t is calculated, using the *actual area of the specimen* instead of dividing the load by the original area A_0, the upper portion of the stress-strain diagram becomes almost a straight line. The curve does not decrease after the maximum load is reached. That is, when based on the actual area of the metal available to withstand the load, the metal continues to increase in strength right up to the point where it is pulled apart.

The engineering strain is the length change compared with the original length; for *true strain* ε_t, the incremental length changes are related to the *immediately preceding length*, and then summed for the length, all the way to fracture. That is,

$$\varepsilon_t = \int \frac{dL}{L} = \ln \frac{L}{L_0}$$

where L is the length at any time and L_0 is the original gage length.

The curve that results from these redefinitions of stress and strain is seen in Figure B-5. In the elastic region the changes in area and length are very slight; therefore the "engineering" and the "true" diagrams are essentially the same for that part of the diagram.

When plastic deformation begins to be significant, the reduced area used in the denominator of the true stress calculation begins to be a factor and the curves begin to separate. The slope of the upper, or plastic, portion of the true stress-strain diagram is referred to as the **modulus of strain-hardening** and is a measure of how fast a metal will work-harden, or strengthen, as it is strained.

If the true stress-strain diagram gives a more valid picture of the actual strength of a metal, why are these values not used more? Because, for most

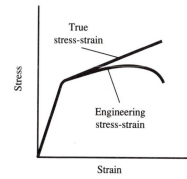

FIGURE B–5

Engineering and true tensile stress-strain curves compared for a ductile metal

engineering work, the whole purpose for design is to prevent ever loading a metal to the point of failure. When designs are done, values determined should be such that the load does not go beyond the elastic region.

True stress-strain data are of some use in making calculations for metalworking, that is, in making calculations where the purpose is, for instance, to bend, extrude, or roll the metal.

B.5   OTHER TESTING CONSIDERATIONS

Assumptions. Whether using data from engineering or true stress calculations, engineers often make a number of assumptions in applying the data: that the metal is homogeneous, that its properties are equal in all directions, and that the stress is uniformly distributed over the area.

The homogeneity of the metal is largely determined by the level of impurities in the original ore and the cleanliness attained in molten-metal processing. The methods and practices used in fabricating the solid metal have a major influence on the directional properties of the metal. The uniformity of stress distribution can be adversely affected by nonuniform loading, end conditions, and stress concentrations. The means of attaching the load may mean that uniform distribution is not achieved until some distance away from the applied load.

Compression Testing. The load-carrying capability of many structural members loaded in compression is determined by their ability to withstand buckling rather than by the strength of the material itself. Building columns, piston rods, and spars in the upper portion of an aircraft's wing are of this type. Under these conditions, load-carrying ability is determined by the moment of inertia and the modulus of elasticity rather than by the strength of the material itself. To determine the actual **compressive strength** of a material, short, relatively stubby specimens must be used, and this complicates the test because of what are termed *end conditions*.

When a short specimen is compressed between flat plates, the material will

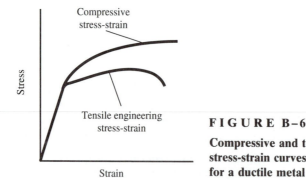

FIGURE B–6

Compressive and tensile stress-strain curves compared for a ductile metal

attempt to move radially in a plane perpendicular to the direction of the applied force. The resulting frictional forces between the specimen and the plates will hinder this radial movement and affect the results. If the metal is ductile, these end conditions will cause the specimen to become barrel-shaped. Because of these difficulties, compressive testing is not commonly done.

One other difference between the compression and tension tests is that in the former there is no localized necking or reduction of area; the plastic part of the curve does not turn down, but just keeps going up. Ductile metals will literally be "squashed flat" (like a penny on a railroad track); brittle materials will crumble at failure. Figure B-6 illustrates a compressive stress-strain curve for a ductile metal.

B.6   ENERGY ABSORPTION

By using data obtainable from the stress-strain diagram it is possible to get measures of a metal's ability to absorb energy. Energy involves forces and the rate at which they are applied; i.e., time is a factor. Although the tensile test and the resulting stress-strain diagram do not theoretically involve time, moving a force through a distance does in fact involve time. So there are two energy measures that can be obtained from the stress-strain curve, one involving the whole curve and the other only the portion in the elastic range.

Modulus of Toughness. The area under the total stress-strain diagram is the **modulus of toughness (M of T)**; it is a measure of toughness, or the *ability to absorb energy if loaded to fracture*. The modulus of toughness cannot be used for design purposes but is a guide in comparing or selecting materials.

This area has the units of stress times strain, or psi × in./in. To gain the concept of toughness do *not* cancel the dimensionless strain units, but continue the multiplication:

$$\frac{\text{Pounds}}{\text{Inch}^2} \times \frac{\text{Inch}}{\text{Inch}} = \frac{\text{Inch-pounds}}{\text{Inch}^3}, \text{ or energy per volume}$$

The concept is shown in Figure B-7.

Often actual stress-strain diagrams are not available for metals being compared, but areas can be approximated from data normally available. Usual handbook data include the tensile strength (σ_u), yield strength (σ_y), and elongation (ε_L). As can be seen in Figure B-7 the elongation provides the horizontal measure for the area calculation, but the vertical height is complicated by the curve in the diagram. However, the average of the tensile and yield strengths provides a stress value that is a reasonable approximation of the metal's strength throughout the range of strain. Thus the modulus of toughness can be *estimated* by the expression

$$\text{M of T} \approx \frac{\sigma_u + \sigma_y}{2} \times \varepsilon_L, \frac{\text{in.-lb}}{\text{in.}^3}$$

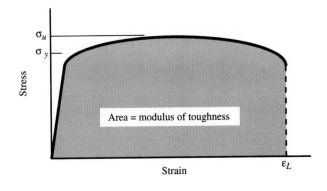

Area = modulus of toughness

$$M/T \approx \frac{\sigma_u + \sigma_y}{2} \times \varepsilon_L, \frac{\text{in.-lbs}}{\text{in.}^3}$$

F I G U R E B–7

Modulus of toughness

This calculation will include the small triangle of area to the left of the elastic region shown in Figure B-7. In this figure the elastic strain is overstated so the elastic region can be seen; the amount of that small area will have little effect on the estimate.

Modulus of Resilience. Although the toughness of a metal all the way to fracture can be important, design work is usually done in the elastic region, so it is often more important that the metal absorb energy without plastic deformation. The area under only the elastic portion of the curve is an indicator of a metal's ability to do this and is termed the **modulus of resilience (M of R).**

This area under the modulus line can be found by taking one-half of the product of the proportional limit (σ_P) times the strain at that stress value (ε_P). This is shown on Figure B-8.

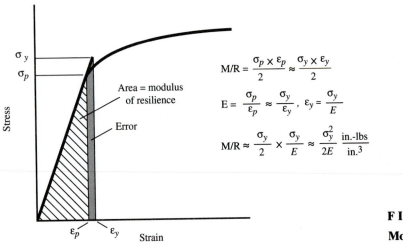

Area = modulus of resilience

Error

$$M/R = \frac{\sigma_p \times \varepsilon_p}{2} \approx \frac{\sigma_y \times \varepsilon_y}{2}$$

$$E = \frac{\sigma_p}{\varepsilon_p} \approx \frac{\sigma_y}{\varepsilon_y}, \quad \varepsilon_y = \frac{\sigma_y}{E}$$

$$M/R \approx \frac{\sigma_y}{2} \times \frac{\sigma_y}{E} \approx \frac{\sigma_y^2}{2E} \frac{\text{in.-lbs}}{\text{in.}^3}$$

F I G U R E B–8

Modulus of resilience

Since the stress-strain diagram is usually not available and handbooks do not normally give the proportional limit, another method must be found to calculate the modulus of resilience. Handbooks do provide values for the yield strength and modulus of elasticity. The yield strength is not the proportional limit, but it is close enough to allow the calculation of a reasonable estimate. Figure B-8 shows the error caused by the use of the yield strength rather than the proportional limit. As with the modulus of toughness, the modulus of resilience cannot be used for design purposes, but it is useful for comparison and selection of materials. In making comparisons the consistent use of the yield strength will affect results very little.

If the modulus of elasticity is used, the strain value at the yield strength can be found, and the modulus of resilience becomes

$$\text{M of R} \approx \frac{\sigma_y^2}{2E} \frac{\text{in.-lb}}{\text{in.}^3}$$

Toughness vs. Resilience. If a vehicle is being designed to withstand a high-speed collision, then the material chosen for the outer shell and structure should be evaluated on the basis of toughness;[§§] i.e., it should have a large area under its stress-strain curve. If, on the other hand, a machine part is being designed for shock loading and is not supposed to deform permanently, then the selection should be based on resilience.

Note that the resilience relationship has the modulus of elasticity E in the denominator. Thus more resilient metals will have a low modulus of elasticity. This is logical since the modulus of elasticity is a measure of stiffness, and stiff members are not resilient.

So, *a resilient material will have a high yield strength but a low modulus of elasticity.* Or, a material that is "strong" and "flexible" (not stiff) will be more resilient. Thus, for a car bumper that is to remain undamaged after a low-speed impact, aluminum with the same yield strength as steel will have 3 times the resilience since its modulus of elasticity is 10×10^6 psi versus 30×10^6 psi for steel. *Note that elongation, the strain at rupture, has no effect on a metal's resilience.*

On the other hand, a bumper will provide better protection in a high-speed collision if made out of low-carbon steel rather than aluminum; the low-carbon steel will have comparable strength, an elongation 2 or 3 times that of aluminum, and therefore a higher toughness.

It is important to recognize that both moduli are calculated on the basis of the volume, not the weight, of the specimen, since the units of the stress-strain diagram involve area and length. Thus more cubic inches of a ligher, though not as strong, material may provide more energy absorption than a smaller volume of a heavier material; i.e., wood, aluminum, or magnesium may absorb energy better than steel on a weight basis.

[§§]It's hoped that the portion of the vehicle around the occupants can be designed not to go to destruction.

The volume of the part participating in the loading is also important. For example, in threaded fasteners the first thread under the nut bears most of the load. If the shank of the bolt or stud is undercut to a diameter just less than the thread root diameter, more of the member will participate in absorbing loads. Connecting rod bolts and engine head studs, which undergo cyclic loading, often have reduced-body diameters. These concepts are also discussed in Chapter 11.

SUMMARY

1. We can now review some definitions:
   - Stress σ is load divided by area; units are psi or ksi.
   - Strain ε is change in length divided by original length; units are in./in. or percent.
   - Modulus of elasticity E is change of stress divided by change of elastic strain; units are psi, usually in the million range (for metals).
   - Yield strength σ_y is the stress at an offset, usually 0.002 in./in.
   - Ultimate tensile strength σ_u is the maximum load divided by original area; units are psi or ksi.
   - Elongation ε_L is the strain at fracture, i.e., the total change in length divided by original length usually expressed as a percent.
   - Reduction of area, R of A, is the total change in area divided by original area; units are percent. (The ratio of the change in area to the original area can be calculated any time in the deformation process as long as the specimen hasn't started to neck; it is given as percent cold work.)

2. Ductile metals have σ-ε curves with larger plastic regions than brittle metals do.

3. Uniform strain exists until the maximum load is reached; the strain or deformation will be distributed along the length of the specimen. The product of length and area will be a constant in this region.

4. Figure B-4 shows the division of strain into elastic, uniform plastic, and non-uniform plastic regions.

5. True stress-strain uses the immediate values of area and length to calculate stress and strain, respectively. These data are helpful in making calculations involving metalworking operations but have no advantage over the simpler engineering values for most design purposes.

6. When σ-ε data are used, assumptions are often made that the metal is homogeneous and has the same properties in all directions and that the stress is uniformly distributed over the area. These are not always valid assumptions.

7. The area under the σ-ε curve can be used as a comparative indictor of the energy-absorbing capability of the metal. The area under the whole curve is a measure of toughness (modulus of toughness) and the area under the elastic portion only is a measure of resilience (modulus of resilience). The units are in.-lb/in.^{3}, or energy per volume.

REVIEW QUESTIONS

1. A brass rod 0.200 in. in diameter and 10 in. long is loaded in tension with 800 lb. Assume this load is within the elastic region. How long will the rod be while under load?

2. A 12-in.-long square bar, 0.50 in. on a side, is made of a steel that has the mechanical properties of σ_u = 90,000 psi, σ_y = 55,000 psi, and ε_L = 25%. How long will the bar be while under a load of 15,000 lb? Can you solve this problem? If not, why not? Can you estimate the answer?

3. Sketch a typical stress-strain curve and identify the location of σ_u, σ_y, ε_L, E, uniform strain, nonuniform strain, elastic strain, plastic strain, modulus of resilience, and modulus of toughness.

4. The gage length of a tensile specimen was 2 in. When the specimen broke and the two ends were placed back together, the distance between the gage marks was $2\frac{7}{32}$ in. What is the elongation?

5. Before testing, a round specimen had a diameter of 0.505 in.; after failure the diameter was 0.455 in. What is the reduction of area?

6. A long piece of rod is drawn from a diameter of 0.400 in. to 0.375 in. (That is, there is no necking; the deformation is uniform.) What is the reduction in area? What is the amount of tensile strain?

7. A tensile test is conducted on a specimen whose cross-sectional dimensions are 0.062 in. thick by 0.515 in. wide. The slope of the elastic region of the tensile-test plot is measured; between two points on the vertical axis the difference in load is 1750 lb. The corresponding change in the strain axis is 0.003 in./in. What is the modulus of elasticity of this material?

8. You are designing the enclosure of a race car, and you're trying to decide whether to use an aluminum alloy or an alloy steel.

    Aluminum: σ_u = 80,000 psi, σ_y = 70,000 psi, ε_L = 12%.
    Steel: σ_u = 190,000 psi, σ_y = 170,000 psi, ε_L = 15%.

    Assume the steel density is 3 times that of the aluminum. If weight is one of your design concerns, which material would you use and why?

9. A structure is to help absorb the energy of a fluctuating tensile load without being permanently deformed. You need to select among aluminum, brass, and steel. Which material would you use if the constraint was on the cross-sectional size of the part? Which material would you use if the weight of the part was the constraint? Make sure you can explain your answers. Assume the yield strengths of the three metals are the same and that steel and brass have the same density, which is 3 times that of aluminum.

FOR FURTHER STUDY

Annual book of ASTM standards. Philadelphia: American Society for Testing and Materials, vol. 03.01, Standard E8.

Avner, Sidney H. *Introduction to physical metallurgy* (2d ed.). New York: McGraw-Hill, 1974, secs. 1.30–1.32.

DeGarmo, E. Paul, J. Temple Black, and Ronald A. Kohser. *Materials and processes in manufacturing* (7th ed.). New York: Macmillan, 1988, 37–44.

Dieter, George E. *Mechanical metallurgy* (3d ed.). New York: McGraw-Hill, 1986, chaps. 1 and 8.

Flinn, Richard A., and Paul K. Trojan. *Engineering materials and their applications* (3d ed.). Boston: Houghton Mifflin, 1986, secs. 3.2, 3.6, and 3.7.

Harris, J. N. *Mechanical working of metals*. New York: Pergamon, 1983, chap. 1.

APPENDIX C

Hardness Testing

OBJECTIVES

After reading this appendix (and portions of Chapter 2) and (preferably) participating in a lab exercise, you should be able to

- explain the bases of the Brinell, microhardness, and Rockwell tests and how they differ
- explain the advantages and disadvantages of each test and be able to select which one to use in a described situation
- perform the tests and determine the hardness reading
- explain the cold-working nature of these tests
- estimate tensile strengths from hardness data

382

DID YOU KNOW?

The tensile strength of a metal is an absolute value; it is expressed in units that can be used in quantitative calculations. Hardness, on the other hand, is a relative value. All hardness measurements merely compare the metal tested against a scale, and the value obtained varies with the method used. In spite of these seemingly negative factors, hardness testing is one of the most often used methods for selecting materials, verifying that metal processes are in control, and checking metal quality. It is also used as a tool to evaluate research and development efforts.

A PREVIEW

Our discussion of the general concepts of hardness testing is followed by descriptions of the most commonly used hardness-testing methods: Brinell, microhardness, and Rockwell. We'll review the test procedure, the method of determining the hardness number, and the advantages and disadvantages of each method. Some general guidelines for hardness testing, a hardness-comparison chart, and information about other testing methods conclude our discussion.

C.1 GENERAL CONCEPTS OF HARDNESS TESTING

The most often used hardness-testing methods are *plastic deformation* or *cold-working processes*. Performing a Brinell, microhardness, or Rockwell test involves a small-scale compression test. In each of these tests the ability of an **indenter** to penetrate, or compress, the surface of the metal is what is being determined. Depending upon the method used, the hardness number is related to either the area of the indentation or the depth of the indentation created under specified standard conditions.

When a compression test is performed, the whole specimen is subjected to the load. In a hardness test a much smaller area is loaded, but the affected area will be larger than just the area of the impression. The metal in the vicinity of the impression will also be erupted, i.e., plastically deformed. If impressions are observed under the microscope it will be seen that the edge of an impression and the original surface of the metal cannot both be kept in focus. That is, they are at different heights; the hardness test has deformed the metal both under the indenter and in the adjacent area.

In Appendix B it was pointed out that the very act of making the tensile test causes the yield strength of the metal specimen to increase. In a somewhat similar fashion, the plastic deformation caused by the hardness test changes the hardness of the metal. That is, performing the test alters the original metal.

C.2 THE BRINELL HARDNESS TEST

Test Procedure. The Brinell hardness testing method uses a standard-sized ball that is pushed into the test piece under a standard load; the length of time the load is applied is also standardized.* The diameter of the indentation or "hole" is then measured with a Brinell microscope that has a special eyepiece. *The Brinell hardness (HB) is the load in kilograms divided by the area of the impression in millimeters squared.*

Determining the Brinell Hardness Number. The area used in this calculation is *not* the area of the circle whose diameter d is read with the microscope, i.e., $\pi d^2/4$. Instead, the area is the surface area that is in contact with the ball when the impression is made. This area is represented by the denominator in the following equation, which is used to calculate the Brinell hardness:

---

*ASTM Standard E10 (Standard Test Method for Brinell Hardness of Metallic Materials), volume 03.01, should be consulted for more information. The standard test conditions call for a 10-mm ball of either hardened steel or tungsten carbide and a 3000-kg load; the load must be applied for 10 to 15 sec. Under these conditions the test is recorded as a number followed by HB, e.g., 155 HB. If other conditions are used, they are listed after the hardness number, e.g., 155 HB 10/500/30, which indicates a 10-mm ball, 500 kg, applied for 30 sec.

$$HB = \frac{P}{(D - \sqrt{D^2 - d^2})\,\pi D/2}$$

where P = load in kilograms, D = diameter of the ball in millimeters, usually 10 mm, and d = diameter of the impression in millimeters. The small chart mounted on the front of the tester usually eliminates the need to make this calculation; it converts the impression diameter to HB for a 10-mm ball and the more commonly used loads.

The Brinell hardness is a load divided by an area. For any given load, the larger the diameter measured, and therefore the larger the area, the softer the metal; conversely, harder metals would allow smaller impressions that would have smaller areas and hence would generate higher hardness numbers. For harder metals the 3000-kg maximum load is often used; for the softest metals the 500-kg minimum load is used. Other nonstandard loads, in increments of 500 kg, can be used if desired.

The size of the load used will affect the Brinell hardness number; that is, tests of the same metal made with different loads will not usually give the same number. Therefore when stating a Brinell hardness, give the load used if it is other than standard.

Advantages and Disadvantages.   Some significant disadvantages of the Brinell method center around its use of a ball indenter. When metals are very hard, i.e., about 500 HB, the steel ball doesn't remain spherical, but begins to flatten, and accurate readings are not obtained. For the extreme condition, visualize a ball being turned into an ellipsoid by being squashed between two surfaces that are harder than it is. In the hardness range of 450 to 650 HB the steel ball should be replaced with one made of tungsten carbide. A safety caution: The ball or specimen may shatter if metals that are too hard are tested.

In the opposite direction, when very soft metals are tested the impression starts to approach the diameter of the ball. The closer the impression gets to the ball diameter, the less its diameter changes with the vertical travel of the tester and the sensitivity of the test is lost. This can be seen at the "soft end" of a hardness-comparison table (see Table C-1); some Rockwell scales have two or three whole number readings between successive Brinell readings.

The size of the Brinell impression has both advantages and disadvantages. Since the impression diameter will normally be in the range of 3 to 5 mm, the test will cover enough of the metal to be very representative, e.g., for quality control of foundry products. However, the large impression will probably be detrimental to the appearance and function of the part. That is, the Brinell test is probably a destructive test than can only be performed on sample specimens rather than on a finished part.

Because of the large impression, the surface to be tested requires only minimal preparation, e.g., removal of dirt and excessive oxide or rust. The surface must be reasonably perpendicular to the tester axis to get a uniform impression.

If standard or known loads are used, once the impression is made it can always be reread.

Since the only role of the Brinell tester itself is to create an impression in the metal (it doesn't have to measure anything), the tester is a very rugged piece of equipment and requires very little attention, for instance, for calibration.

Summary. The Brinell test is a load-over-area hardness test, gives a reading very representative of the metal, but leaves a large impression. Because of its round indenter, it is best suited for metals that are neither too hard nor too soft.

C.3 THE MICROHARDNESS TEST

The microhardness test[†] is also a load-over-area hardness measurement. As in the Brinell method, a known load causes an indenter to penetrate the metal and the area of the resulting impression is determined. The test differs from the Brinell in a number of ways:

1. The loads are much smaller (1 kg maximum, and often ½ kg or less).
2. The impressions are much smaller; they usually can't be seen with the naked eye and must be measured with a microscope.
3. The indenters used are pyramids rather than a ball, so the area is determined using a diagonal or diagonals rather than a diameter.
4. The small size of the impression requires that the surface be prepared as if for metallographic examination.[‡]

The equipment used for the microhardness test has two stations and a "stage," which supports the specimen and moves it to either of the stations. One station contains a microscope, the other the indenter and load-applying mechanism.

Test Procedure. To perform the test, the specimen is positioned on the stage and then moved under the microscope. By viewing the specimen through the microscope, and shifting the specimen with X-Y coordinate micrometer screws mounted on the stage, the observer selects the area to be tested. The stage is then shifted under the indenter (proper adjustment of the machine ensures that the area seen through the microscope is under the indenter). The timed load cycle is initiated electrically. At the completion of the cycle, the stage and specimen are moved under the microscope and the diagonal(s) of the impression measured with the special micrometer eyepiece.

Two indenters are used to make these tests—the Vickers and the Knoop indenters.

[†]The prefix "micro" indicates that the impressions made by this test are very small or that they must be read with a microscope; it does not mean that the test is used on materials with low hardness values, i.e., very soft materials.

[‡]ASTM Standard E384 (Standard Test Method for Microhardness of Materials), volume 03.01, should be consulted for further information. Loads are to be from 1 to 1000 g; they are to be applied for 10 to 15 sec unless specified otherwise.

Determining the Microhardness Number.   The Vickers indenter is a square pyramid with 136° angles between opposite sides. The impression seen through the
microscope is a square with diagonals running between the corners. The average
length of the two diagonals is used to determine the area of the surface impressed
into the metal. Because of the standard angle, the constants of the formula convert
the diagonal reading to the *impressed area.*

The Knoop indenter is also a pyramid, but it is not square; instead it is ground
to a long diamond shape with a ratio of long diagonal to short diagonal of approximately 7 : 1. The impression seen in the microscope is a diamond with two diagonals, one of which is much longer than the other. Only the length of the long
diagonal is used to determine the area of the surface impressed into the metal. As
with the Vickers indenter, a constant converts the diagonal reading into the *impressed area.* Figure C-1 shows the impressions made by the Vickers and Knoop
indenters.

The Vickers hardness number, or HV,[§] and Knoop hardness number, or HK,
are found from the following formulas:

$$HV = 1854.4P/d^2$$

$$HK = 14229P/d^2$$

where P = the load in grams and d = the diagonal length in micrometers (μm),
i.e., meters × 10^{-6}.[‖]

Loads used in the microhardness test are usually a few hundred grams, e.g.,
300 to 500 g, but can be a kilogram; for very soft material much lower loads of just
a few grams can be used.[#]

[§]Hardness values from the Vickers indenter are often identified as DPH, or diamond-pyramid hardness
numbers.
[‖]If the units of kilograms and millimeters are used, the formulas become $HV = 1.854P/d^2$ and $HK =$
$14.229P/d^2$.
[#]The Vickers indenter can also be used with loads of 1 to 120 kg. This type test is described in ASTM
Standard E92, vol. 03.01.

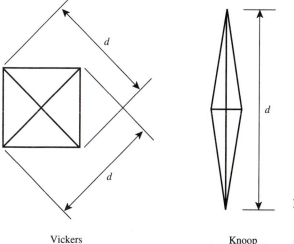

Vickers   Knoop

FIGURE C–1

**Impressions made by Vickers
and Knoop indenters**

Advantages and Disadvantages.   The advantage of the microhardness test is that it can be used to investigate variations in hardness in very small areas or products, or through very small thicknesses. For example, this test is used to investigate the hardness gradient of case-hardened gear teeth. The case of the tooth may be only a millimeter or two thick, so a number of indentations must be made in a very small distance to detect variations. The number of indentations that can be placed in an area depends on their size, i.e., the load used, but under typical circumstances they can be placed within 100 μm (0.1 mm) of each other fairly easily.

The test can also be used to measure the hardness variation across the thickness of a razor blade, or of individual grains in a microstructure, or of other parts of a microstructure that are of interest.

The small size of the indentation might be an advantage since it probably will not cause rejection of a finished part; however, because of the surface preparation and equipment required, usually only specimens rather than finished products are tested.

It has been reported that the Knoop hardness number is unaffected by the load size for loads greater than 300 g.** This is an advantage over Brinell since it means readings taken over a range of loads can be compared directly. Since the Vickers indenter is also a straight-sided pyramid this *may* be true for that indenter also.

Summary.   To gain its advantages, the microhardness test necessitates the disadvantages of requiring extensive surface preparation, delicate equipment, and readings not necessarily representative of the whole part. Surfaces are polished and often etched to remove any cold-worked layer left from the polishing operation. The microscope and associated equipment are delicate and require special care and maintenance. The microscopic nature of the test means the readings obtained will represent a very small part of the specimen.

C.4   THE ROCKWELL HARDNESS TEST

The Rockwell hardness test differs from those just discussed in that the *depth* of the indentation or penetration is measured rather than the *area*. The Rockwell tester applies a standard load to the specimen through a standard **penetrator** and then measures the depth of penetration. The Rockwell hardness is read directly from a dial mounted on the face of the machine; the hardness reading will be from 0 to 100. The Rockwell hardness is not calculated by load and area, but is a relative measure of how difficult it is to push a penetrator into the metal.

The Rockwell test has numerous scales, which are made possible by the

---

**L. P. Tarazov and N. W. Thibault, *Trans. ASM, 38,* 331–53, as reported in George E. Dieter. *Mechanical metallurgy* (2d ed.). New York: McGraw-Hill, 1974, 292.

combination of different loads and penetrators.[††] The more common combinations involve loads of 60, 100, and 150 kg with indenters that are a diamond cone (brale), a 1/16-in.-diameter ball, and a 1/8-in.-diameter ball. The scales are identified by letter and the common scales range from A through K, with I and J left out. Unfortunately, the scales are not in an order that relates to load or penetrator, but a metal tag affixed to the pedestal of the tester identifies the load and penetrator for each scale.

The indicator dial has two sets of numbers, one in black, the other in red. The black numbers are used to read the scales that use the diamond cone (A, C, and D), and the red numbers are used for both ball penetrators. The setpoints are the same, but the red scale is displaced 30 units from the black.

Test Procedure. The test is begun by placing the specimen on an anvil under the indenter and, using a large adjusting nut, forcing the anvil upward so the specimen pushes on the penetrator. This upward movement is continued until the dial indicator points to "Set" on the dial face. At this location the machine has applied a **preload,** or **minor load,** of 10 kgf to the specimen. This minor load ensures that the specimen is firmly held in the machine, that the penetrator is well seated against the specimen, and that any minor surface films that may be on the specimen are broken through.

To initiate the test a lever is tripped, applying the **major load.** This load will first stress the specimen through the elastic region and then cause plastic deformation, i.e., cause a permanent cavity to be formed. For most metals, as soon as the needle stops moving the major load may be removed and the hardness read from the dial.

In the hardness reading the numbers are followed by HR and the letter of the scale used; e.g., 63 HRC indicates a reading of 63 obtained with 150-kg load and diamond-cone penetrator.

Determining the Rockwell Hardness Number. The measuring device of the Rockwell tester is actually a dial indicator or depth gage, each division of which is equal to 0.00008 in. (Thus the rotation of the needle through 360° would equal a vertical movement of the indenter of 0.008 in.) But if the depth of penetration was used as the *hardness* reading, then the softer materials would have the higher numbers; the number would be a "softness" reading rather than a hardness reading. The mechanical linkage of the Rockwell tester solves this problem.

Inspection of the tester dial shows that the numbers go around the face in the normal *clockwise* fashion. However, close observation when the major load is applied shows that the indicator rotates in a *counter*clockwise direction. On the black scale it starts at zero, which is also 100, and proceeds through 90 to 80 to 70,

[††]ASTM Standard E18 covers the Rockwell and Rockwell superficial hardness tests. In addition to the penetrators mentioned above, 1/4- and 1/2-in.-diameter balls are also used with the same loads. The superficial test uses loads of 15, 30, and 45 kg with the diamond cone and 1/16-in. ball penetrators for scales identified as "N" and "T," respectively, e.g., 15-N, 45-T.

etc. In effect, the indicator is saying that the hardness is at 100 when the test begins; if the major load was applied and the needle never moved the material would be 100 units hard, as hard as it could get for that Rockwell scale. As the needle moves counterclockwise, it is showing that the material is softer than the original 100, but the harder materials would still have the higher number, so the number indicates hardness, not softness.[‡‡]

Unusual Features of the Test.   *When the reading is taken, the minor load is still on the penetrator.* This is enough load to keep the indenter in the bottom of the indentation so the "depth indicator" can make its reading, but not so much load as to greatly stress the specimen elastically. That is, since the major load has been removed, the indentation being measured is due to plastic deformation only. It is not reading any of the elastic deformation that would have been caused by the full load. Note that when the major load is removed, the needle always moves clockwise to a higher number. That is, the depth of penetration being measured by the tester *goes down* as the load is removed, a very logical result.

The combination of load and penetrator makes certain scales more appropriate for certain materials. Hardened steel is tested using the C scale, 150 kg and a diamond-cone penetrator; for brass the B scale is often used, 100 kg and a $1/16$-in. ball; the E scale, 100 kg and a $1/8$-in. ball, is often used with aluminum. The harder metals are tested with the higher loads and sharper penetrators, the softer metals with lower loads and larger penetrators.

It is important to recognize that the diamond-cone penetrator does not have a sharp point; it is not a true cone. It really is a 120° cone with a portion of a sphere of 0.200-mm radius at its tip. Thus for very hard materials it will tend to do the same thing as the Brinell ball; it will start to deform or "squash." The result is that for harder metals the test is less sensitive.

For example, on a hardness-comparison chart (Table C-1) it can be seen that in the range of hardness where Rockwell C numbers go from 58 to 68 the Vickers numbers go from 655 to 942. That is, over the same range of hardness the Rockwell scale uses about one-sixth of its total range while the Vickers uses about one-third. If the two tests are plotted, the HRC curve will be flatter (less sensitive) than the HV scale for the same range of hardnesses.

A general rule: It is desirable to keep the Rockwell readings between 20 and 80 when selecting the scale, i.e., the load and penetrator.

Advantages and Disadvantages.   The Rockwell hardness testing method has some advantages that make it one of the most commonly used in industry:

- It is a direct reading and therefore is fast; no secondary operations are necessary to get the reading.
- The indentation is small enough that it is not injurious to the appearance or function of many products and parts.

[‡‡]For the scales read in black, the depth of penetration can be calculated by subtracting the hardness reading from 100 and multiplying by the dial indicator constant of 0.00008 in./unit.

- Although more surface preparation is required than for the Brinell test, surfaces prepared by ordinary hand grinding or filing give good results; polishing is not needed.
- Compared with the Brinell method, the number of scales enable the testing of a wider range of hardness with greater sensitivity.
- As will be discussed shortly, the thickness of specimen that can be hardness-tested is related to the indenter size and testing load. Compared with Brinell, the Rockwell can test thinner specimens; usually a Rockwell scale can be found to test most engineering materials in the thicknesses commonly used.

The disadvantages of the Rockwell tester are few:

- Compared with the Brinell, the machine is not quite as rugged and may require occasional adjustment.
- Since the impressions are relatively small, they are not as representative as a Brinell reading; a good general rule is to take at least two Rockwell readings and then a third if the first two are not reasonably close.
- Erroneous readings result if the indicator rotates more than one revolution.

C.5   GENERAL GUIDELINES FOR HARDNESS TESTING

To get accurate, reliable, and representative hardness measurements there are some general guidelines that must be followed when using any of the testing methods just discussed.

- Surface preparation must be appropriate for the test to be performed. For steels all scale or rust must be removed.
- The surface to be tested must be reasonably perpendicular to the axis of the indenter. Where possible, curved surfaces should have a flat ground on them, although there are correction tables for the Rockwell testing of specimens less than 1 in. in diameter.
- The piece to be tested must be supported so it will not move when the load is applied. Particularly in the Rockwell test, the contact between the anvil and specimen must be such that the specimen does not deflect under the load. A nonflat thin specimen can behave as a leaf spring when supported by a large anvil; for these type specimens it is often better to use a small anvil that directly supports the specimen only under the penetrator.
- If the specimen is tested too close to its edge, the indenter will not be supported and will penetrate further than it should, giving an indication that the metal is softer than it actually is. To avoid this "edge effect" the general rule is that the centers of impressions should be made at least 2½ diameters from the edge.
- In a similar fashion, indentations that are too close to each other can also give false measurements. (Remember that hardness-testing methods are cold-working operations.) The rule is to keep the centers of indentations a minimum of 2½ to 3 diameters apart.

• Specimens should be thick enough that the test does not cause a bulge on the bottom surface. The general rule is that specimen thickness should be at least 10 times the depth of the indentation.

There are two possible consequences of testing too thin a specimen. First, the indenter will read through the specimen and measure, at least in part, the hardness of the anvil, which would normally be harder than the specimen; this is called "anvil effect." The other would be due to an anvil in poor condition. When someone incorrectly performs a Rockwell test on the anvil itself, a cavity is left. Under load a thin specimen might not bridge this cavity, which would allow the indenter to travel too far, giving the indication that the metal was softer than it really was.

C.6   PRACTICAL INSIGHTS INTO
         THE TESTING METHODS

If for some reason only one hardness reading can be taken, the Brinell method is the best choice since its large impression will give a reading most representative of the metal. If speed, flexibility, and ease of operation are major concerns, the Rockwell test will be the most desirable. If it is necessary to measure a hardness gradient across a very short distance, or to measure the hardness of very small parts, or to investigate the hardness of various parts of a microstructure, then the microhardness test will be the most appropriate.

C.7   HARDNESS-COMPARISON CHARTS

As stated earlier, the hardness test is a relative measurement that does not measure a well-defined property of a metal. Hardness readings can only be directly compared with data from the same test. However, because of the limitations, advantages, and disadvantages of the various methods, all hardness measurements cannot be made with the same test, so it is advantageous to be able to compare them. Table C-1 enables comparisons of some frequently used tests and scales.

Many texts, handbooks, and ASTM publications[§§] provide charts which compare one hardness test or scale to another. These charts can be very useful, but it is important to recognize that they are based on hardness data from tests on

---

[§§] ASTM Standard E140 (Standard Hardness Conversion Tables for Metals), vol. 03.01, is one such source.

specific metals. Other metals with different elastic and/or strain-hardening characteristics might respond differently to the various indenters and loads.

C.8   RELATIONSHIP OF HARDNESS TO TENSILE STRENGTH

Both the tensile test and the commonly used hardness tests involve plastic deformation or slip; it is normally expected that if the result of one test goes up, the other will also. Therefore it is reasonable that the two tests would have some relationship to each other.

For normalized and slow-cooled plain carbon and medium-alloy steels, it has been found empirically that their tensile strengths will be approximately 500 times their Brinell hardness. This is a useful relationship that can be used to estimate either value if the other is known. Although this constant is for steels, it will often come close to estimating tensile strengths of some nonferrous metals in various conditions of heat treatment.

C.9   OTHER HARDNESS TESTING METHODS AND NEW DEVELOPMENTS

The scleroscope hardness test is a means of measuring the "elastic hardness" of a metal.[‖] The test consists of dropping a standard "hammer" from a specified height onto the surface to be tested; the hardness number is related to how high the hammer bounces. The test is really a measure of the resilience of the metal. The advantages of the method are that the test is nondestructive, and the tester is portable; e.g., it can be used to check finished parts to see if they have been heat-treated or to check shafts or other equipment in place. The disadvantages of the method are that it is neither as sensitive nor as representative as the other methods discussed.

Methods of mechanizing and automating testing equipment have been and are being developed. Some Rockwell testers are now motorized and others are automated so they don't need an operator.

One of the newest developments[##] is the automating of the microhardness test by converting it from a load-over-area test to a depth-of-penetration test. In the load-over-area mode the diagonals must be measured, a chore that is difficult to automate. For the depth-of-indentation mode the geometry of the simple, straight-sided indenter is used to advantage. That is, if the depth of indentation is

[‖]ASTM Standard E448, vol. 03.01, has details on this testing method.
[##]Sue A. Bruskin. Automation gains acceptance with microhardness testers. *Heat treating, XXI*(5), May 1989, 28.

TABLE C–1 Comparison of hardnesses measured by various tests and scales*

HRC (150 kg, brale)	HRA (60 kg, brale)	HV (10 kg)	HK (≥500 g)	HB (3000 kg, 10 mm)	Tensile Strength (ksi)	HRB (100 kg, 1/16 in.)	HRA (60 kg, brale)	HRE (100 kg, 1/8 in.)	HRK (150 kg, 1/8 in.)	HK (≥500 g)	HB (500 kg, 10 mm)	HB (3000 kg, 10 mm)	Tensile Strength (ksi)	HRB (100 kg, 1/16 in.)	HRA (60 kg, brale)	HRH (60 kg, 1/8 in.)	HRE (100 kg, 1/8 in.)	HRK (150 kg, 1/8 in.)	HK (≥500 g)	HB (500 kg, 10 mm)
80	92.0	1865				100	61.5			251	201	240	116	50			87.0	64.5	107	
79	91.5	1787				99	61.0			246	195	234	112	49			86.5	63.5	106	83
78	91.0	1710				98	60.0			241	189	228	109	48			85.5	62.5	105	82
77	90.5	1633				97	59.5			236	184	222	106	47			85.0	61.5	104	81
76	90.0	1556				96	59.0			231	179	216	103	46	35.0		84.5	61.0	103	80
75	89.5	1478				95	58.0		100	226	175	210	101	45	34.5		84.0	60.0	102	79
74	89.0	1400				94	57.5		99.5	221	171	205	98	44	34.0		83.5	59.0	101	78
73	88.5	1323				93	57.0			216	167	200	96	43	33.5		82.5	58.0	100	77
72	88.0	1245				92	56.5			211	163	195	93	42	33.0		82.0	57.5	99	76
71	87.0	1160				91	56.0			206	160	190	91	41	32.5		81.5	56.5	98	75
70	86.5	1076	972			90	55.5		98.5	201	157	185	89	40	32.0		81.0	55.5	97	74
69	86.0	1004	946			89	55.0		98.0	196	154	180	87	39	31.5		80.0	54.5	96	73
68	85.5	942	920			88	54.0		97.0	192	151	176	85	38	31.0		79.5	54.0	95	72
67	85.0	894	895			87	53.5		96.5	188	148	172	83	37	30.5		79.0	53.0	94	
66	84.5	854	870			86	53.0		95.5	184	145	169	81	36	30.0		78.5	52.0	93	
65	84.0	820	846			85	52.5		94.5	180	142	165	80	35	29.5		78.0	51.5	92	71
64	83.5	789	822			84	52.0		94.0	176	140	162	78	34	29.0		77.0	50.5	91	70
63	83.0	763	799			83	51.0		93.0	173	137	159	77	33	28.5		76.5	49.5	90	69
62	82.5	739	776			82	50.5		92.0	170	135	156	75	32	28.0	100	76.0	48.5	89	
61	81.5	716	754			81	50.0		91.0	167	133	153	74	31	27.5	99.5	75.5	48.0	88	68
60	81.0	695	732	614		80	49.5		90.5	164	130	150	72	30	27.0	99.0	75.0	47.0		
59	80.5	675	710	600		79	49.0		89.5	161	128	147		29	26.5	98.5	74.0	46.0		67
58	80.0	655	690	587		78	48.5		88.5	158	126	144		28	26.0	98.0	73.5	45.0	85	
57	79.5	636	670	573		77	48.0		88.0	155	124	141		27	25.5	97.5	73.0	44.5		66
56	79.0	617	650	560		76	47.0		87.0	152	122	139		26	25.0	97.0	72.5	43.5		
55	78.5	598	630	547	301	75	46.5		86.0	150	120	137		25	24.5	96.5	72.0	42.5		65
54	78.0	580	612	534	291	74	46.0		85.0	147	118	135		24	24.0	96.0	71.0	41.5		
53	77.5	552	594	522	282	73	45.5		84.5	145	116	132		23	23.5	95.5	70.5	41.0	82	64
52	77.0	545	576	509	273	72	45.0		83.5	143	114	130		22	23.0	95.0	70.0	40.0		63
51	76.5	528	558	496	264	71	44.5	100	82.5	141	112	127		21	22.5	94.5	69.5	39.0		62

50	76.0	513	542	484	255
49	75.5	498	526	472	246
48	74.5	485	510	460	237
47	74.0	471	495	448	229
46	73.5	458	480	437	221
45	73.0	446	466	426	214
44	72.5	435	452	415	207
43	72.0	424	438	404	200
42	71.5	413	426	393	194
41	71.0	403	414	382	188
40	70.5	393	402	372	182
39	70.0	383	391	362	177
38	69.5	373	380	352	171
37	69.0	363	370	342	166
36	68.5	353	360	332	162
35	68.0	343	351	322	157
34	67.5	334	342	313	153
33	67.0	325	334	305	148
32	66.5	317	326	297	144
31	66.0	309	318	290	140
30	65.5	301	311	283	136
29	65.0	293	304	276	132
28	64.5	285	297	270	129
27	64.0	278	290	265	126
26	63.5	271	284	260	123
25	63.0	264	278	255	120
24	62.5	257	272	250	117
23	62.0	251	266	245	115
22	61.5	246	261	240	112
21	61.0	241	256	235	110
20	60.5	236	251	230	108

70	44.0	99.5	81.5	139	110	125		20	22.0		68.5	38.0		
69	43.5	99.0	81.0	137	109	123		19	21.5	94.0	68.0	37.5	79	61
68	43.0	98.0	80.0	135	107	121		18		93.5	67.5	36.5		
67	42.5	97.5	79.0	133	106	119		17	21.0	93.0	67.0	35.5		60
66	42.0	97.0	78.0	131	104	117		16	20.5		66.5	35.0		
65		96.0	77.5	129	102	116		15	20.0	92.5	65.5	34.0	76	59
64	41.5	95.5	76.5	127	101	114		14		92.0	65.0	33.0		
63	41.0	95.0	75.5	125	99	112		13			64.5	32.0		58
62	40.5	94.5	74.5	124	98	110		12		91.5	64.0	31.5		
61	40.0	93.5	74.0	122	96	108		11		91.0	63.5	30.5	73	
60	39.5	93.0	73.0	120	95	107		10		90.5	62.5	29.5		57
59	39.0	92.5	72.0	118	94	106		9			62.0	29.0		
58	38.5	92.0	71.0	117	92	104		8		90.0	61.5	28.0	71	
57	38.0	91.0	70.5	115	91	103		7		89.5	61.0	27.0		56
56		90.5	69.5	114	90	101		6			60.5	26.0		
55	37.5	90.0	68.5	112	89	100		5		89.0	60.0	25.5	69	55
54	37.0	89.5	68.0	111	87			4		88.5	59.0	24.5		
53	36.5	89.0	67.0	110	86			3		88.0	58.5	23.5		
52	36.0	88.0	66.0	109	85			2			58.0	23.0	68	54
51	35.5	87.5	65.0	108	84			1		87.5	57.5	22.0		
50	35.0	87.0	64.5	107	83			0		87.0	57.0	21.0	67	53

*HRC = Rockwell, 150-kg load, diamond brale penetrator.
HRA = Rockwell, 60-kg load, diamond brale penetrator.
HRB = Rockwell, 100-kg load, 1/16-in.-diameter ball penetrator.
HRH = Rockwell, 60-kg load, 1/8-in.-diameter ball penetrator.
HRE = Rockwell, 100-kg load, 1/8-in.-diameter ball penetrator.
HRK = Rockwell, 150-kg load, 1/8-in.-diameter ball penetrator.
HB = Brinell, 3000- or 500-kg loads, 10-mm-ball indenter.
HK = Knoop, 500 g or more load, long diamond-shaped pyramid indenter.
HV = Vickers, 10-kg load, square pyramid indenter.
Tensile strengths, ksi, are inexact estimates for steel.

measured and the geometry of the indenter is known, the area can be calculated electronically and the hardness read directly.

SUMMARY

1. Tensile strength is an absolute measure; it can be used in quantitative calculations. Hardness, on the other hand, is a relative value.

2. In spite of being a relative measure, the hardness test is one of the most often used methods for selecting materials, controlling processes (especially heat-treating), checking quality, and evaluating research and development efforts.

3. Most hardness testing methods plastically deform the metal in a small-scale compression test. The effect of the plastic deformation goes beyond the indentation itself.

4. The hardness number is related to either the area of the indentation or the depth of the indentation, created under specified standard conditions.

5. In the Brinell and microhardness tests the area used in the calculation is the surface area that is in contact with the indenter when the impression is made.

6. The Brinell test is a load-over-area hardness test, gives a reading very representative of the metal, leaves a large impression, and because of its round indenter is best suited for metals that are neither too hard nor too soft.

7. The microhardness test is also a load-over-area hardness measurement using loads and indenters that, compared with Brinell or Rockwell, are very small.

8. In the microhardness test the diagonals are measured and substituted in formulas whose constants convert them to the impressed area of the indentation.

9. The microhardness test can investigate variations in hardness in very small areas or products, or across very small thicknesses.

10. The Rockwell tester uses standard loads and penetrators and directly converts the depth of penetration to a hardness reading.

11. Rockwell hardnesses have values between 0 and 100; the test is not load-over-area but a relative measure of how difficult it is to push a penetrator into the metal.

12. The Rockwell's general advantages are that it is a direct reading and therefore fast, the indentation is fairly small, relatively little surface preparation is required, it can cover a wide range of hardness, and it can test reasonably thin specimens.

13. Perhaps the biggest disadvantage of the Rockwell is that the small indenters used test a small area, so more readings may have to be taken to get a representative measure of hardness.

14. Hardness readings taken by various methods can be compared with reasonable accuracy using comparison charts; comparisons need to be made among similar metals.

15. Tensile strength and hardness are directly related. Estimates of tensile strength often show on hardness-conversion charts; for hot-worked steels the ratio is $500 \times HB \approx$ estimated tensile strength in psi.

REVIEW QUESTIONS

1. A standard Brinell test results in an impressed area of 8.57 mm^2. What is the Brinell hardness number of the metal?

2. Using a load of 500 g, a microhardness test using the Vickers indenter gives an impressed area of 0.87 μm^2. What is the Vickers hardness number?

3. With a load of 450 g in a microhardness test, the Knoop indenter creates a diagonal 93 μm long. What is the Knoop hardness number?

4. In measuring a hardness of 75 HRA, how deep into the metal did the penetrator go?

5. Comparing the three hardness-testing methods discussed in this appendix, identify three advantages each has over the others.

6. If the hardness of a hot-worked low-carbon steel is 190 HB, estimate its tensile strength.

7. You need to compare the hardness of some very soft metals. You use a Brinell tester with a 500-kg load but use a 20-mm-diameter ball indenter. You conduct the first test, allowing the load to stay on the specimen only 5 sec, and find the diameter of the impression to be 12 mm. What is your hardness number and how would you report the result?

FOR FURTHER STUDY

Annual book of ASTM standards. Philadelphia: American Society for Testing and Materials, vol. 03.01, Standards, E10, E18, E384, and E92.

Avner, Sidney H. *Introduction to physical metallurgy* (2d ed.). New York: McGraw-Hill, 1974, secs. 1.21–1.27.

Bruskin, Sue A. Automation gains acceptance with microhardness testers. *Heat Treating, XXI*(5), May 1989, 28.

DeGarmo, E. Paul, J. Temple Black, and Ronald A. Kohser. *Materials and processes in manufacturing* (7th ed.). New York: Macmillan, 1988, 45–50.

Dieter, George E. *Mechanical metallurgy* (3d ed.). New York: McGraw-Hill, 1986, chap. 9.

Flinn, Richard A., and Paul K. Trojan. *Engineering materials and their applications* (3d ed.). Boston: Houghton Mifflin, 1986, sec. 3.11.

Haga, L. J. Measuring metal hardness: Using the Brinell method. *Heat treating, XXI*(5), May 1989, 22.

Haga, L. J. The Rockwell method: Measuring penetration depth for hardness. *Heat treating, XXI*(6), June 1989, 24.

Harris, J. N. *Mechanical working of metals*. New York: Pergamon, 1983.

APPENDIX D

Calculation of Cold Work

OBJECTIVES

After studying this appendix and portions of Chapter 2 and (preferably) participating in a lab exercise involving cold-working, you should be able to

- recognize the nature of deformation (slip in lattices) as cold work in tension and compression processes
- use area and/or length relationships to calculate percent cold work
- solve typical problems in cold-working and apply the annealing process to a sequence of cold-working operations so cold-working can be continued
- demonstrate an understanding that cold-working accumulates by correctly using the *annealed* size to make cold-work calculations, and being able to differentiate between reduction per operation and cold work.

DID YOU KNOW?

Metal cold-working operations, as opposed to hot-working operations, are used commercially to achieve a good surface finish, hold dimensions to close tolerances, and/or achieve a desired level of properties. This appendix deals with the last of these purposes.

Commercial metal-forming and fabricating processes are based on the ability of metals to slip. That is, these operations involve plastic deformation. When this occurs at temperatures below the recrystallization region, the metal's tensile and yield strengths and hardness go up, and its plasticity goes down. Many common products rely on this characteristic. For example, the draw and iron process used to produce thin-wall steel or aluminum beverage containers, or "pop cans," gives them the strength needed to withstand the pressure of carbonated beverages.

If cold-working is continued, most metals eventually become so brittle that they cannot be worked any further. To predict the eventual mechanical properties of the metals, or to know when metals have been worked "too much," requires the ability to calculate the amount of cold work performed in any process. Commercial operations such as cold-rolling, drawing, swaging, and extruding depend on the ability to make calculations of cold work. They also rely on the annealing process so that cold-working processes may be continued. Annealing is dealt with in Chapter 3, but for this discussion it is sufficient that you know that there are means to soften the metal and return the level of cold work to zero.

A PREVIEW

In this appendix we'll examine some of the basic methods of calculating cold work, develop equations, and do some example calculations. We'll use graphs of properties versus cold work to relate the size of a cold-worked product with its properties. We'll distinguish the calculation of cold work from the calculation of percent reduction of area, or simply reduction, and stress the importance of calculating percent cold work from the annealed, or zero cold-work, base. Along the way, we'll give example calculations for drawing and rolling operations.

D.1   DEFINITIONS AND TERMINOLOGY

The measure of plastic deformation is the change in the area of the metal part due to the deformation process, compared with the area of the part before deformation, or *change in area divided by the original area*; **cold work** can thus be expressed as

$$C.W. = \frac{\Delta A}{A_a}$$

In this expression A_a is the area before deformation, i.e., when the metal has zero cold work, or when it is in the annealed condition.

In Appendix B this measure was introduced as the reduction of area,

$$R \text{ of } A = \frac{A_0 - A_x}{A_0} = 1 - \frac{A_x}{A_0}$$

which compares the area before deformation with the area at some later point x. If the area before deformation, A_0, is also the area of the metal when it was annealed, A_a, then the reduction of area is also the cold work; if A_a is not A_0, then the calculation determines the reduction of area.

If point x occurs where the deformation is uniform, then the volume equation holds true ($L_0 \times A_0 = L_x \times A_x$) and can be rewritten as

$$\frac{A_0}{A_x} = \frac{L_x}{L_0} \quad \text{or} \quad \frac{A_x}{A_0} = \frac{L_0}{L_x}$$

and

$$C.W. = 1 - \frac{A_x}{A_0} = 1 - \frac{L_0}{L_x}$$

Thus as long as deformation is uniform, the cold work can be found by either area or length relationships. Note that since the products of terms are involved the area and length subscripts are inverted. Also note that cold work is *not* the same as tensile strain ε, which was discussed in Appendix B.

D.2   RELATIONSHIP OF PROPERTIES TO ACCUMULATED COLD WORK

The relationship between mechanical properties and cold work can be experimentally determined by performing tensile and hardness tests on specimens containing a range of cold work. The specimens are initially annealed so they contain zero cold work, and then deformed by whatever means is available, e.g., a small rolling mill, a tensile tester, or a drawing machine. Figure D-1 shows plots of typical mechanical-property data versus percent cold work. The cold work axis begins at zero, i.e., the percent cold work is based on when the metal was annealed. Note

F I G U R E D – 1 **Mechanical properties vs. percent cold work for a typical metal**

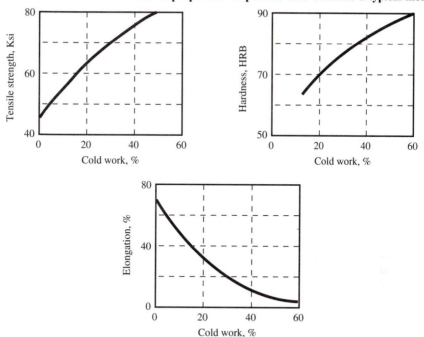

that the strength and hardness curves vary directly with the cold work, and that elongation (ductility) varies indirectly.

The experimental relationships of Figure D-1 can be used to determine either the amount of cold work a piece of metal has already received, or the cold work required to impart a specified level of properties. If a piece of the metal has been cold-worked and has a hardness of 70 HRB, it is a reasonable conclusion, reading from the hardness curve of Figure D-1, that the amount of cold work present is 20%. In the reverse fashion, if a customer specified a product to have a hardness of 70 HRB and/or a tensile strength of 64 ksi, that could be achieved by giving the metal a *total* of 20% cold work.

Total is emphasized because when the cold work is calculated it must be the *accumulation of all cold work since the part was annealed*. For example, assume that the metal had an area of A_a when it was annealed and then was cold-worked, first to area A_1, and then to area A_2. After the first operation the percent cold work would be

$$\% \text{ C.W.} = \left(\frac{A_a - A_1}{A_a}\right) \times 100$$

The *total* cold work after the second operation would be

$$\% \text{ C.W.} = \left(\frac{A_a - A_2}{A_a}\right) \times 100$$

Note that this last expression takes into account the total change in area from A_a to A_2, and that the expression cannot be obtained by adding the two operations separately. That is, the ratio $(A_a - A_2)/A_a$ does not equal

$$\frac{A_a - A_1}{A_a} + \frac{A_1 - A_2}{A_1}$$

The second term of the above expression, $(A_1 - A_2)/A_1$, equals the reduction done in that second pass; it is *not* the amount of cold work. Sometimes the sum of the reductions per pass will *approximate* the total cold work of multiple operations, but mathematically it can never equal the cold work.

D.3   CALCULATION OF COLD WORK FOR A TWO-STEP OPERATION

To see that the sum of reductions will *never* equal the accumulated cold work, consider the following example.*

Problem.   Assume that a metal rod, with the cold-work characteristics shown in Figure D-1, is to be drawn to a smaller size in two operations. We want to estimate what the properties will be. Assume that the rod has a starting (annealed) diameter of 0.400 in. (d_a) and is to be drawn down first to a 0.350-in. diameter (d_1) and then to a 0.300-in. diameter (d_2). (See Figure D-2.)

*The calculations are similar and use the same basic formula, but the meanings of the two terms are different. The cold work is accumulating internal deformation; the reduction is concerned with the change per operation.

F I G U R E D–2 Drawing of a rod in a two-step operation

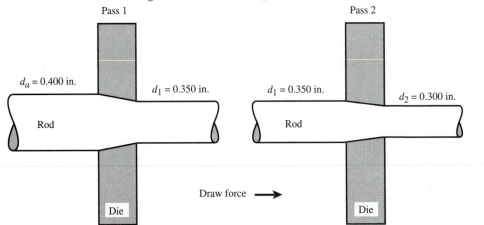

Solution. The area of a circle is given by $\pi d^2/4$. Since the cold-work expression is area divided by area, the $\pi/4$ terms will cancel out and the expression becomes

$$\% \text{ C.W.} = \left(\frac{A_a - A_2}{A_a}\right) \times 100 = \left(\frac{d_a^2 - d_2^2}{d_a^2}\right) \times 100$$

By substituting the values for d_a and d_2 the *total* cold work can be found:

$$\% \text{ C.W.} = \left(\frac{0.400^2 - 0.300^2}{0.400^2}\right) \times 100 = \left(\frac{0.160 - 0.090}{0.160}\right) \times 100 = 43.8\%$$

Reading from the curves of Figure D-1, you can see that this rod should have properties of approximately 77-ksi tensile strength, 84 HRB hardness, and 10% elongation.

If the *reductions* of the two operations were calculated independently,

$$\% \text{ R of A} = \left(\frac{d_a^2 - d_1^2}{d_a^2}\right) \times 100 = \left(\frac{0.400^2 - 0.350^2}{0.400^2}\right) \times 100 = 23.4\%$$

$$\% \text{ R of A} = \left(\frac{d_1^2 - d_2^2}{d_1^2}\right) \times 100 = \left(\frac{0.350^2 - 0.300^2}{0.350^2}\right) \times 100 = 26.5\%$$

their sum, almost 50%, would be greater than the actual cold work of 43.8%.

D.4   ANNEALING TO CONTINUE COLD-WORK OPERATIONS

If a piece of metal is extensively cold-worked, say by rolling, hammering, drawing, or bending, and the working continues, the metal eventually gets so hard that it will crack or break. If working is stopped early enough, and the metal annealed, cold work can be continued. Also, the strength, hardness, and ductility desired can be ensured by the amount of annealing or cold-working the metal receives in the final operations.

Most metalworking processes involve a number of steps, but the sequence of the final operations determines, for the most part, the final properties. For example, aluminum foil can be produced to thicknesses as small as 0.00025 in. using a rolling process that starts with a slab many inches thick. Some rolling passes are done by hot-rolling rather than cold-rolling where no work-hardening is done. Other passes, although done at elevated temperatures, still work-harden the metal. (Hot-working and working at elevated temperatures are discussed in Chapter 3.)

Eventually, to get the surface finish and tolerances required for foil, the metal is cold-rolled, but regardless of the processes, when work-hardening gets too severe the metal is annealed so the process can continue.

If the last cold-working operation is followed by an anneal, as it is in making household foil, the product will be very soft and ductile, and not very strong. If, however, a cold-working operation comes after an anneal, as it might if making a structural member, then the part will be harder, stronger, and less ductile.

D.5   COLD-WORK CALCULATIONS INVOLVING AN ANNEAL

For the example problems that follow, various limits have been assumed for the maximum cold work possible between anneals and for the maximum reduction possible in one operation. In practice, these limits would be determined experimentally for the metal and equipment being used in the given mill.

Problem.   Assume an annealed metal rod of 0.370-in. diameter, with the mechanical properties versus cold-work characteristics shown in Figure D-1, is to be drawn to a 0.250-in. diameter with properties of 60,000 psi minimum tensile strength and 20% minimum elongation. Hardness is not specified.

For this metal it is assumed that cold work can accumulate to 70% before an anneal is required. This does not mean that 70% cold work can necessarily be done in one operation or pass. Since drawing involves tensile forces in the metal exiting the die, a very large reduction might cause the metal to pull apart. If the metal is being rolled, the rolls might not "bite," or grip, the metal if too much reduction is attempted, or the mill might stall. Similar to the limit on the amount of cold work per anneal, a limit on the amount of reduction per pass can also be found experimentally for the metal and equipment used. For this example it is assumed to be 30%.

Solution.   The first step is to find out how much cold work is required to develop the properties required by the finished product. Using Figure D-1 it can be determined that to develop 60-ksi tensile strength the metal must be cold-worked 14%; since 60 ksi is the *minimum* specified, the metal must be cold-worked 14% or more; the *cold work cannot be less than 14%*.

From the elongation graph, to develop 20% minimum elongation the cold work cannot be *greater than* 30%[†] Remember, elongation varies *indirectly* with cold work; the greater the amount of cold work, the *less* the elongation.

Thus the limits of cold work have been determined; the cold work must be somewhere between 14 and 30%. The choice of a target cold-work figure will depend on the consistency or repeatability of the process, but one approach would be to use the midpoint of these values, 22%. This amount of cold work will develop properties within the limits desired, so 22% cold work must be given the rod *after* the last anneal.

Let d_f be the diameter of the final product that has the properties specified. The diameter at which annealing will be done to achieve those properties is d_P. The diameter at which the metal was annealed *before* the production sequence starts will be d_a, and subsequent operations will be d_1, d_2, etc.

[†]Percent elongation and percent cold work are not the same thing. Elongation is an indication of the metal's ductility and is the strain at tensile failure; it is a ratio of the change in length to the original length. Cold work is a ratio of the change in area to the annealed area.

Starting with the basic equation for cold work,

$$\% \text{ C.W.} = \frac{\Delta A}{A_a} \times 100 = \left(\frac{d_P^2 - d_f^2}{d_P^2}\right) \times 100$$

Substituting the values given and solving,

$$22\% = \left(\frac{d_P^2 - 0.250^2}{d_P^2}\right) \times 100$$

we find d_P to be 0.283 in. Drawing the rod from 0.283-in. diameter to 0.250-in. diameter gives 22% cold work, which should enable the rod to meet the properties specified.

The cold work to reduce the original rod, with 0.370-in. diameter, to this intermediate or anneal size, 0.283-in. diameter, must now be determined. Using the same equation,

$$\% \text{ C.W.} = \left(\frac{d_a^2 - d_P^2}{d_a^2}\right) \times 100 = \left(\frac{0.370^2 - 0.283^2}{0.370^2}\right) \times 100 = 41.5\% \text{ is required}$$

Since the starting material is annealed,[‡] and 41.5% is less than the 70% limit on cold work per anneal, the metal can be drawn from a 0.370-in. diameter to a 0.283-in. diameter, annealed, and cold-worked the remaining 22% to get the required size, 0.250-in. diameter, and the desired properties. *Notice that the cold-work percentages are always based on, or calculated from, an annealed size where the cold work is assumed zero.*

Although 41.5% is less than the limit on cold work per anneal, it is not less than the limit per pass of 30%. Thus another draw or draws will be required. As a first approximation the number of 30% reductions required to draw the 0.370-in. diameter to 0.283-in. diameter is found. Starting from the original annealed size, 0.370-in. diameter, the cold-work relationship is used to determine a diameter that has 30% less area; note the formula can be rewritten for easier use:

$$\% \text{ C.W.} = \frac{\Delta A}{A_a} \times 100 = \left(\frac{d_a^2 - d_1^2}{d_a^2}\right) \times 100 = \left(1 - \frac{d_1^2}{d_a^2}\right) \times 100$$

where d_x is any intermediate diameter. Substituting the 30% limit and the starting or annealed diameter, we find that the diameter of the maximum draw possible is

$$0.30 = 1 - \frac{d_1^2}{0.370^2} \qquad d_1^2 = 0.70 \times 0.370^2 \qquad d_1 = 0.310\text{-in. diameter}$$

[‡] If the starting material was *not* in the annealed condition, but already contained cold work, a preliminary step would have to be taken before the calculations as described here could be done. As stated above, accumulated cold work must be measured from an annealed size. The cold work in the starting material must be either known or determined. If data like those given in Figure D-1 are available, and if the hardness or strength is known, then the cold work in the starting size can be determined and the larger annealed size calculated. The calculations can then proceed from the size where the metal was annealed.

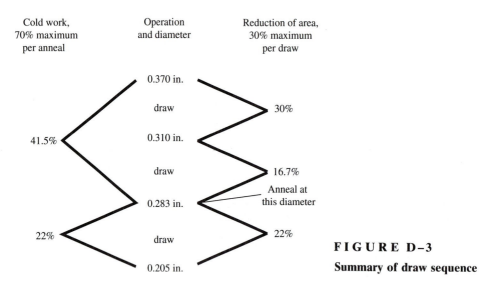

Cold work, 70% maximum per anneal	Operation and diameter	Reduction of area, 30% maximum per draw

0.370 in.

draw   30%

41.5%   0.310 in.

draw   16.7%

Anneal at this diameter

0.283 in.

22%   draw   22%

0.205 in.

FIGURE D–3

Summary of draw sequence

The rod can be drawn from 0.370-in. diameter to an intermediate diameter of 0.310 in. at a 30% reduction. It must now be determined whether the reduction from 0.310-in. diameter to the property annealing diameter of 0.283 in. is within 30%; if it is not, then another intermediate draw will be necessary:

$$\% \text{ R of A} = \left(1 - \frac{d_P^2}{d_1^2}\right) \times 100 = \left(1 - \frac{0.283^2}{0.310^2}\right) \times 100 = 16.7\%$$

This is within the 30% limit, so only two draws are required to get from the starting size of 0.370-in. diameter to the anneal size of 0.283-in. diameter. In actual practice, a producer might choose an intermediate size larger than 0.310 in. and distribute the reductions more evenly between the draws.

A summary of this analysis is shown in Figure D-3. This shows the draw sizes, the percent reduction between draws, and the cumulative percent cold work between anneals.

Note that the cold work to the anneal, 41.5%, is less than the percent reduction of the two draws added together, 30% + 16.7% = 46.7%.

When a large number of reductions are needed to achieve the size required, the production sequence can get lengthy. Two or more anneals might be distributed among dozens of draws or passes. In such cases the calculation process becomes a reiterative one that requires first approximations and later refinements.

D.6   CALCULATING COLD WORK IN OTHER TYPE OPERATIONS

In this appendix we've looked at examples of drawing a solid rod. If the metal to be drawn is instead a square solid, a rectangular solid, or a tube, the relationships are the same. Areas are calculated differently, but reductions are still the change

$$\% \text{ R/A} = \frac{\Delta A}{A_1} \times 100 = \left(1 - \frac{t_2 \times w_2}{t_1 \times w_1}\right) \times 100 \qquad \text{If } w_1 = w_2 \text{ then}$$

$$\% \text{ R/A} = \left(1 - \frac{t_2}{t_1}\right) \times 100 = \frac{\Delta t}{t_1} \times 100 \qquad \text{If } t_1 = t_a \text{ then}$$

$$\%\text{C.W.} = \left(1 - \frac{t_2}{t_a}\right) \times 100$$

$$\text{or, } \%\text{C.W.} = \frac{\Delta t}{t_a} \times 100$$

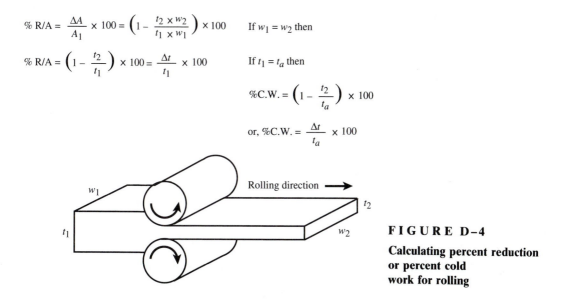

FIGURE D–4

Calculating percent reduction or percent cold work for rolling

in area through an operation divided by the starting area. Cold work is the change in area since the metal was annealed divided by the original area.

The calculation of reductions in a rolling operation is done on the same basis as in a drawing operation. However, as shown in Figure D-4, it is assumed that the width of the sheet does not increase as the metal progresses through the rolls. The frictional force between the rolls and metal inhibits almost all the movement of the metal in anything but the direction of rolling. Within practical limits this is a good assumption and is commonly made.

Thus the calculation of percent reduction or percent cold work is an easy one—the difference in thickness is divided by the prior thickness for reduction or the annealed thickness for cold work.

D.7   OTHER METHODS OF CONTROLLING MECHANICAL PROPERTIES

In this appendix we have studied the method of changing mechanical properties by cold working. Starting with metal that had zero cold work (annealed), or considering an anneal at a strategic point in the process, we saw how the metal can be given the correct amount of cold work to achieve the properties desired.

In a similar fashion the *amount of annealing* can be adjusted. That is, the metal can be worked down to its final size, the amount of cold work calculated, and then an anneal of the right temperature and length of time can be used to get the properties desired.

There are two other methods of altering the mechanical properties of metals that are unrelated to cold-working and annealing—alloying and heat-treating—which we have discussed elsewhere.

SUMMARY

1. Cold work is the change in area divided by the area of the part in the annealed condition, i.e., before any cold work has been done, or

$$\text{Cold work, or C.W.} = \frac{\Delta A}{A_a}$$

The percentage amount of cold work can only be calculated in reference to an annealed size of the metal.

2. If the cold-working is done uniformly, as in drawing, then the volume of the part equals its area times its length, and percent cold work can also be found from a ratio of the lengths, or

$$\% \text{ C.W.} = \frac{\Delta A}{A_a} = 1 - \frac{A_x}{A_a} = 1 - \frac{L_a}{L_x}$$

3. Reduction of area is the change in area divided by the area of the part through any cold-working operation, whether the part was in the annealed state or not, or

$$\text{Reduction of area, R of A} = \frac{A_0 - A_x}{A_0} = 1 - \frac{A_x}{A_0}$$

4. Since the starting size for the reduction calculation changes with each operation, the total cold work done through a series of operations will not equal the sum of the reductions.

5. The relationship between percent cold work and mechanical properties can be found experimentally and used to evaluate cold-work levels or to estimate the amount of cold work needed to achieve a desired level of mechanical properties.

6. Cold-working will increase strength and hardness measures and reduce ductility. Annealing can be used to soften the metal so that cold-working can be continued, or it can be used to produce a soft, ductile, finished product.

7. In many cold-working operations (e.g., drawing, extruding, or swaging) both dimensions of the cross-sectional area are changed by the cold work. In calculating cold work for these operations, both dimensions must be taken into account. In cold-rolling, the friction between the roll face and the metal prevents much widening of the metal, so cold working in rolling is calculated as the ratio of the change in thickness to the original thickness.

REVIEW QUESTIONS

1. A solid rectangular bar is drawn from an original size of 1.000 in. × 0.875 in. to 0.900 in. × 0.788 in. If the bar was annealed to begin with, what percent cold work was done? If the bar *wasn't* annealed before the draw, what have you calculated?

2. It takes a number of operations to draw a piece of tubing from 2.154 in. O.D. × 0.250 wall to 1.660 × 0.140 wall. If the metal was annealed to begin with, and if no anneals were done along the way, what percent cold work was required?

3. A 1.000-in.-diameter metal rod, whose properties versus cold-work characteristics are shown in Figure D-1, has a hardness of 20 HRB. If the rod is drawn to 0.750-in. diameter, what will its hardness be?

4. If the metal in the preceding question was a piece of 1.000-in.-thick plate (rather than rod), and was *rolled* to a thickness of 0.750 in., what would its hardness be?

5. A piece of sheet metal is reduced in successive cold-rolling passes from an annealed starting thickness of 0.250 in. to 0.160 in. to 0.085 in. to 0.050 in. What is the reduction for each pass and what is the total percent cold work done?

6. A piece of plate, of a metal whose properties versus cold-work characteristics are shown in Figure D-1, is to be rolled down to produce a finished product that meets the following specifications: tensile strength of 70-ksi min. to 75-ksi max.; hardness of 75 HRB min.; elongation of 5% min. The starting material is 1.125 in. thick and has a hardness of 77 HRB. The desired finish size is 0.125 in. thick. This metal can accumulate 70% cold work before annealing is required, and the limit on the reduction per pass is 35%. Design a rolling and annealing schedule that produces the product in the fewest anneals and draws. (Anneals cost more than two or three draws.)

FOR FURTHER STUDY

DeGarmo, E. Paul, J. Temple Black, and Ronald A. Kohser. *Materials and processes in manufacturing* (7th ed.). New York: Macmillan, 1988, chap. 17, on forming processes in general, and chap. 19, on cold-working processes.

Dieter, George E. *Mechanical metallurgy* (3d ed.). New York: McGraw-Hill, 1986, chaps. 1, 17, and 19.

Flinn, Richard A., and Paul K. Trojan. *Engineering materials and their applications* (3d ed.). Boston: Houghton Mifflin, 1986, 97–100.

Harris, J. N. *Mechanical working of metals*. New York: Pergamon, 1983.

Hardenability: Testing and Application

OBJECTIVES

After studying this appendix, you should be able to

- differentiate between hardness and hardenability and explain the factors that determine each
- describe the performance of the Jominy end-quench test, the nature of the data collected, and how it is plotted
- explain how Jominy curves and hardness readings can be used to establish a cooling rate
- use Jominy and cooling-rate data to select steels to meet hardness specifications

DID YOU KNOW?

In the 1930s Walter Jominy and A. L. Boegehold proposed a measure of hardenability that has come to be known as the Jominy end-quench test. Prior to the use of this test, the hardening ability, or hardenability, of steels could only be determined by direct comparisons among steels. A typical method compared hardness gradients on rods of different diameters quenched in a common medium.

Although the maximum diameter that would quench to all martensite was objective data, it gave little information that could be used to design parts requiring certain hardnesses at different locations. That is, there was no direct link between the cooling rates and the results. Jominy's method solved this problem; it put on one diagram the relationship between hardening ability and cooling rate.

A PREVIEW

In Chapter 7 we discussed the concepts and principles of hardenability, the Jominy test, and H-band steels. This appendix takes those concepts and principles and shows how they can be used to select steels to meet practical specifications.

After a brief review of hardenability, we'll discuss the Jominy test in detail, stressing the relationships among positions on the Jominy specimen, cooling rate, and hardness.

We'll look at an example of selecting a steel to meet a specification for an irregularly shaped part and then use the same example to explore the use of H-band steels. Finally, we'll discuss the use of quench charts to determine cooling positions for regularly shaped parts like rods.

E.1   PRINCIPLES OF HARDENABILITY

In Chapter 7 *hardenability* was defined as the ability to convert austenite to martensite at slower cooling rates; when comparing two steels, we found that the one capable of forming martensite at the slower cooling rate had the higher hardenability. Since the hardness of martensite is largely dependent on its carbon content, for any given chemical composition the hardness depends on the cooling rate, that is, on whether or not the austenite has the opportunity to transform to all martensite or to martensite and/or some softer product(s), e.g., bainite, pearlite, or ferrite. Thus if the hardness of martensite of a particular carbon content is known, any hardness less than that value indicates that other, softer products are present.

These general principles have some practical consequences. For example, if there are two pieces of steel with the same composition, we know that the harder one was quenched faster, and conversely, if two pieces of steel are quenched identically, we know that the harder one has more martensite.*

The concept of the Jominy test is that if a specimen of a standard size is heated to austenite and cooled by spraying water on one end, cooling rates will vary along the bar from fairly fast at the water-cooled end to very slow at the other. It is a good assumption that the composition of the specimen is uniform; thus the cooling rate is the only variable influencing the hardness. It is also a good assumption that the heat conductivity of steel is essentially a constant, regardless of composition. Thus one steel can be compared with another using the Jominy end-quench hardenability test.

E.2   THE JOMINY TEST

Standard Conditions. The purpose of the Jominy test is to quench specimens under standardized conditions so that the two consequences stated above—the quench rate and the amount of martensite—can be used to compare the hardenability of steels of different compositions. This requires that two major items be standardized—the method and media used for quenching and the size and shape of the specimen. The standardization used in the Jominy test results in controlling all the variables mentioned in Section 7.10—agitation of the quenchant, quenchant medium, quenchant temperature, and surface-to-volume ratio.

The specimen used in the Jominy test is 1 in. in diameter by 3 to 4 in. long. One end of the specimen has a collar larger than 1 in. in diameter so the specimen can be hung vertically. (See Figures 7-8 and E-1b.)

The quenching medium is water, specified to have a temperature of 40 to 85°F (5 to 30°C) and supplied through a vertically oriented outlet. The volume of water and the amount of agitation are controlled by the size of the water outlet (½-in. inside diameter), the vertical free height of the unobstructed water fountain (2½

---

*We assume here, of course, that the hardness is measured at comparable points on the specimens.

F I G U R E   E – 1   Control of the variables for the Jominy test

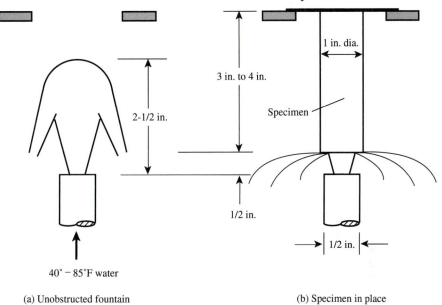

(a) Unobstructed fountain   (b) Specimen in place

in.), and the distance between the water outlet and the end of the specimen when it is in place (½ in.). The arrangement of the equipment, with the unobstructed fountain (a) and the specimen in place (b), is shown in elevation in Figure E-1.

Heating the Specimen.   The specimen, of normalized steel, is heated to a temperature within the austenite region. To reduce scaling or decarburization, a controlled atmosphere furnace should be used; if such a furnace is not available, the specimen can be suspended in a fixture made from pipe (see Figure E-2); a layer of charcoal or graphite in the bottom of the fixture will help reduce scaling. With the furnace at temperature, such a fixture and specimen take about 1¼ h to ensure transformation to austenite.

Performing the Test.   Once the specimen is removed from the furnace, and from the fixture if used, it should be placed in the quenching apparatus and the water applied within 5 sec; care should be taken that the water strikes only the bottom of the specimen.

When removed from the furnace, the specimen, at approximately 1600°F (870°C), will be orange-red in color; almost immediately after the water is turned on the lower end will become black (at about 1200°F, or 650°C), and in a minute or two the whole specimen will be dark.

The specimen should be left in the quench unit for a minimum of 10 min and then the whole specimen may be immersed in water to cool it so it can be handled.

Hardness Readings.   To accurately measure the hardness along the length of the specimen, a minimum of 0.015 in. of outer surface must be ground off to remove

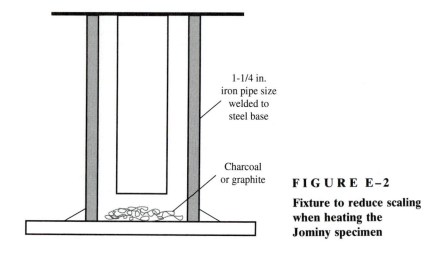

1-1/4 in.
iron pipe size
welded to
steel base

Charcoal
or graphite

F I G U R E E–2

**Fixture to reduce scaling
when heating the
Jominy specimen**

scale and any decarburized layer. Ideally this will be done on a surface grinder at two locations 180° apart. Lacking these facilities, the grinding can be done by hand if done carefully. The grinding must be done without overheating the specimen, since this will change its properties.

The specimen is then hardness-tested, with readings being taken ideally every $\frac{1}{16}$ in. for 2 in. along the length of both of the ground flats. Except in rare cases, the readings are taken using the Rockwell C scale.

Because of the relationship between the quench rate and the ability to form martensite, the hardness readings decrease as they move away from the quenched end. However, if readings are taken much beyond 2 in. from the quenched end, the hardness readings may go up; this means heat is being extracted faster by the quenching apparatus and/or the fresh air available to the top of the specimen than it is being conducted to the lower end to be removed by the water.

Maximum Hardness. It should not be assumed that the highest hardness shown on the Jominy curve for any particular steel is the highest hardness possible for that steel. A small slice, say $\frac{1}{2}$ in. long, cut from a 1-in. rod can often be plunged in water and achieve a higher hardness than that achieved on the Jominy specimen cut from the same rod. The level of agitation at the end of the Jominy specimen is really not very high. Also, at the quenched end the heat is being removed through the 1-in. diameter end; that surface-to-volume ratio is not as high as it is for a small slice of the rod.

E.3   INTERPRETING THE RESULTS

The Jominy end-quench test results are plots of hardness versus distance from the quenched end and will be similar to those shown in Figure E-3. As discussed in Chapter 7, the vertical location of the curve is primarily determined by the carbon

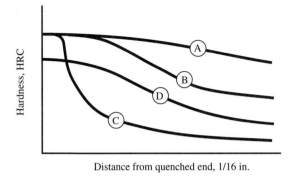

FIGURE E-3

Typical Jominy end-quench curves

Distance from quenched end, 1/16 in.

content, and the shape and/or "flatness" of the curve is determined by the alloy content.

In Figure E-3 curves A and B represent alloy steels with the same carbon content; steel A's curve is flatter than B's because it is more highly alloyed and thus has higher hardenability. Steel C represents a plain carbon steel that has the same carbon content as steels A and B,[†] but lacks hardenability; at the slower cooling rates, i.e., farther from the quenched end, steel C is not capable of generating hardnesses as high as steels A and B.

Steel D represents a steel with less carbon content than the other three steels, but with about the same alloy content as steel B; thus its curve is shaped like steel B, but is lower because it has less carbon.

E.4   DETERMINING THE RATE OF COOLING

The principles and concepts of the Jominy end-quench test can be investigated by considering an application of the test data to an example.

Relation of Rate of Cooling to Hardness. Suppose that a steel must be selected to produce the unusually shaped part shown in Figure E-4(a), and that for various reasons the steel is to be quenched in oil from an austenite temperature. Also suppose that a trial piece of steel is used to make a sample part *and* a Jominy test bar. After the part is heat-treated, its hardness is measured at six different locations, with the results shown in Figure E-4(a).

The Jominy specimen (made from the same piece of steel used to fabricate the sample part) is given the end-quench test and generates the data plotted in Figure E-4(b). The six hardness readings taken on the sample part are 59, 51, 45, 59, 51, and 58 HRC, going from top to bottom. On the Jominy plot of Figure E-4(b), horizontal dotted lines have been drawn from these same values to the curve,

---

[†]Steels A, B, and C have the same hardness at the fast-quench or martensite end of the specimen; since they were quenched the same and have the same hardness they must have the same carbon content.

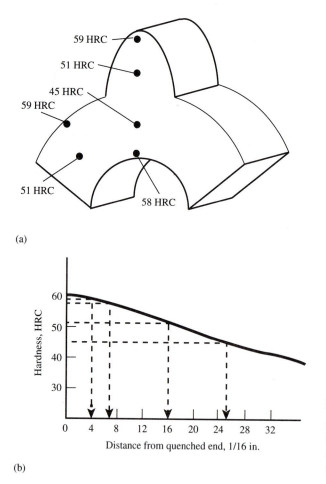

(a)

(b)

F I G U R E E–4

(a) Steel part and hardnesses achieved with oil quench; (b) the Jominy curve for the steel used to make the part

where intersecting vertical lines with arrowheads indicate the corresponding cooling positions 4, 7, 16, and 25, respectively. That is, for the composition of steel[‡] used to make the part and the Jominy specimen, developing a hardness of 58 HRC requires that the steel part cool at the same rate as the cooling position $7/16$ in. on the Jominy specimen; to develop a hardness of 45 HRC on the part requires that it cool at the same rate as the cooling position $25/16$ in., etc.

Role of the Quenchant. This method of determining the rate of cooling is dependent on the fact that for a given composition of austenite its eventual hardness is dependent on how fast it was cooled. Note that the part was quenched in oil while the Jominy standard quench was done with water. At first this may seem confusing, but the key is that, regardless of the medium used to quench the part, its hardness relates to a position on its Jominy specimen, and that position is related to the rate of cooling.

[‡]Technically, the austenite grain size must also be similar.

Rate of Cooling in Terms of Position on the Jominy Specimen. Note that the rates at which the steel cools are not expressed in degrees Fahrenheit per second, but in terms of positions on the Jominy specimen. Actually, knowing a cooling rate in terms of degrees per second is of little value since, as discussed in Section 7.10, the quench rate, the slope of the temperature-versus-time curve, varies during the quenching process; to use such data for comparing steels, the actual cooling curves would have to be plotted, related to hardness, and then compared on some basis. It is far more useful to have data on cooling rates in terms of positions on the Jominy test specimen; if tests are run according to standard, the curve of hardness versus position of one steel can be compared directly with another.

Meeting a Specification. Now suppose that the specification for the part shown in Figure E-4(a) calls for a hard outer surface and a soft, tough interior. The sample part produced does have a hard outer surface, 58 to 59 HRC, but the interior is between 45 and 51 HRC, which would not be considered soft and tough.

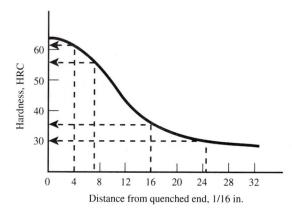

(a)

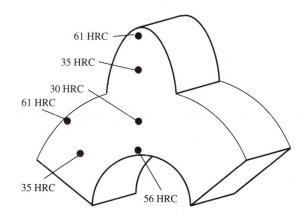

(b)

FIGURE E–5

(a) Jominy curve for proposed steel; (b) part made from proposed steel, with hardnesses achieved by oil quench

Figure E-5(a) shows the Jominy curve for a proposed steel of a different composition. The cooling positions for the areas of the part that are of interest are already known—positions 4, 7, 16, and 25. Vertical dotted lines have been drawn from these values to the curve, where intersecting horizontal dotted lines with arrowheads indicate the corresponding hardness that will be achieved by quenching the part identically to that of Figure E-4. These values have been transferred to the drawing in Figure E-5(b).

The steel of Figure E-5 has less hardenability than that of Figure E-4, but enough carbon that its surface can quench to martensite while the interior transforms to a softer product. Thus the Jominy curve has been used to select a steel composition that, given a certain quench, will meet the hardness range specified for an unusually shaped part.

E.5   USING H-BAND STEELS TO MEET SPECIFICATIONS

H-Band Steels.   Hardenability-band (H-band) steels are commercially available steels with guaranteed hardenability; i.e., limits are placed on the variability of their Jominy curves. Because of the varying influence of alloying elements on hardenability, and the variations in chemistry that occur in commercial production, the hardenability of steels of the same grade can vary appreciably. The concept of H-band steels is that, regardless of composition, their Jominy curves will fall within maximum and minimum limits.[§] H-band steels are also discussed in Section 7.9.

Figure E-6 is the H-band for the AISI/SAE 50B44H steel.[∥] The upper curve represents the Jominy results when the composition is on the top side of its range, and the lower curve when it is toward the bottom.

The dotted lines of Figure E-6 help to describe the two basic ways this figure can be read or used—cooling this H steel at the rate of Jominy position $^{16}/_{16}$ in. is likely to produce hardnesses between 26 and 48 HRC—or a hardness of 48 HRC is likely to be obtained by cooling the steel at rates between positions $^{7}/_{16}$ and $^{16}/_{16}$ in. Although these values have seemingly large ranges, H-band steels can be used to ensure that various areas of steel parts will have the relative hardness and/or strength desired.

Establishing a Specification in Maximum and Minimum Hardnesses.   To produce the part described in Figure E-5, with the hardnesses shown, the Jominy curve of the steel used would have to be *identical* to that of Figure E-5(a). Thus to produce a large number of these parts is impractical. However, if the specifications desired can be translated to conform to an H-band steel, then commercially available

---

[§]The chemistry of H-band steels is also specified, but to wider limits than the non-H-band steel of the same designation.

[∥]50B44H is a chromium-alloy steel with 0.42–0.49% C, 0.65–1.10% Mn, 0.15–0.35% Si, 0.30–0.70% Cr, and 0.0005–0.003% B added.

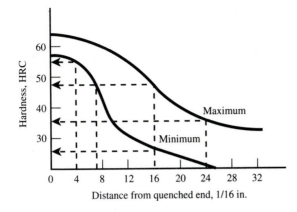

FIGURE E-6

Hardenability band for SAE/AISI Steel 50B44H, showing maximum and minimum hardnesses of a *possible* **solution**

steels can be used. By using the results shown in Figure E-5 and making some assumptions a specification can be determined:

1. The outer surface, which will cool at the rate of position 4/16 in., needs to be about 60 HRC and at least *more than* 55 HRC.
2. The portion of the inner circle, which cools at the rate of position 7/16 in., does not need to be as hard as the outer surfaces, but should be *more than* 40 HRC.
3. The inner portion of the "tooth," which cools at the rate of position 16/16 in., should be softer and tougher and therefore *less than* 40 HRC.
4. The central portion, which cools at the rate of position 25/16 in., should be even tougher and therefore should be *less than* 30 HRC.

To select a commercial H-band steel these specifications must be superimposed on the Jominy curves of a steel, for example, 50B44H in Figure E-6. The cooling positions for the above specification are the vertical dotted lines shown; they translate to the hardnesses indicated at the arrowheads of the horizontal intersecting dotted lines. By analyzing the hardnesses it is seen that specifications 1 and 2 are satisfied; at cooling positions 4/16 and 7/16 in. the *minimum* hardnesses that would be achieved are 55 and 48 HRC, respectively, which meet the 55 and 40 HRC minimums of the specification.

However, at the slower cooling rates, positions 16/16 and 25/16 in., the hardnesses can be as high as 48 and 36 HRC, respectively, or higher than desired. That is, this steel has too much hardenability; at slower cooling rates it produces hardnesses that are too high.

A Solution That Meets the Specifications. Figure E-7 shows the H band for 15B48H.# The vertical dotted lines are again the cooling positions of interest, and the resulting minimum or maximum hardnesses are shown at the arrowheads. The minimum hardnesses for positions 4/16 and 7/16 in. are 55 and 44 HRC, respectively, and meet specifications 1 and 2.

#15B48H is a carbon-manganese steel with 0.43–0.53% C, 1.00–1.50% Mn, 0.15–0.35% Si, and 0.0005–0.003% B.

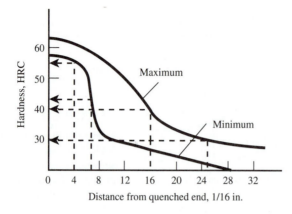

FIGURE E–7

Hardenability band for SAE/AISI Steel 15B48H, showing maximum and minimum hardnesses of a solution

At the slower cooling rates the maximum curve of this steel drops enough that at cooling positions $^{16}/_{16}$ and $^{25}/_{16}$ in. the hardnesses are less than 40 and 30 HRC, respectively, which satisfies specifications 3 and 4.

Thus the part of Figure E-5 with the hardness ranges as specified can be produced from an AISI/SAE 15B48H steel quenched in the same oil as used for the original trial part of Figure E-4.

E.6   SPECIFYING STEELS

From the discussion above it can be realized that if Jominy bands can be developed for steels, then steel characteristics or properties can be specified relative to those curves. In the problem just described if the specification was reduced to just the hardnesses at the surface and the interior, then the steel required must be able to develop 55 HRC minimum at J $^4/_{16}$ in. (Jominy distance) and 30 HRC maximum at J $^{25}/_{16}$ in.; this can be written in shorthand as

Combination of any maximum with any minimum hardness	55 HRC at J $^4/_{16}$ min.
	30 HRC at J $^{25}/_{16}$ max.

Other combinations of hardnesses and distances that can be used in standard specifications are shown below; the data are from Figure E-7, but are merely examples of specifications and have nothing to do with the prior example.

Minimum and maximum hardnesses at a cooling distance	36 to 40 HRC at J $^{16}/_{16}$
One hardness value at minimum and maximum distances	30 HRC at J $^{10}/_{16}$ min.
	30 HRC at J $^{25}/_{16}$ max.
Two maximum hardnesses at two positions	50 HRC at J $^{12}/_{16}$ max.
	40 HRC at J $^{25}/_{16}$ max.
Two minimum hardnesses at two positions	55 HRC at J $^4/_{16}$ min.
	44 HRC at J $^7/_{16}$ min.

E.7   QUENCH CHARTS FOR SYMMETRICAL SHAPES

Review of the Process for an Irregularly Shaped Part.   The process of selecting a steel to produce the part of Figure E-5 began by determining the "cooling rates," in terms of Jominy distances, for various areas of the unusually shaped part. To do this, a Jominy test was conducted for the steel used to produce the part. The hardness of the part was measured at certain points; comparing those hardness readings with the Jominy curve defined the cooling position that was required to achieve that hardness.

Developing Relationships for Symmetrical Parts.   However, if parts are to be quenched that have more symmetrical geometries, e.g., rods, squares, and hexagons, then relationships can be developed between the size of the part and Jominy cooling positions for a specific quenchant and composition of steel.

Assume that a number of rods with diameters between, say 0.25 and 4.00 in., are given the same austenitizing treatment, quenched in the same medium, and hardness-tested at their centers. The resulting data would show that the smaller diameters have the harder centers and the larger diameters are softer. This is logical, since the smaller diameters would cool faster and the larger would cool slower.

If a Jominy test was also run on this same steel, then the hardness readings from the centers of the rods could be used to equate the rate of cooling of the center of each size rod with a position on the Jominy specimen. The resulting curve might look like that of Figure E-8.

If these same rod diameters were quenched in other quenchants, e.g., agitated water and agitated oil, curves like those of Figure E-9 could also be generated.

If the hardness of these test rods was also measured at other locations, e.g., at the surface and at ¾ of the radius from the center, then these other locations could also be related to the Jominy positions.

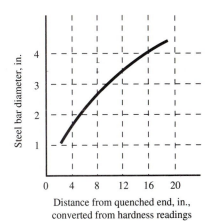

FIGURE E-8

Jominy cooling positions for the centers of rods, converted from center-hardness readings

Steel bar diameter, in.

Distance from quenched end, in., converted from hardness readings

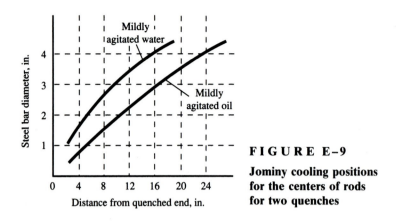

F I G U R E E–9

Jominy cooling positions for the centers of rods for two quenches

Quenching Charts for Steel Rods. Figure E-10 shows hypothetical examples of the relationships just described—between rod diameter and its Jominy cooling position, at specific locations on the rod, for two types of quench. The top chart is for steel rods quenched in mildly agitated water, the bottom for rods quenched in mildly agitated oil. The three curves on each chart relate the cooling positions at the center of the rod, at ¾ of the radius from the center, and on the surface of the rod. Obviously, the center of a rod cools more slowly than the surface, so its curve is farther to the right (slower cooling position) than the surface curve.

For the agitated-water quench, the curve labeled "Center" equates the cooling rate at the center of a rod with its Jominy cooling position; for example, the center of a 4-in.-diameter rod cools at the rate of the ¹⁶⁄₁₆-in. Jominy position, or J ¹⁶⁄₁₆. The oil quench, being a quench of less intensity or severity, cools the same 4-in.-diameter rod more slowly—at the rate of J ²³⁄₂₄.

The "¾ radius" location is ¾ of the way from the center of a rod to its surface. For a 1-in.-diameter rod this would be ⅛ in. under the surface; for a 2-in.-diameter rod this would be ¼-in. under the surface. If the 2-in.-diameter rod was quenched in mildly agitated water, the ¾ radius location cools like J ³⁄₁₆; in oil this same location cools like J ⁸⁄₁₆.

Using the Charts. Knowing how a particular part of a rod will quench, i.e., the Jominy position, enables a designer to use Jominy curves to select steels to meet hardness specifications. For example, assume that you want to quench a 3-in.-diameter steel shaft in mildly agitated oil so that it achieves a *minimum* hardness of 50 HRC on the surface and a *maximum* hardness of 40 HRC at the center. The lower chart of Figure E-10 gives the key cooling positions of J ⁶⁄₁₆ for the surface and J ¹⁷⁄₁₆ for the center.

These specifications can be compared with Jominy curves that have been run for specific steels or compared with charts from handbooks. From the Jominy curves given in this text, steel 15B48H, shown in Figure E-7, and the specific steel

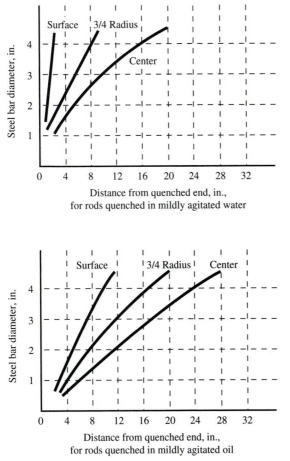

FIGURE E-10

Quenching charts for a quenching system

shown in Figure E-5 both meet the specification. That is, at J ³⁄₁₆ the hardness will be 50 HRC or greater, and at J ¹⁷⁄₁₆ the hardness will be less than 40 HRC.

Practical Limitations. Curves similar to those seen in Figure E-10 are often found in handbooks and other sources. They are helpful in understanding the relationships that exist between the size of a part, the quench rate, and the eventual hardness, but their usefulness in practice is somewhat limited because of the difficulty in comparing quenching systems, i.e., quenching media and levels of agitation. However, these charts can be developed for specific quench systems by running the heat-treat/hardness/Jominy tests described above. The charts can then be used to predict the hardness that will be achieved when different-sized rods of known steels are quenched.

Similar charts can also be developed for symmetrical shapes like squares, hexagons, and flats.

SUMMARY

1. Hardenability is the ability to achieve martensite at slower and slower cooling rates of austenite.

2. The Jominy end-quench test is a standardized way to achieve a range of cooling rates and measure their effect on the hardness.

3. The Jominy test standardizes the quenchant, the agitation of the quenchant, the temperature of the quenchant, and the surface-to-volume ratio of the specimen.

4. Other quenching methods (media, agitation level, temperature, and/or surface-to-volume ratio) may produce hardnesses greater than those at the quenched end of the Jominy specimen.

5. The distance from the quenched end is essentially a cooling rate. If the Jominy curve is known for a steel, a hardness reading of that steel tells the position at which it cooled.

6. That cooling position can be used to select another steel that, if cooled at the same rate, would have properties that are more desirable.

7. H-band steels are commercially available steels with guaranteed hardenability; Jominy curves for H-band steels are guaranteed to fall within certain limits. That is, that for any cooling position the hardness of an H-band steel will fall within specified limits.

8. H-band steels enable the hardness of a part, or a location on it, to be specified as a range or as a maximum or a minimum.

9. Cooling positions of irregularly shaped parts can be found by quenching the part and comparing its hardness with that of a Jominy specimen of the same steel.

10. Cooling positions of more regularly shaped parts, like round rods, flat bars, or hexagonal bars, can be found by using charts constructed for the quenching system used.

REVIEW QUESTIONS

1. What factors influence the hardness of a steel?

2. Define hardenability. What is the difference between hardness and hardenability?

3. Describe the factors responsible for the hardenability of a steel.

4. What factors are standardized by the Jominy test? Describe how they are standardized.

5. Explain how the Jominy test is related to the CT diagram.

6. A part of unusual shape is fabricated from the steel used in Figure E-5. The part is heat-treated and its hardness is measured as 45 HRC. The part is then fabricated from 50B44H, the steel used in Figure E-6. After heat-treating, using the same quench as before, what range of hardness is likely to be found?

7. A gear tooth is specified to be hard on the surface (> 50 HRC) and soft and tough in the interior (< 35 HRC). Describe the process you could use to select a steel for this purpose.

8. A 2½-in.-diameter shaft of 15B48H steel (Figure E-7) is hardened by quenching in water; it develops a center hardness of 45 HRC and, unfortunately, cracks. If the same size shaft is made from the same steel and quenched in the mildly agitated oil of the quench system of Figure E-10, what will be its likely hardness?

FOR FURTHER STUDY

Annual book of ASTM standards. Philadelphia: American Society for Testing and Materials, vol. 01.05, Standard A255.

Avner, Sidney H. *Introduction to physical metallurgy* (2d ed.). New York: McGraw-Hill, 1974, secs. 8.20 and 8.21.

DeGarmo, E. Paul, J. Temple Black, and Ronald A. Kohser. *Materials and processes in manufacturing* (7th ed). New York: Macmillan, 1988, 112.

Flinn, Richard A., and Paul K. Trojan. *Engineering materials and their applications* (3d ed.). Boston: Houghton Mifflin, 1986.

Grossman, M. A., and E. C. Bain. *Principles of heat treating*. Materials Park, OH: ASM International, 1933, 1964.

Jominy, Walter E. Hardenability tests, in *Hardenability of alloy steels* (1938 symposium). Materials Park, OH: ASM International, 1939, 66–94.

SAE handbook. Warrendale, PA: Society of Automotive Engineers, vol. 1, Standards J406 and J1268.

APPENDIX   F

Impact Testing

OBJECTIVES

After studying this appendix, you should be able to

- describe the operation of an impact testing machine, including the safety procedures to be followed
- identify the information that can be obtained in the impact test and describe how it can be used
- describe the general configurations of the three Charpy impact test specimens and how they are put into the tester
- differentiate between the Charpy and Izod tests and specimens
- describe how the appearance of the fractured surface of ferrous metals changes with temperature
- define transition temperature, describe the methods used to determine it, and explain its significance

DID YOU KNOW?

The impact test originated in the early 1900s and evolved to the Charpy and Izod tests, which became ASTM standards in the 1930s.* The impact test played a major role in resolving the dilemma of the catastrophic failure of the Liberty ships described in Chapter 12.

Although the impact test is used primarily to investigate the transition temperatures of ferrous metals, it is also helpful in understanding some of the differences between ferrous and nonferrous metals and the significance of notches in creating the complex stress distributions discussed in Chapters 11 and 12.

A PREVIEW

We'll first look at the purpose and performance of the impact test in a general way. The impact test measures toughness or the ability of a metal to absorb energy to failure; more details of the test will come to light as we study the means of measuring this energy. Next we'll discuss the types of specimens used in the test, with emphasis on the three Charpy-type specimens. We'll outline step-by-step procedures for conducting the test.

Once the test is conducted, there is the matter of determining and interpreting the results. This can be done three ways: using the energy measurements taken during the test, looking at the amount of distortion of the fractured specimens, and examining the fracture surface of the specimens.

Although this appendix can be read by itself, it is anticipated that you'll be reading it along with Chapters 11 and 12. It will be especially helpful to use the equipment or see it demonstrated.

*Metals handbook (8th ed.). Materials Park, OH: ASM International, 1961, vol. 1, 225.

F.1 <u>**GENERAL PURPOSE OF THE IMPACT TEST**</u>

The impact test, as its name suggests, applies a load at a very high rate of speed and uses specimens designed to create complex stress distributions. This combination of high strain rate and stress concentration enables the impact test to determine the energy-absorbing capabilities, or toughness, of specimens; i.e., it measures the energy absorbed by specimens as they are fractured rather than measuring, for example, their strength as determined in the tensile test. Commonly, specimens of BCC metals are tested over a range of temperatures to determine their ductile-brittle transition temperature.

The apparatus, procedure, and specimen are standardized so that results can be duplicated by others. Figure F-1 is a schematic representation of the progressive steps of the test. The Charpy specimen, a horizontally oriented simple beam, receives the impact blow from a weighted pendulum. The weight or hammer, connected to the pendulum arm, is raised to an elevated position (1) and held there with a locking device. The specimen is placed at position 2. The hammer is released from position 1, goes through the specimen at position 2, traveling at a high rate of speed,[†] and rises to some height, position 3.

Although the hammer provides the weight for the potential energy, it is not what actually strikes the specimen. The hammer, often of an inverted U shape,

---

[†]Approximately 5.2 m/sec (17 ft/sec).

F I G U R E F–1 Schematic drawing of an impact test

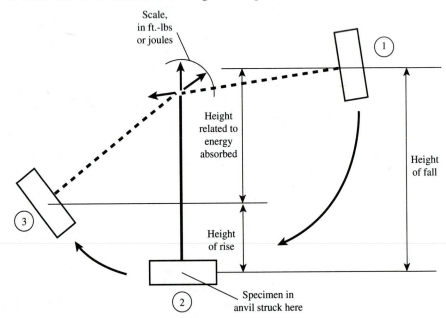

has a small, bullet-shaped striker tucked up inside the U that contacts the specimen.

The Charpy simple-beam specimen (Figure F-2) is commonly used in the United States. In the Charpy test (see Figure F-3) the specimen is struck on the side opposite from the notch. The Izod specimen (Figure F-4) is a cantilever beam oriented vertically that is favored in the United Kingdom.[‡] In the Izod test the specimen is held in a vise with the notch just even with the top of the vise and facing the hammer. (This orientation is opposite from the Charpy.) The hammer strikes the specimen a specified distance above the notch. Both tests measure the same properties of metals, but the data are not interchangeable.

F.2   MEASURING ENERGY ABSORBED

At its elevated height, position 1 of Figure F-1, the hammer represents potential energy. As the hammer descends, this energy becomes kinetic energy. At the bottom of its arc, just before the hammer strikes the specimen, position 2, its velocity is at a maximum, and except for windage and bearing losses, its kinetic energy equals the original potential energy. When the hammer strikes the specimen, energy is absorbed. Assuming the tester breaks the specimen, the specimen may fly some distance away, and the hammer swings through and rises to some height, position 3. This last height is a measure of the potential energy that *remains;* i.e., it is the portion *not* used in breaking the specimen. Thus the energy absorbed by the specimen is indicated by the *difference* between the starting and finishing heights of the hammer; this height is identified in Figure F-1.

If no specimen is placed on the anvil at position 2, the hammer will swing through to a height that, except for losses, equals the starting height; no energy is absorbed or expended since the difference in the height before and after is zero. On the other hand, if a specimen that the tester cannot break is placed on the anvil, the hammer will go no farther than position 2, and the full amount of the original potential energy will be absorbed by the specimen, tester, and its anchors.

To determine test results easily, a pointer and curved scale are provided that directly indicate the energy absorbed. As just described, zero is the point on the scale to which the pendulum swings when no specimen is present; the maximum-energy end of the scale is where the pendulum hangs vertically, just before it would strike the specimen. Between these two points, the scale is calibrated to convert the difference in heights to energy absorbed. This scale is indicated in Figure F-1.

The numerical results of the impact test are in energy units, i.e., foot-pounds (English) or joules (SI).[§] These numerical results can be used to compare the

[‡]George E. Dieter. *Mechanical metallurgy* (3d ed.). New York: McGraw-Hill, 1986, 472.
[§]The conversion is foot-pounds $\times$ 1.356 = joules.

energy-absorbing characteristics of metals (or other materials) and to specify or select metals; however, they cannot be used for design, as a yield strength is.

A plot of these energy data versus test temperatures is one of the methods used to determine the ductile-brittle transition temperature for BCC metals.

F.3   SPECIMENS

Figure F-2 illustrates the size and configuration of three types of Charpy impact specimens. The specimen is basically a bar, 10 mm (1 cm) square by approximately 2 in. long, with a notch or slot at the midpoint of one face. The purpose of this stress concentration is to create a complex stress distribution in the specimen which makes it more difficult for a ductile failure. Or, in the case of BCC specimens, it *raises* the temperature required for it to be a ductile failure.

The ASTM Standard E23 (volume 03.01) specifies in detail how specimens are prepared. From a practical standpoint, since the test is primarily a comparative one, other stock sizes can give repeatable results. For example, the U.S. standard ⅜-in.- (0.375-in.-) square steel bar is close to the specified 10 mm (0.394 in.), readily available and inexpensive for laboratory purposes.

Although three types of specimens are covered in the ASTM standard, the V-notch is the most popular. In order to form the V-notch or rounded-slot specimens, milling cutters with the required contours are required. The keyhole slot, on the other hand, can be generated by drilling a hole of the required diameter and then sawing to the hole with a band saw. However, the keyhole and machined-slot configurations can be *approximated* with a saw cut of proper depth; the depth of

F I G U R E F-2   Three types of Charpy impact test specimens

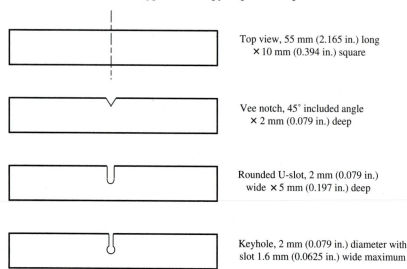

Top view, 55 mm (2.165 in.) long
× 10 mm (0.394 in.) square

Vee notch, 45° included angle
× 2 mm (0.079 in.) deep

Rounded U-slot, 2 mm (0.079 in.)
wide × 5 mm (0.197 in.) deep

Keyhole, 2 mm (0.079 in.) diameter with
slot 1.6 mm (0.0625 in.) wide maximum

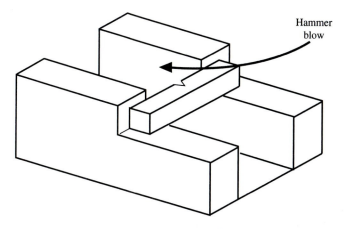

FIGURE F-3

Charpy impact specimen on the anvil, with notch or slot to the left

cut can be closely duplicated using a jig with a band saw. The impact data resulting from such specimens are repeatable, making it an inexpensive way to demonstrate the principles involved.

Figure F-3 shows a Charpy specimen placed on the anvil. The requirements of the anvil are rather straightforward—to support the specimen so the hammer can strike it and continue its swing unimpeded. However, when the specimen breaks, and either folds around the end of the striker or separates into two pieces, it may jam between the sides of the anvil and the hammer. Therefore a shroud, not shown in Figure F-3, is added to allow enough space for the specimen to deform, if that is its mode of failure, or if it separates, to keep it inside the anvil until the striker has passed through.

Figure F-4 illustrates the Izod specimen in position. The notch for the Izod specimen has the same dimensions as for the Charpy V.

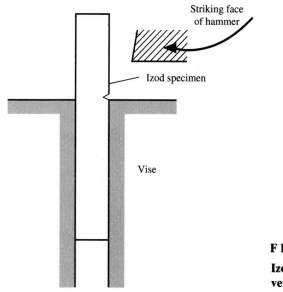

FIGURE F-4

Izod impact specimen in its vertical position in a vise

F.4   CONDUCTING THE TEST

The Charpy test is conducted in the following sequence:

1. Select the specimen and record the appropriate data.
2. Place the specimen on the anvil with the notch or slot away from the hammer and in the center of the anvil; the hammer should strike the back side of the specimen immediately opposite the notch or slot.
3. Before the specimen is placed on the anvil, make sure the hammer is locked and cannot swing through the anvil; an easily added safety device is a block of wood placed across the entry end of the anvil.
4. Place the indicator or pointer at the maximum end of the energy scale.
5. Stand out of the way of the hammer and release the hammer.
6. After the specimen is broken, catch and hold the hammer on the back swing, so someone is not struck by the swinging hammer.
7. Read the indicator before latching the hammer in the upper position, since the latching may cause the indicator to move, changing the reading.
8. Record the energy reading, retrieve the specimen, and note the condition of the fractured surface.

The impact test is often performed with specimens at differing temperatures, in which case steps 2, 3, and 4 must be well coordinated. The ASTM standard allows five seconds to remove the specimen from the temperature-controlling device, locate the specimen on the anvil, and release the hammer.

F.5   DETERMINING AND INTERPRETING RESULTS

Numerical Values. The energy data collected indicate toughness, or energy-absorbing capability, in a rather straightforward manner. The significant thing here is how the data vary with respect to test temperature. Although plain carbon steels are normally considered to be ductile, when impact-tested at room temperature, they usually fail in a brittle mode.

Figure F-5 is the typical result of plotting the energy data, in foot-pounds or joules, versus the test temperature, for plain carbon steels. At higher temperatures, where impact strengths are high, ductile failure occurs. At lower temperatures, where impact strengths are low, brittle failure occurs. The transition between these two types of failure occurs over a range of temperatures. At temperatures above and below this transition, the impact strengths are at two decidedly different levels; within the transition range the impact strength ramps down rapidly as temperature decreases.

Transition Temperature: Two Definitions. The temperature at which the transition in strength occurs is defined as the **transition temperature.** Unfortunately, there are (at least) three ways it is defined; two of those make use of the energy

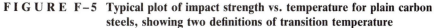

FIGURE F–5 Typical plot of impact strength vs. temperature for plain carbon steels, showing two definitions of transition temperature

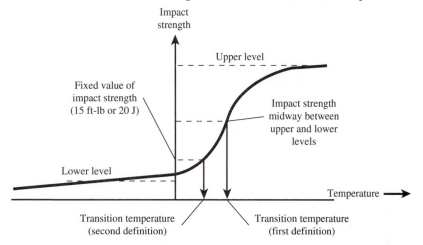

data recorded during the impact test and are discussed here; the third will be discussed below.

The first definition of transition temperature is the temperature at which the impact strength is halfway between the upper and lower levels; this midpoint is shown on Figure F-5. The second definition is the temperature that occurs at a specified value of impact strength. In Figure F-5 the specified value is 15 ft-lb or 20 J; other values can be used but this criterion was used in the investigation of the ship failures of World War II described in Chapter 12.

Distortion. In addition to using numerical data to interpret the ductile-brittle transition, the appearance of the specimen itself can be used as an indicator. Brittle behavior is characterized by lack of distortion. The two halves of impact specimens that have failed by complete cleavage can be placed together and the specimen will be almost straight, indicating that very little distortion occurred during the fracture.

On the other hand, a severely bent ductile bar gets narrower at the outer surface of the bend, and wider at the inner surface (see Figure F-6). In the same way, ductile Charpy specimens become narrower on the notch or slot side and wider on the struck side. The extent of this narrowing/widening is an indication of the amount of ductility, which can be measured and stated as a percentage of the original specimen width for each test temperature.

Fracture Surface. In brittle failure, the metal cleaves on crystallographic planes. Overall the resulting fracture surface is relatively flat without large undulations or gross irregularities. However, the cleaved surface sparkles and has a rough texture; for a steel specimen it is somewhat like looking into a bowl of light-gray sugar.

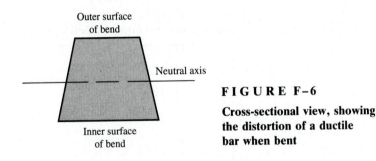

Outer surface
of bend

Neutral axis

Inner surface
of bend

F I G U R E F–6

**Cross-sectional view, showing
the distortion of a ductile
bar when bent**

Ductile failure occurs because of the presence of shear forces which enable slip to occur; final failure occurs because basic atomic forces holding the metal together have been exceeded. The resulting distortion gives the surface of a ductile failure contours that may contain gross irregularities. Often the shear forces cause projections on the failed surface at the traditional maximum shear-stress angle of 45°. In a steel specimen the ductile area has a rough, dark-gray appearance.

In the temperature range where BCC metals are in transition between ductile and brittle failures, the surface of the specimen has an appearance that is a combination of the two conditions just described. That is, at some high temperature the specimen fails ductilely, and the fracture surface is distorted and has a dark-gray color; as the test temperature is lowered, a small amount of light-gray, cleavage-type failure begins to appear, approximately in the center of the fracture surface. As the temperature is further lowered, the amount of the light-gray surface increases and the dark gray becomes a border around the edge. Eventually a temperature is reached where the light gray covers the whole fractured surface; i.e., there is complete brittle failure. Figure F-7 is a schematic representation of the fracture surface of a specimen intermediate between ductile and brittle.

Figure A-40 shows the fracture surface of steel specimens at six temperatures. As the temperature increases, distortion and the percent of the surface that is dark gray increases; both of these are indications of increased ductility.

The sawed or milled surfaced that makes up
approximately 20% to 50% of the thickness,
depending on the type of specimen.

Shear area; dark gray outer ring,
indicating ductile failure.

Cleavage area; sparkling, shiny, light gray
center area, indicating brittle failure.

Impact force
applied here

F I G U R E F–7

**Schematic representation
of the cleavage and shear
areas of a fractured
impact specimen**

Transition Temperature: A Third Definition. By a third definition, transition temperature is that temperature where the fracture surface is half ductile and half brittle. This requires that the brittle and ductile fracture areas be determined in order to determine the transition temperature. These areas can be found by four different methods:

1. Actually measure the size of the light-gray area with a scale or vernier and express it as a percentage of the original area of the specimen (minus the area taken up by the notch or slot).
2. Compare the specimen to a fracture-comparator chart similar to that available in the ASTM standards, Standard E23, volume 03.01.
3. Magnify the surface and use an overlay chart.
4. Photograph the fracture surface at some magnification and then use a planimeter to measure area.

SUMMARY

1. The impact test is used primarily to determine the ductile-brittle transition temperature for ferrous metals, but it is also useful in studying the notch sensitivity of metals in general.
2. The Charpy test is most often used in the United States. Three specimen types are standard: V-notch, rounded slot, and keyhole slot. The V-notch is the most popular.
3. The data that results from the test of the specimen give the impact strength, in energy units, of that specimen.
4. Ductility or lack of ductility can also be measured by the extent of distortion of the specimen.
5. Test results can also be interpreted by examination of the fracture surface. Ductile failure is indicated by a dark-gray surface; brittle failure is indicated by a crystalline-looking light-gray surface.
6. In the region of the transition temperature, the fracture surface has a dark-gray border with a light-gray center.
7. The transition temperature can be determined as
   • the temperature that occurs midway between the ductile and brittle impact-strength levels
   • the temperature where the impact strength reaches a certain fixed minimum value, e.g., 15 ft-lb
   • the temperature where the fracture surface of the specimen is half ductile (dark gray) and half brittle (crystalline white)

REVIEW QUESTIONS

1. What is the purpose of the notch or slot in the Charpy specimen?
2. What data or information can be obtained from the impact test? What is the data used for?

3. What is the general purpose of the impact test? Why is it used?

4. Describe the appearance of a ductile failure in a ferrous specimen. Describe the appearance of a brittle failure. Describe a combination ductile-brittle failure.

5. Sketch the curve of impact strength vs. temperature for a low-carbon steel, an alloy steel, and a nonferrous metal. (See Chapter 12.)

FOR FURTHER STUDY

Annual book of ASTM standards, volume 03.01. Philadelphia: American Society for Testing and Materials, Standard E23.

Avner, Sidney H. *Introduction to physical metallurgy* (2d ed.). New York: McGraw-Hill, 1974, secs. 1.33 and 17.3.

DeGarmo, E. Paul, J. Temple Black, and Ronald A. Kohser. *Materials and processes in manufacturing* (7th ed.). New York: Macmillan, 1988, 51.

Dieter, George E. *Mechanical metallurgy* (3d ed.). New York: McGraw-Hill, 1986, chap. 14.

Flinn, Richard A., and Paul K. Trojan. *Engineering materials and their applications* (3d ed.). Boston: Houghton Mifflin, 1986, sec. 3.15.

Reed-Hill, Robert E., and Reza Abbaschian. *Physical metallurgy principles* (3d ed.). Boston: PWS-Kent, 1992, 827–33.

G L O S S A R Y

Å Angstrom unit, 10^{-8} cm or 10^{-10} m.

A_1 On the iron–iron carbide phase diagram, the designation used for the lower temperature-transformation line (eutectoid temperature) for hypoeutectoid steels. Under slow-cooling, austenite transforms to ferrite and cementite at this temperature.

A_3 On the iron–iron carbide phase diagram, the designation used for the upper transformation line for hypoeutectoid steels. It is the limit of saturation of ferrite in austenite; in cooling across the A_3 line, ferrite precipitates from austenite.

$A_{3,1}$ On the iron–iron carbide phase diagram, the designation used for the lower temperature-transformation line (eutectoid temperature) for hypereutectoid steels. Under slow-cooling, austenite transforms to ferrite and cementite at this temperature.

A_{cm} The designation used for the upper transformation line for hypereutectoid steels. This is a solvus line and is the limit of saturation of cementite in austenite; in cooling across the A_{cm} line, cementite precipitates from austenite.

abrasive wear A type of wear caused by moving intersurface elements (particles) that remove metal.

acicular Needlelike.

active The opposite of passive; a condition whereby the metal surface is not protected from corrosion.

adhesive wear See galling.

ADI See austempered ductile iron.

aging or age-hardening See precipitation-hardening.

alclad A coating applied, especially to aluminum alloys, that is anodic to the base alloy and thus protects it from corrosion.

allotropy An ability of some metals to change from one crystal structure to one or more other crystal structures and then change back again. See polymorphism.

alloy A combination of two or more elements, one of which is metallic.

alloy steels In general, steels that have alloying elements added to improve hardenability, hardness, toughness, etc., beyond what would be found in carbon steels; the AISI/SAE series of alloy steels.

alpha brass Brass containing less than about 35% Zn whose microstructures are the alpha solid solution.

alpha iron BCC form of iron below 1666°F (908°C); also called *ferrite*.

alternating stresses   A general term to describe stresses that vary in intensity and/or direction and that can do so in a number of different patterns.

annealing   In general, a thermal treatment intended to soften or remove residual stresses from metals. See recovery, recrystallization, grain growth stages of annealing.

anodic   Descriptive of the more easily corroded metals at the top of the electromotive force and galvanic series; a metal is said to be *anodic* to those below it in the list.

anodizing   An electrochemical process that forms an oxide coating on the surface of a metal, especially aluminum alloys.

artificial aging   After solution heat treatment, a precipitation-hardening process requiring additional heating to reach optimum properties. Contrast with natural aging.

asperities   In wear, the peaks or projections from the surface of metal that cause its roughness.

atom diameter   The distance between the centers of the nuclei of atoms when they are at their minimum energy, or closest-approach distance from each other; the atoms are assumed to be hard balls.

atom lattice   A three-dimensional repeating crystalline or geometrical structure in which atoms are located at specific points.

austempered ductile iron (ADI)   Ductile or nodular cast iron heat-treated to form a bainitic matrix, which gives it high strength and high toughness.

austempering   Cooling austenite such that it transforms to bainite, usually by quenching to a temperature above M_s and holding; produces steel with a good combination of strength, hardness, and toughness, without the internal stresses of martensite.

austenite   An interstitial solid solution of carbon in FCC-gamma iron; exists above 1333°F (723°C) under normal conditions and can dissolve 2.0% C at 2065°F (1130°C). All steels freeze as austenite.

austenitizing   Heating steel to a temperature above the A_3/A_{cm} lines and holding it long enough that it all transforms to austenite.

bainite   A steel microstructure; nonequilibrium transformation product of austenite to a mixture of ferrite and cementite occurring at temperatures below those where pearlite forms and above those where martensite forms. At higher temperatures it has a feathery appearance, at lower temperatures it is acicular, or needlelike.

basal plane   The hexagonally shaped faces of the hexagonal lattice, or the {0001} planes.

base   Used to describe the more easily corroded (anodic) metals in the upper portion of the electromotive or galvanic series; used in the sense of "low-grade."

BCC   Body-centered cubic atom lattice; eight atoms form a cube with another atom in the center of the cube.

BCT   Body-centered tetragonal atom latice; similar to BCC except one dimension is greater so it is a rectangular rather than cubic solid.

beach marks   The pattern left by a fatigue crack as it propagates through the metal; the failed surface has a pattern similar to that left on a beach by wave action.

binary   Generally meaning "two"; an alloy of two metals, e.g., iron and carbon, is a binary alloy.

blue brittleness   A loss of toughness caused by tempering medium- or high-carbon steels in the range of 450 to 700°F (230 to 370°C).

body-centered cubic lattice   See BCC.

body-centered tetragonal lattice   See BCT.

bonding   Coming together of atoms to form salts, gases, and metals, requiring the ionic, covalent, and metallic bonds, respectively. Molecular, or van der Waal's, bonding is a weak bond used by inert gases and molecules with completed outer shells.

brale In Rockwell hardness testing the diamond penetrator used with the C, A, and D scales.

brass Alloys of copper, with zinc as the major alloying element; usually identified by the ratio of the two metals, e.g., 70-30, or by a name, e.g., cartridge brass.

brittle Descriptive of metals that do not undergo deformation before final failure or rupture; the opposite of ductile.

bronze Alloys of copper, with major alloying elements other than zinc, the major alloying element is usually a prefix in its name, e.g., tin bronze and aluminum bronze.

carbon steels See plain carbon steels.

carbonitriding A process raising the carbon and nitrogen content of the surface of a ferrous alloy by holding it at a temperature above A_1 in the presence of a medium containing carbon and nitrogen; parts are later quenched.

carburizing A process raising the carbon content of the surface of a ferrous alloy by holding it at a temperature above A_1 in the presence of a medium containing carbon; subsequent quenching converts the high-carbon surface to hard martensite while the center stays softer.

case hardening A general term for processes used to increase the hardness of the surface, or case, of a ferrous alloy.

cathodic Descriptive of the more corrosion-resistant metals toward the bottom of the electromotive force or galvanic series; a metal is said to be cathodic to those above it in the list.

cathodic protection The coupling of an otherwise anodic metal to a more anodic metal, i.e., one higher on the galvanic series, in order to convert it to a cathode and prevent corrosion.

cavitation corrosion A wear and corrosion phenomenon due to the collapse of vaporous bubbles against a metal surface in the presence of a liquid electrolyte.

cavitation erosion See cavitation corrosion.

cell See galvanic cell.

cementite An interstitial compound, Fe_3C-iron carbide, with a composition of 6.67% C.

chipping abrasion Abrasive wear in a metal brittle enough that the abrading particle causes the metal to chip away.

close-packed hexagonal lattice See CPH

closest-approach distance See atom diameter.

cold-short A brittle condition of some metals that occurs at temperatures near room temperature; phosphorus makes steels cold-short.

cold work In deformation processes, e.g., rolling or drawing, the change in area caused by the process divided by the original area at zero cold work, i.e., when the metal was annealed; often expressed as a percentage.

cold-working A plastic deformation process performed at temperatures and work rates such that the metal's strength and/or hardness are higher after working than before.

columnar grains In solidification, long grains or crystals which form because of very strong directional cooling such that the grains have the appearance of columns.

complete (solid) insolubility The solubility condition that exists when different metals are not soluble in any amount in the solid state.

complete (solid) solubility The solubility condition that exists when different metals in any combination or composition can form solid solutions; that is, their solution never becomes saturated.

complex stress distribution Stresses on more than one axis.

compound A homogeneous, i.e., single-phase, substance containing discrete ratios of atoms of two or more elements.

compressive strength The stress equal to the maximum compressive force required to shatter a specimen divided by its original area. Metals that readily deform do not have a compressive strength; they just keep deforming until squashed flat.

contact fatigue In wear, fatigue that occurs because of the cyclical application of a moving (usually rolling) load.

contact wear A type of wear caused when surface and countersurface are in intimate contact and have relative motion.

continuous yielding behavior In the loading of a metal, as the stress approaches the elastic limit the stress continues to increase; contrast with discontinuous yielding.

cored crystal A grain or crystal whose composition varies from the center, or core, to its outer portions because of nonequilibrium freezing. See coring.

coring A variation in composition from the center, or core, of a crystal to its outer portions, caused when time is not available during the freezing process for complete diffusion to occur. See cored crystal.

countersurface In wear, the surface that is in relative motion with the surface being worn.

coupled In corrosion, metals in electrical contact in the presence of an electrolyte. See galvanic cell.

covalent bonding Atomic bond involving the sharing of valence electrons to complete outer electron shells; commonly used to form gaseous and liquid molecules, e.g., NH_3, H_2O.

CPH Close-packed hexagonal atom latice, where atoms are in close-packed layers stacked *ABAB*.

creep Slow plastic flow that occurs under constant stress; most metals must be at elevated temperatures to exhibit creep.

creep curve A plot of strain (deformation) versus time at a particular temperature, the data resulting from a creep test.

creep strength The stress required to produce a specified rate of creep at a given temperature; typical units are psi for 0.1% strain/1000 h, or psi for 0.01% strain/1000 h.

critical cold work For a given metal, the minimum amount of cold work stored in the lattice that will enable recrystallization to occur at any temperature.

critical cooling rate In steel heat-treating, the slowest cooling rate that enables austenite to just reach the M_s temperature without transforming to a softer product.

critical lines See transformation lines.

critical resolved shear stress The component of shear stress, resolved in the direction of slip, necessary to initiate slip in a grain; a constant for a metal.

crystal A repetitive geometric structure; in metals a structure of atoms. See crystalline, grain.

crystal lattice See atom lattice.

crystalline Having a geometrical structure like a crystal. See atom lattice.

crystallization The process by which metallic atoms take their crystalline form, that is, solidification.

CT diagram A diagram showing the nonequilibrium transformations of austenite as a result of continuous cooling from austenite rather than transformation at one temperature. See TTT diagram.

cube face The outer surfaces of the cubic form of atom lattice, or the {100} planes.

cutting abrasion An abrasive wear caused by chips being removed from the surface by a particle.

cyaniding   A process raising the carbon and nitrogen content of the surface of a ferrous alloy part by holding it above A_1 in the presence of a medium containing carbon and nitrogen; parts are later quenched.

cyclical stresses   See alternating stresses.

decarburizing   Removal of carbon from the surface of ferrous alloys by heating in the presence of a medium that reacts with the carbon, e.g., oxygen.

deleted steels   AISI/SAE alloy steels that have been "deleted" from a list of approved alloys because their annual production fell below a benchmark; the steel may be difficult to purchase.

dendrite   A treelike atom crystal structure formed during the solidification of metals, especially if the rate of cooling during solidification is slow.

deoxidizer   An element added to a molten metal to react with oxygen and remove it, i.e., deoxidize it; Mn, Si, and Al are often added to steels for this purpose.

diffusion   In general, the spreading of one material throughout the whole; used here to describe the movement of the atoms of one element into or through another, usually at a high temperature.

directional cooling   Descriptive of the situation where heat is extracted from a solidifying metal in such a way that it travels in a particular direction.

discontinuous yielding   The condition that occurs, especially in tensile testing of hot-worked or annealed steels, when the strength, after reaching the yield point, fluctuates before increasing to maximum load.

dislocation   A region of misaligned atoms existing between otherwise properly aligned atoms.

dissimilar metals   Metals that are separated on the electromotive force series or galvanic series; the greater the separation, the greater will be the corrosion of the metal that is more anodic.

drawing   The process of pulling metal wire, bar, or tubing through a die, or forming a metal with a punch and die to create a container with a bottom, e.g., a cooking pot; also, "shop language" for tempering.

dual-phase steel   Steel, typically hypoeutectoid, cooled or quenched from an intercritical temperature, resulting in a two-phase structure of ferrite and martensite.

ductile   Descriptive of metals that undergo deformation before final failure or rupture; the opposite of brittle.

ductile cast iron   See nodular cast iron.

ductility   The ability to be plastically deformed without rupture or fracture in a tensile test.

edge dislocation   A series of dislocations that extend across a grain or crystal such that they form an "edge."

elastic deformation   Physically deforming a metal by a stress (or force) such that upon its removal the metal returns to its original size; the stress is below the elastic limit and the deformation is not permanent.

elastic limit   The highest stress that, when removed, allows the tensile specimen to return to its original length.

electrochemical   Involving the electrical phenomenon of chemicals; neutral atoms of chemicals form ions that have electrical charges.

electrode   An electrical conductor that is part of a cell.

electrolyte   A chemical, usually a liquid, that will ionize, allowing the transfer of ions and electrons.

electromotive force series   Elements and alloys ranked by the voltage (electromotive force) generated when tested in a standard cell containing a salt of the metal tested, with hydrogen as a reference.

electron   An atomic particle of negative charge and very little mass that lies in energy shells about the nucleus; behaves both as a particle and a wave.

elongation   In the tensile test the strain at failure or fracture; usually expressed as a percent; used as an indicator of ductility.

endurance limit   In fatigue testing, the applied stress below which the metal can withstand an infinite number of stress cycles.

equiaxed grains   Grains or crystals formed during freezing with little or no directional cooling such that they freeze "equal on the axes"; contrast with columnar grains.

equilibrium   Heating or cooling such that whatever phase changes will occur have time to do so, especially in interpreting phase diagrams.

erosive wear   A type of wear characterized by fluid flow; may involve corrosion.

eutectic   A reversible reaction where metals soluble in the liquid state freeze as two (insoluble) phases; the reaction occurs at a eutectic temperature and a eutectic composition. See Raoult's law.

eutectoid   A reversible reaction, occurring at one temperature, where a solid phase forms two different solid phases upon cooling.

extension   The change in length or "stretch" of the specimen in a tensile test.

face-centered cubic lattice   See FCC.

fatigue   A failure process dependent on alternating or cyclical stresses, not on age or passage of time.

fatigue strength   In fatigue testing, the applied stress required to cause failure in a specified number of stress reversals.

FCC   Face-centered cubic atom lattice; atoms are located at the eight corners of a cube, with one atom located in the center of each of the six faces; the atoms are stacked *ABCABC*.

ferrite   An interstitial solid solution of carbon in BCC alpha-iron; has a very low solubility for carbon. It is the normal form of iron at room temperature.

ferrous   Referring to metals or alloys whose composition is primarily iron.

film stage   The first stage of the quenching process; the hot metal converts the quenchant into a vapor or film on the surface of the metal, hindering heat transfer so the rate is usually slower than the second, or high-transfer, stage.

final anneal   An annealing treatment used at the end of cold-working operations to achieve completely annealed properties or an intermediate level of properties. See partial anneal.

flame hardening   One of the methods used to case-harden steels; a flame is used to austenitize the surface of the part and subsequent quenching transforms it to martensite.

fretting corrosion   Corrosion that occurs on surfaces that have very little relative motion, e.g., a gear press-fit on a shaft.

galling   Also termed *adhesive wear*, where interlocking asperities cause metals to adhere or transfer from one surface to another.

galvanic   Pertaining to an electromotive force (voltage) and/or a flow of electrons produced by the interaction of chemically different metals.

galvanic cell   Dissimilar metal electrodes coupled and capable of generating a voltage and current flow.

galvanic series   Elements and alloys ranked by the voltage generated when each is electrically connected to dissimilar metals while both are in the same electrolyte, usually seawater.

galvanizing Coating steel with zinc, anodic to the steel, which protects the steel from corrosion. See sacrificed.

gamma-prime A compound that forms between nickel and aluminum and/or titanium, e.g., Ni_3Al, Ni_3Ti, or $Ni_3(Al, Ti)$; one of the phases that strengthens the superalloys.

grain An individual crystal of a polycrystalline metal; except for interstitials, dislocations, and vacancies, atoms are in their geometric or crystalline alignment.

grain growth stage The highest temperature stage of annealing; individual metal crystals or grains grow when metals are held at high temperatures for a period of time.

gray cast iron A ferrous alloy usually containing 2.0 to 4.0% C plus Si whose free carbon is present as graphite flakes; its fracture has a gray appearance.

grooving abrasion Wear caused when sharp particles abrade the surface and cause grooving.

groups The vertical columns of the periodic table; elements in the same group tend to have the same number of valence electrons and similar properties.

H steels Steels produced within chemical specifications that also produce Jominy end-quench results within published limits; the producer is guaranteeing a certain level of hardenability. An "H" is placed after the alloy number, e.g., 4340H.

Hadfield's steel Steel containing about 12% Mn, which retards the austenite-to-martensite transformation; deformation of the steel at room temperature causes martensite to form, hardening the steel.

hard The industry standard for the cold-working of a metal or alloy. Variations from "full-hard" are expressed as fractions of hard, e.g., quarter-hard, half-hard, etc. See temper.

hard-ball diameter See atom diameter.

hardenability The property of a steel that enables it to form martensite at slower cooling or quenching rates; a TTT diagram that is moved to the right or a flatter Jominy curve indicates hardenability.

high-speed steels Tool steels typically containing chromium, vanadium, and tungsten; used because of their ability to be operated at high speeds and resist accompanying high temperatures. Some versions contain molybdenum and cobalt.

high-transfer stage The second stage of the quenching process, characterized by agitation provided by boiling but with the liquid in contact with the metal's surface; has the highest rate of heat transfer.

homogenization A thermal process primarily intended to make the chemical composition of metals more uniform by reducing or removing the segregation caused during the freezing process; will also remove residual stresses formed during the freezing process.

hot-short A brittle condition that exists in the hot-working temperature range; sulfur in steels can promote hot-shortness.

hot-working A deformation process performed at a temperature and rate of working such that the strength and/or hardness are not increased.

HSLA steel High-strength, low-alloy steel, a family of steels that in general do not get their improved properties by quenching or alloying but by microalloying and controlled cooling to refine grain size.

hydrodynamic slipper A hydrodynamic wedge formed between the surface and counter-surface, with a lubricant as an intersurface element, separating the surfaces at some combination of viscosity, load, and speed.

hypereutectic   Descriptive of an alloy that contains a eutectic structure whose percent composition is greater than the composition of the eutectic for the alloy system.

hypereutectoid   See hypereutectic; same relationship but in the solid state.

hypoeutectic   Descriptive of an alloy that contains a eutectic structure whose percent composition is less than the composition of the eutectic for the alloy system.

hypoeutectoid   See hypoeutectic; same relationship but in the solid state.

impressed current   In a corrosion cell, electrons in the external circuit flow from anode to cathode; a battery or other source of DC current can impress a current on the cell so electrons are supplied to an anodic metal, converting it to a cathode, and thus protecting it.

incipient melting   Descriptive of the melting of eutectic composition metal that occurs when a rapidly quenched alloy is solution-heat-treated at a temperature above the eutectic.

incubate   The action that initiates recrystallization; the strain energy (from the cold work) and the heat energy must accumulate for a period of time before nucleating sites appear.

indenter   In hardness testing, the tool used to press against and into the metal tested.

induction hardening   One of the methods used to case-harden steels; electrical induction is used to austenitize (heat) the surface of the part and subsequent quenching transforms it to martensite.

inoculate   The addition of a material to molten metal, usually just before pouring or casting and usually to promote the formation of nuclei during crystallization or solidification.

insoluble impurities   Gases, oxides, and the like, which do not become part of the metal's atom lattice but can provide sites for nucleation.

intercritical anneal   Annealing done by slow-cooling steels from temperatures between the critical lines, i.e., usually just above A_1 or $A_{3,1}$. The original proeutectoid ferrite or cementite is not transformed; only the portion that was pearlite transforms to austenite before cooling.

intercritical heat-treating   Heat-treating done by quenching steels from temperatures between the critical lines, i.e, just above A_1 or $A_{3,1}$. The original proeutectoid ferrite or cementite is not transformed; only the portion that was pearlite becomes austenitized and can be quenched to martensite. See dual-phase steel.

interface   A common boundary between adjacent regions; in solidification, the interface is the junction of the solid and liquid phases.

intermediate anneal   An annealing treatment done between cold-working operations. See process anneal.

intermetallic compounds   Compounds formed in discrete ratios by elements that are separated in the periodic table so that they have electrical attraction for each other. See valency compounds.

interstitial   Between the spaces; in atom lattices, the space(s) between the solvent atoms making up the lattice. Thus an interstitial particle or alloying element resides between the atoms forming the lattice.

interstitial compound   A compound formed by a transition element and elements that have small atom diameters, e.g., B, O, N, C, and H.

intersurface element   In wear, an element between the surface and countersurface, e.g., a lubricant, an abrasive particle, or an oxide.

ionic bonding   Atomic bond involving the donating of valence electrons from one atom to the other so each becomes an ion; commonly used to form compounds between elements that are separated on the periodic table so there is a high electrostatic attraction.

IT diagram   Isothermal transformation diagram. See TTT diagram.

Jominy test   A measurement of hardenability; under standard conditions a steel specimen is quenched at one end and the hardness measured relative to that end. The Jominy curve (plot of hardness vs. the position on the specimen) is useful in metal selection since it indicates the steel's hardenability. Usually done for steels, but other metals may be tested also.

journal bearing   A sleeve-type bearing, usually made of a soft metal, that encloses a portion of a shaft and relies on a hydrodynamic slipper to reduce friction and prevent excessive wear.

killed steel   A steel sufficiently deoxidized by metals such as silicon or aluminum that no reaction occurs between carbon and oxygen during solidification.

kinetic energy   Energy dependent on the motion of a body or mass.

ksi   Abbreviation for thousands of pounds per square inch; an English unit for stress.

latent heat of fusion   The heat that must be added or removed just to change between the liquid and solid states.

lattice parameter   The length along an axis or edge of an atom lattice; a measure of the size of the atom lattice, usually given in angstrom units (Å).

ledeburite   The eutectic of the iron—iron carbide alloy system; occurs at 4.3% C and 2065°F (1130°C).

Liberty ships   Cargo ships and tankers constructed to ferry large numbers of personnel and supplies overseas during World War II.

liquid state   The condition of matter that is not constrained but free to flow; in metals the condition where atoms are not in a crystalline form but can move in a random manner.

liquid-cooling stage   The third stage of the quenching process; liquid comes in contact with the metal's surface but with little agitation unless provided by circulation of the quenchant; it has the lowest rate of heat transfer.

liquidus   On a phase diagram the line representing the temperatures at which the various compositions begin to freeze on cooling, or finish melting on heating.

localized plastic deformation   See nonuniform plastic deformation.

M_f temperature   The temperature at which all the austenite has transformed to martensite; in higher-carbon steels this temperature can be below room temperature.

M_s temperature   In steel heat treatment, the temperature at which austenite begins to transform to martensite.

major load   In Rockwell hardness testing, the total load applied on the penetrator during the hardness test; standard loads are 60, 100 and 150 kg.

malleability   The ability to be plastically deformed by compressive forces, e.g, by hammering.

malleable cast iron   A ferrous alloy usually containing 2.0 to 4.0% C and low Si produced by tempering white cast iron, i.e., holding at an elevated temperature, which converts the cementite to temper carbon, or carbon in an irregular shape or clump.

maraging steel   Steel containing large amounts of nickel that forms a soft martensite, which is hardened by precipitation of intermetallic compounds created by elements such as Al, Mo, Ti, and Co; has high strength, with the final thermal treatment done at about 900°F.

marquenching   See martempering.

martempering   The quenching of austenite, interrupted at a temperature above M_s to allow the surface and core of the part to equalize in temperature; the transformation to martensite thus occurs more uniformly with time, reducing the level of internal stresses.

martensite   A microstructure of steel; a supersaturated solid solution of carbon in BCT iron, which has a hard needlelike, or acicular, microstructure. See tempered martensite.

matrix   A background phase or alloy in which another material is present; e.g., steel is the matrix for the graphite flakes of gray cast iron.

metallic bonding   The bond typical of metals involving the attraction of atom nuclei and inner electron shells (with a net positive charge) to a cloud of valence electrons (with a negative charge).

microalloying   The process in which small amounts of alloying elements such as Nb, Al, V, and Ti are used to control the grain size of steels, especially HSLA, during hot-processing.

Miller indices   A system of numerical notations and brackets used to identify planes and directions in crystal lattices. The reciprocal of the intercepts of a plane in a unit cell identifies that plane; the direction perpendicular to that plane is of the same number but has different brackets.

minor load   In Rockwell hardness testing, a 10-kg preload to establish the reference point for the depth measurement. It remains on during the reading of the hardness to ensure the penetrator is seated in the impression.

modulus of elasticity   The slope of the linear portion of the stress-strain diagram, equal to the change in stress divided by the change in strain for the elastic portion of the stress-strain diagram.

modulus of resilience (M of R)   A measure of the ability to absorb energy and remain elastic, indicated by the area under the elastic portion of the tensile stress-strain curve; units are psi/in.^{3}

modulus of strain-hardening   The slope of the plastic portion of the true stress-strain curve; it measures the ability of the metal to be strengthened by plastic deformation or strain-hardening.

modulus of toughness (M of T)   A measure of the ability to absorb energy if loaded to fracture, indicated by the area under the tensile stress-strain curve; units are psi/in.^{3}.

monel   An alloy of 70% nickel-30% copper; has good corrosion resistance and good strength.

multiaxial stress distribution   See complex stress distribution.

natural aging   After solution heat treatment, a precipitation-hardening process that occurs "naturally," i.e., at room temperature, especially in aluminum alloys. Contrast with artificial aging.

necking   See nonuniform plastic deformation.

neutron   An atomic particle of neutral charge and high mass contained in the nucleus.

nitriding   A process raising the nitrogen content of the surface of a ferrous alloy part by holding it below A_1 in the presence of a medium containing nitrogen. The purpose is to form hard nitrides; no subsequent heat-treating is required.

noble   Descriptive of the more corrosion-resistant (cathodic) metals in the lower portion of the electromotive or galvanic series; used in the sense of "high grade" or "important."

nodular cast iron A ferrous alloy usually containing 2.0 to 4.0% C, plus Si and inoculants, whose carbon is present as graphite nodules or spheroids; also referred to as *ductile cast iron*.

nonferrous Referring to metals or alloys whose composition is primarily of a metal other than iron, though iron may be an alloying element in a nonferrous alloy.

nonmetallic A general term used to describe materials that have properties unlike metals; e.g., metals generally have plasticity; materials described as nonmetallic tend to be hard and brittle and lack plasticity.

nonuniform plastic deformation In the tensile test the condition that occurs after maximum load is reached; the specimen area reduces locally, not uniformly along its length; i.e., necking takes place.

normalizing Air cooling of a steel from austenite, i.e., from a temperature above A_3/A_{cm}; this cooling is typical of the treatment received by many hot-worked steels that are worked at higher temperatures as austenite and then allowed to air-cool.

nucleating sites Sites or locations about which grains or crystals begin to form; they occur during solidification (crystallization) and during the second stage of annealing (recrystallization).

nucleus The positively charged central portion of the atom containing protons and neutrons; has most of the atom's mass.

octahedral planes The major slip planes for the FCC lattice (the {111} planes), so called because of the eight corners in a cube and thus eight {111} planes in a cubic lattice.

offset The strain value used in determining the yield strength; usually 0.002 in./in. (0.2%). Other values used are 0.001 in./in. and 0.005 in./in.

orange peel A rough surface condition, normally undesirable, developed during the forming of metals with large grain size.

orbit See shell.

ordering The condition by which some alloys form solid solutions having discrete ratios of the atoms of the metals, e.g., 1:1, 2:1, etc., and the atoms repeat in the atom lattice in this ratio; for example, at 1:1 the atoms of two metals would alternate, i.e., every other one, throughout the lattice.

overaged temper After solution heat treatment, an artificial aging treatment applied especially to high-strength aluminum alloys, that uses a higher aging temperature to achieve "overaging." The treatment improves corrosion resistance, with some reduction in strength.

overaging After solution heat treatment, performing the artificial treatment at too high a temperature or for too long a time; optimum mechanical properties are not achieved.

oxidation Combining a substance with oxygen, e.g., heating a metal in air. Also, a chemical reaction in which a substance loses electrons.

packing factor A measure of the density of the atom lattice structure; equal to the volume of the atoms in the unit cell divided by the volume of the cell.

partial anneal A final annealing treatment used to achieve a product with a certain level of properties as opposed to a fully softened product.

partial (solid) solubility The solubility condition where the metals have solid solubility only over a range of compositions, not all compositions; that is, the solution formed is capable of becoming saturated.

passivation A process that forms a nonporous film on metals that serves to protect them from further corrosion.

passive The condition achieved by passivation when the metal is protected from further corrosion, or is passive; the opposite of active.

pearlite   A steel microstructure; under slow-cooling of a carbon steel, the laminated mixture of ferrite and cementite formed at the eutectoid composition (0.8% C) and temperature (1333°F or 723°C).

penetrator   In Rockwell hardness testing, the tool used to penetrate the surface of the metal; the depth of this penetration is an inverse indication of the hardness of the material.

periods   The horizontal rows of the periodic table; they are the same as the shells or principal quantum number.

peritectic   A reversible reaction, occurring at one temperature, where a liquid and a solid phase freeze into one, different solid phase; often occurs when the two metals have a large difference in melting temperature. The single phase can be a solid solution or a compound.

peritectoid   A reversible reaction, occurring at one temperature, where two solid phases form one, different solid phase on cooling.

phase   A homogeneous, uniform, separate, and distinct metallic structure; metallic single phases are pure metals, solid solutions, and compounds.

pickling   The removal of surface contaminants (oxides or scale) by applying an acid or other solution; especially used on steels.

pipe   A shrinkage cavity, formed especially in ingots, where the surface metal freezes and does not permit molten metal to feed the center of the ingot.

pitting corrosion   Localized corrosion due to variations in electrical potential caused by nonhomogeneity of the metal's surface.

plain carbon steels   Steels whose principal alloying element is carbon, though other elements (e.g., Mn, Si) may be present to control sulfur and oxygen.

plastic deformation   Physically deforming a metal by a stress (or force) above the elastic limit, such that upon removal of the stress the metal does not return to its original condition and deformation is permanent.

plasticity   The ability to be plastically deformed without failure or rupture.

plowing abrasion   A form of abrasive wear where the particle makes a furrowlike groove without removing any metal.

polarization   In corrosion, the condition that results when electrons flowing in the external circuit of a galvanic cell are not removed from the cathode; the accumulation of electrons on the cathode stops both the electron flow and the corrosion, and the cell is said to be polarized.

polycrystalline   Descriptive of the more common form of metals; metallic microstructures containing more than one grain or crystal.

polymorphism   An ability to change from one form to another, but not necessarily change back again.

potential energy   Energy dependent on the location or position of a body or mass.

precipitation-hardening   Hardening that takes place by the precipitation of a supersaturated phase from a solid solution. Because in some alloy systems this can occur with the passage of time at room temperature, it is often termed *aging*.

preferred direction   The specific direction in which a metal prefers to freeze; for the cubic metals it is ⟨100⟩, for hexagonal metals it is ⟨10$\bar{1}$0⟩.

preload   See minor load.

primary arm   The center or "trunk" arm of a dendrite.

primary crystals   The first crystals to form from the melt during solidification, e.g., primary alpha, primary austenite. See primary solid solutions.

primary solid solutions   Solid solutions at the extremes of composition of the alloying metals; on the phase diagram they are immediately adjacent to the pure metals and thus termed *primary*. See primary crystals, terminal solid solutions.

primary stage of creep   The first stage of creep that begins after the initial elastic reaction to the load. Some work-hardening occurs, which slows the rate of deformation, reducing the slope of the creep curve.

prism   A crystalline solid with three or more faces parallel to an axis.

prism face   The nonhexagonal faces of the hexagonal lattice, or the rectangularly shaped faces of the hexagonal lattice, or the $\{10\bar{1}0\}$ planes.

process anneal   An anneal used in a sequence of cold-working steps to soften the metal so further cold work can be done; a recrystallization-type anneal. See intermediate anneal.

proeutectic   During cooling, a phase that forms at temperatures above the eutectic temperature.

proeutectoid   During cooling, a phase that forms at temperatures above the eutectoid temperature.

proportional limit   The stress where the proportionality of the elastic (straight-line) portion of the tensile stress-strain diagram stops.

proton   An atomic particle of positive charge of high mass contained in the nucleus; its charge is opposite and equal to that of the electron.

psi   Abbreviation for pounds per square inch; an English unit for stress and pressure.

purity   The extent to which a metal's atom lattice contains only atoms of that element; solute atoms of another metal in the lattice lower the purity.

quanta   Plural of quantum; the increments by which electron energies vary; quanta are related to the wavelength of the electron.

quenchant   The medium in which metals are quenched during heat-treating, e.g., air, water, oil, or a special polymer.

quenching medium   See quenchant.

Raoult's law   The temperature at which a pure metal freezes will be lowered by the addition of a second metal if the two metals are soluble as liquids and insoluble as solids. Metals that obey Raoult's law form eutectics.

recovery stage   The stage of annealing or heating that occurs at the lower end of the annealing temperature range; it relieves residual stresses and causes little change in mechanical properties.

recrystallization stage   The stage of annealing or heating that occurs at an intermediate temperature range and causes the largest reduction of mechanical properties; atoms recrystallize into new unstrained lattices. Prior cold work is required for recrystallization to occur.

reduction of area (R of A)   In tensile testing, the percent change in area of the specimen at failure; an indicator of ductility. In cold-working, the percent change of area through an operation; if the metal was annealed before the operation, R of A is equal to percent cold work.

residual stress   Elastic or locked-in stress caused by rapid cooling, welding, machining, etc.; can be relieved by recovery or stress-relief anneals.

retained austenite   Particularly in higher-carbon steels the austenite that is not transformed to martensite upon fast cooling and is then retained at room temperature, softening the metal.

sacrificed   In corrosion prevention, purposely connecting an anode (a metal higher in the galvanic series) to protect the metal of concern, which thus becomes a cathode; hence the anodic metal is "sacrificed" to protect the metal of concern.

saturated   The state whereby, for a given temperature and pressure, a solution cannot dissolve any additional solute.

scale   A thick layer of oxidation products formed on the surface of metals at high temperatures.

screw dislocation   A series of dislocations with nonuniform spacing, i.e., a taper, that progresses through the crystal in a spiral like a screw thread.

secondary arm   The portion of a dendrite that branches from the primary arm; the "limbs" of the treelike dendrite.

secondary hardening   Tempering of some steels at high temperatures to develop complex carbides to give the steels a higher hardness than if tempered at a lower temperature; the reverse of what would normally occur.

secondary stage of creep   The second stage of creep; the portion of the creep curve with a uniform slope, i.e., a condition of steady-state strain.

segregation   A nonuniform distribution; a separation; alloying elements, impurities, or phases may become segregated.

shell   The orbit, or energy state, of an electron that is its principal quantum number n. The maximum number of electrons that can be in any shell is given by $2n^2$.

simply loaded   A condition whereby the applied force generates a stress that acts on only one axis.

slip   The sliding of a layer or layers of atoms over each other because of a shearing component of an applied force (or stress); also, the progressive movement of dislocations across a lattice.

slip planes   The planes within the various lattices on which slip occurs. The major planes are {111} for FCC, {110} for BCC, {0001} for CPH.

slip system   The combination of the plane and direction on which slip occurs. The major slip systems are {111} and ⟨110⟩ for FCC, {110} and ⟨111⟩ for BCC, {0001} and ⟨112$\bar{0}$⟩ for CPH.

slipper effect   See hydrodynamic slipper.

slurry erosion   Wear that occurs when a liquid stream carries solid particles against a metal surface; the liquid may also contribute to corrosion.

S-N diagram   A diagram of stress versus number of stress cycles to cause failure; the data are obtained by fatigue testing.

solid solution   Solution of metals when in the solid state; the atoms of one metal (the solute) take up residence in the lattice of another (the solvent), or the metals exist together in a lattice different from either one.

solid-solution strengthening   The strengthening that occurs when interstitial or substitutional solute atoms strain the lattice of a solvent.

solid state   The condition of matter that has shape and volume and is constrained and not free to flow; in metals atoms are in a crystalline form.

solidus   The line representing the temperatures at which the various compositions of the phase diagram finish freezing on cooling or begin to melt on heating.

solute   In a solution the substance dissolved by the solvent.

solute atoms   In a metallic solution the atoms of the metal being dissolved into the lattice of the solvent metal.

solvent   In a solution the substance that dissolves the solute.

solvent lattice   The metal in whose lattice the solute atoms take up residence.

solvus   On a phase diagram the line representing the limit of solid solubility; a line separating solid phases.

spall   Deterioration of a metallic surface by eruptions or craters formed by (1) subsurface cracks induced by contact wear or (2) oxide layers that do not adhere to the parent metal and allow corrosion to proceed.

spheroidizing   Heating and cooling treatments designed to convert cementite to a spheroidal or globular form, usually to improve machinability of steels above 0.5% C. The processes used involve holding for long periods at, or above and below, $A_1/A_{3,1}$ and slow cooling.

stacking   The order or arrangement of atoms in layers within atom or crystal lattices; the layers are designated A, B, C. The CPH lattice is stacked $ABAB$; the pattern repeats every other layer. The FCC lattice is stacked $ABCABC$; it goes through three layers before repeating.

stainless steel   A special variety of steel containing chromium to promote corrosion and heat resistance. There are three major types—ferritic, martensitic, austenitic—which are dependent on other alloying elements and treatments to achieve their properties.

stiffness   Rigidity, an elastic property of metals related to the modulus of elasticity.

strain   In the tensile test the change in the length, or extension, of the specimen divided by the original length, usually 2 in.; expressed as inch per inch or percent.

strain-hardening   See cold-working.

stress   A tensile, shear, or compressive force per area over which the force is applied; English units are psi or ksi.

stress concentration   A surface discontinuity, e.g., sharp angle, notch, change of contour, or internal discontinuity, that raises the level of stress in the local area.

stress-relief anneal   See recovery stage.

stress-rupture strength   The stress to cause rupture in a certain time at a given temperature; typical units are psi for 1000 h.

subcritical anneal   Annealing of steels done at temperatures below the lower critical or transformation temperature; recrystallization occurs because of cold work present and not because of allotropy.

subshell   In atomic structure, a division of the principal shell because of the ranges of electron energies within a shell; these energy ranges are related to the two less significant quantum numbers. See also shell.

substitutional solid solution   A solid solution that results from alloying, where solute atoms substitute into the lattice of the solvent metal; e.g., in alpha brass, zinc atoms take up residence in the FCC lattice of copper; in steels an element such as chromium takes up residence in FCC austenite or BCC ferrite.

superalloys   Alloys based on nickel, nickel-iron, and cobalt used at temperatures above about 1000°F (540°C).

superlattice   In solid solutions, an overall arrangement or ordering of the atoms superimposed on the regular atom lattice. See ordering.

supersaturated   Descriptive of a state wherein a solution contains more solute than it should be capable of holding at that temperature and pressure; supersaturation is usually accomplished by rapid cooling from a high temperature.

S/V ratio   The ratio of the surface area (length^2) of a quenched part to its volume (length^3); thus related to length^{-1} and not a constant. The larger the S/V ratio, the faster a part will quench in a given medium.

swaging   A cold-forging process where dies are driven by a rotating mechanism; commonly used to reduce the size of a round rod or tubing.

temper   The degree to which a metal or alloy has been heat-treated or cold-worked, e.g., heat-treated, heat-treated and aged, or hard, half-hard, etc. See hard.

temper carbon   The carbon formed in tempering white cast iron to produce malleable cast iron; appears as irregular shapes or clumps.

tempered martensite   The microstructure achieved by tempering the product of the transformation of austenite to martensite; martensite is tempered to reduce the stress caused by the volume increase that occurs during the transformation.

temper embrittlement   A loss of toughness caused by tempering medium- to high-carbon steels containing certain alloying elements in the range of 700–1100°F (375–575°C).

tempering   A thermal treatment of steels done at temperatures below $A_1/A_{3,1}$ for the purpose of reducing brittleness caused by previous rapid cooling or quenching processes; usually associated with the tempering of martensite; see tempered martensite.

tensile strength   The stress equal to the maximum load determined during the tensile test divided by the original specimen area.

terminal solid solutions   See primary solid solutions, which are also called terminal solid solutions, since they occur at the ends of the range of compositions.

tertiary stage of creep   The third stage of creep, which leads to separation; the slope of the creep curve goes up steeply.

three-body abrasion   Abrasion involving the surface, countersurface, and intersurface element(s).

tool steels   In general, steels of high quality used for tooling of all types; the AISI/SAE series of tool steels. Tool steels often have compositions that enable large amounts of hard carbides to be formed.

tramp elements   Unwanted elements present in metals; usually carried over into the metal from the ore or picked up from scrap as metals are recycled (remelted).

transformation lines   On phase diagrams, the lines indicating the temperatures where metals change phase as they are heated and cooled.

transition elements   Those elements in the center of the periodic table (the B groups of Table I-1) that add electrons to an inner shell rather than an outer shell as the atomic number increases.

transition temperature   The temperature of BCC metals, determined in the impact test and defined three ways: (1) the temperature at the impact strength that is midway between the upper and lower levels of impact strength, (2) the temperature that occurs at a specified level of impact strength, and (3) the temperature at which the fracture surface of the impact specimen is half ductile and half brittle.

tribology   The study of rubbing, or the study of wear in its broadest scope.

true strain   Strain as calculated by dividing the incremental change in length by the immediately preceding length, or ln L/L_0.

true stress   Stress as calculated using the actual area under load rather than the original load, as done in determining engineering stress.

TTT diagram   A time-temperature transformation diagram showing the nonequilibrium transformations of austenite at one temperature; therefore also termed an isothermal-transformation (IT) diagram.

twin or twinning   A distortion of the atom lattice of a grain or crystal, resulting in a geometric distortion that is a mirror image, or twin, of the original structure. Twinning changes the relationship of atoms within their lattice but major deformation does not occur; atoms retain the same neighbors, but at different distances.

two-body abrasion   Abrasion caused by a projection from the countersurface that abrades the surface; no intersurface element is present.

ultimate tensile strength   See tensile strength.

undercooling   A condition where, under actual cooling conditions, the metal temperature goes below the normal temperature of a phase transformation.

uniform corrosion   Corrosion occurring more or less uniformly over the surface of a metal.

uniform elastic deformation   In the tensile test, the deformation that occurs during elastic strain; the change in area of the specimen is approximately uniform along the test length.

uniform plastic deformation   In the tensile test, the deformation that occurs during plastic strain before maximum load is reached; the change in area of the specimen is approximately uniform along the test length.

unit cell   The number of atoms that make up the repeating cell of an atom lattice, found by summing the fractional atoms that make up or belong to any one lattice; in CPH and BCC it is 2; in FCC it is 4.

UNS   Unified numbering system used to identify metals and their alloys; can be found in various handbooks and publications, e.g., the SAE handbook and the ASTM series of standards.

unsaturated   A condition whereby, for a given temperature and pressure, a solution can dissolve additional solute.

vacancy   An unoccupied atom site in a crystal lattice.

valence electrons   The electrons located in the outermost, i.e., highest, energy shell of the element which interact with the electrons of other metals in the metallic bond.

valency compound   Compounds formed by following the normal valence rules as identified in the periodic table. See intermetallic compounds.

vapor-transport stage   See high-transfer stage.

white cast iron   A ferrous alloy, usually containing 2.0 to 4.0% C and low Si, whose carbon is present as cementite; its fracture surface has a white appearance.

X-ray diffraction   Electromagnetic energy of short wavelength diffracted by the atoms of crystalline materials; by measuring the angle of diffraction, the separation of the atoms, or hard-ball atom diameter, can be measured.

yield point   In a tensile test, the first stress where strain occurs without an increase in stress; typical of annealed and hot-worked steels. The stress may actually decrease after (discontinuous) yielding starts, so upper and lower yield points may be identified.

yield strength   The stress at the intersection of a line parallel to the elastic portion of the stress-strain diagram at a strain value equal to a specified offset.

I N D E X